The publisher and the University of California Press Foundation gratefully acknowledge the generous support of Lorrie and Richard Greene.

The Jew Who Would Be King

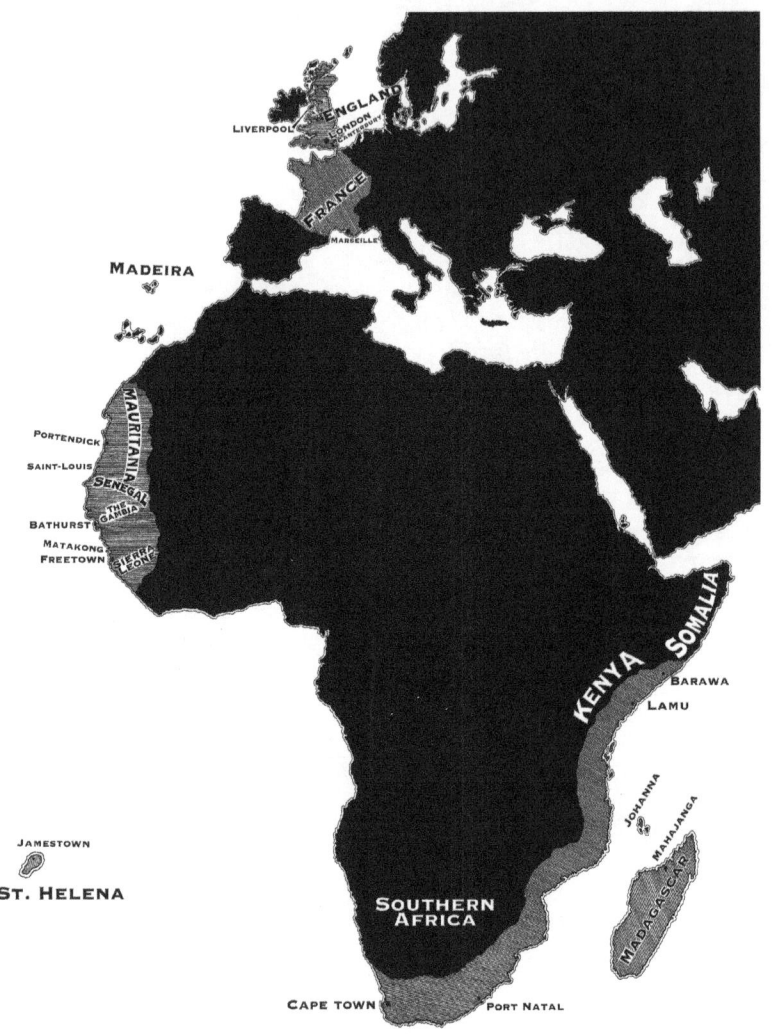

MAP 1. Nathaniel Isaacs' world.

The Jew Who Would Be King

A True Story of Shipwreck, Survival, and Scandal in Victorian Africa

Adam Rovner

UNIVERSITY OF CALIFORNIA PRESS

University of California Press
Oakland, California

Library of Congress Cataloging-in-Publication Data

Names: Rovner, Adam, author.
Title: The Jew who would be king : a true story of
    shipwreck, survival, and scandal in Victorian Africa /
    Adam Rovner.
Description: [Oakland, California] : University of
    California Press, [2025] | Includes bibliographical
    references and index.
Identifiers: LCCN 2024030674 (print) | LCCN 2024030675
    (ebook) | ISBN 9780520403000 (cloth) | ISBN
    9780520403017 (ebook)
Subjects: LCSH: Isaacs, Nathaniel, 1808–1872. |
    Explorers—Biography. | Imperialism—19th century.
Classification: LCC G246.173 R68 2025 (print) |
    LCC G246.173 (ebook) | DDC 916.04/23092 [B]—dc23
    /eng/20241015
LC record available at https://lccn.loc.gov/2024030674
LC ebook record available at https://lccn.loc.gov
    /2024030675

33   32   31   30   29   28   27   26   25
10   9   8   7   6   5   4   3   2   1

*In memory of Ruth (Kretzmer) Cohen and
Dr. Stanley Cohen, who pursued justice in their
native South Africa and in their adopted homes
in England and Israel*

I want! I want!

—William Blake, *The Gates of Paradise: For Children*, 1793

Contents

Illustrations

MAPS

Prologue

FRIDAY, JUNE 30, 1854

A brightly colored boat built for coastal fishing cut through the still waters of the bay, heading towards King's Yard in the inky predawn. The boat's pilot sang to himself while he perched in the wooden bow. Behind him, men with powerful shoulders rowed for the wharf. The boat's clandestine cargo—two women and one man—huddled together at the rear of the narrow, deep-hulled vessel. Their eyes could not yet make out the waterfront, but they knew Freetown was awake. Voices echoed from the shoreline. Here and there a human shape formed out of the haze that shrouded the coast. Barracah strained to bring the docks into clear view. His wife, Langoh, cast an anxious glance back across the depths they had traversed. Bangah tilted her head heavenward and prayed by the pale light of the crescent moon to reach solid ground. Her empty stomach churned with nausea, and she shivered from the damp chill spreading through her wet clothes.

When the boat bumped against the stone quay, Black hands reached down to lift the bewildered trio up to safety. They were told in English and then again in Mende that they were now free. But no matter the language, what could freedom mean to Barracah, Langoh, and Bangah? They had been taken captive as children during regional wars, sold as slaves in their youth, and then sent to labor in the fortress and fields ruled by a white man and his African mistress. One thing that freedom meant, they would soon learn, was that their stories, their testimony, might deny their former captor his own liberty. The words of an African could damn as surely as the words of an Englishman. And the voice of the king of Matakong held no dominion in Freetown.[1]

Introduction

Matakong Island rises above the rocky shallows of the coastline of Guinea about fifty-two miles northwest of Freetown, Sierra Leone. To the extent that Americans and Europeans think about West Africa at all, Matakong lies in an imagined torrid zone marked by blood diamonds, child soldiers, and limbs lopped off by machetes. The truth is more mundane, and far more peaceful. Sierra Leone's civil war (1991–2002) spilled north across Guinea's borders, but the fighting did not leave its mark on Matakong. Even for Guineans, the place remains remote. My visit to this outpost required weeks of preparations and a fraught border crossing, then bouncing over dirt roads for hours, changing punctured tires, crossing makeshift log bridges, and racing along a muddy beach against an incoming tide. Most of the island, three miles or so in circumference, is uninhabited and covered by palm forest dotted with patches of waist-high brush. A few hundred villagers live on Matakong much as they have done since the mid-nineteenth century, when Nathaniel Isaacs, the merchant adventurer whose trail I followed here, built his home on Matakong's elevated bluffs.

The men fish from boats painted in riotous colors. They launch these handcrafted pirogues before sunrise. In the heat of the afternoon, they land and untangle their nets while their wives and daughters crouch over piles of reeking fish innards, gutting and sorting the day's catch. The women smoke the silvery fish over open flames on the shoreline at night. From this crowded bank, footpaths branch uphill. Other pathways wind past tumbledown homes, skirting wooded areas. One of these tracks leads from the mosque—Matakong's largest building by far—to a cinder-block school. Children's singsong voices carry from the structure to the fringe of the forest twenty yards beyond.

This isolated patch of ground is an improbable site for modern Jewish history. Yet here, rising from the damp leaf litter, stands a solitary headstone, carved in Hebrew and decorated with a banana plant motif, still intact and legible after more than 165 years. The gravestone marks the final resting place of Benson Isaacs. He was the elder brother of Nathaniel Isaacs, who once ruled Matakong as his personal fiefdom before being brought low by the enslaved people who escaped his grasp, hostile British authorities, and his own overweening ambitions.

Nathaniel Isaacs wanted a crown of his own. From pauper to prince to king, his fate intertwined with those of the imprisoned Napoleon Bonaparte on St. Helena, King Shaka Zulu in South Africa, and royalty in West Africa who fought against—or profited from—the slave trade. His pursuit of wealth and influence propelled him to represent Her Majesty's Government along the serpentine waterways of Sierra Leone and Guinea. But when it suited him, he allied himself with those who opposed Great Britain's interests. As much as Isaacs' story reveals how European forces conspired against Indigenous Africans, his tale presents a case history of the limits of colonial power and demonstrates how Europeans and Africans cooperated for mutual gain, and often at great cost.

Nathaniel Isaacs can be seen as a cross between an orphaned hero from a Charles Dickens novel of Victorian London and a character from an H. Rider Haggard African adventure. In a strange twist, Isaacs' published volumes left their mark on the work of both these famed authors. Yet Isaacs himself may most resemble the protagonists of Rudyard Kipling's "The Man Who Would Be King" (1888). In that story, two brothers-in-arms hatch a scheme to establish themselves as sovereigns of an inaccessible territory. Like Isaacs, Kipling's adventurers seek to transcend their class and abandon their past by lording it over supposedly savage peoples. Also like Isaacs, they sign their names to a dubious contract, train the local inhabitants to use European weapons, and despoil the land they appropriate. Their avarice ultimately proves their undoing. Isaacs also resembles real-life adventure capitalists of the nineteenth century who established themselves as sham monarchs the world over: men like Sir James Brooke (sultan of Sarawak, Borneo), Gregor MacGregor (cazique of Poyais, Honduras), William Walker (president of Baja and later Nicaragua), and Josiah Harlan (the prince of Ghor, in today's Afghanistan), who likely served as the inspiration for Kipling's story. Isaacs' career offers far more nuance, but no less drama, than the lives of these men, or Kipling's fable of imperial overreach.

Isaacs possessed many of the traits of the heroes of the imperial romance genre Kipling mastered. He was an intrepid, resourceful, and ambitious Englishman who kept his head about him when all around him were losing theirs. In this way, Isaacs adheres to the imperialist mythology. But he was also duplicitous, exploitative, self-promoting, and casually violent. He emerges as a real-life kinsman to Harry Flashman, the scoundrel-hero of George MacDonald Fraser's long-running series of novels set in the Victorian era. Moreover, Isaacs was Jewish, and therefore only grudgingly acknowledged as white. This narrative history thus treats Isaacs as an imperial antihero. His chameleon-like ability to adapt to varied circumstances, canny pursuit of profit, and multiple loyalties sit uncomfortably close to the fact of his Jewish heritage and to familiar anti-Semitic slights. Just as my recovery of Isaacs' story complicates efforts to read his life in the context of the imperial romance genre, it challenges the construction of a romanticized Jewish history.

The romance of imperialism has waned, thanks in part to postcolonial historiography, though a stubborn tendency to sentimentalize Africa and its inhabitants persists. In "How to Write about Africa," the Kenyan poet and critic Binyavanga Wainaina charges Western authors with employing a host of demeaning clichés in their work.[1] I have taken Wainaina's words to heart and tried to avoid the worst of the blind spots he identifies. This book features no warm-hearted conservationists, no saintly white saviors, no confident interventionist program, no sotto voce moralism, and no glad tidings from a simpler or more "authentic" world.

Nor do the following chapters condescend to the past by imposing contemporary values on it. Nathaniel Isaacs' life yields little in the way of a redemptive message. Instead, his career reveals equal parts benevolence and malevolence. It is the story of one man who acted with others, European and African, to negotiate sometimes opposed and sometimes parallel interests through the transfer of technologies, goods and resources, information, and systems of control that set the stage for the late-nineteenth-century "scramble for Africa." Isaacs was a man living at the cusp of empire and at its geographic limits. My exploration of his life and times illuminates the origins of many of the issues that continue to divide Europe from Africa: classism, ethnocentrism, racism, military adventurism, and economic imperialism. This microhistory of Nathaniel Isaacs opens a window onto the macrohistory of his world and its relevance today.

My book details the odyssey of a Jewish merchant, mercenary, and would-be monarch, of an explorer and author, a lover and fighter, and a nineteenth-century dreamer who played the role of Joseph in Zululand and of Pharaoh in West Africa. It demonstrates how formal and informal British colonialism in Africa intersected with the rise of global capitalism and the creation of what Hannah Arendt referred to as a "phantom world of race," in which the figure of the Jew participates as a secret sharer.[2] That is, Great Britain's encounter with Africa's Indigenous peoples in the nineteenth century was consistently mediated by the image of the Jew, and it was from this encounter with Black Africans that Jews within British society first emerged as white. Ironically, it was the white Jewish Isaacs who promoted the myth of Shaka Zulu, the king who remains a Black African icon.

Isaacs ably narrated his own life in black and white on the printed page, but his character reveals complex shadings of gray. He was sure of his purpose yet uncertain of the methods to achieve it, curious about the peoples and places he encountered while bound by his era's prejudices, proud of his successes yet embarrassed by his origins. His life was replete with innocence and bloodlust, drama on the high seas, bravery and villainy, freedom and slavery, fortune and calamity, and survival against all odds. Yet his story also emerges from the stubborn materiality of the distant past: leather shoes and cotton cloth, hair dye and ivory, beads and bangles, gunpowder and groundnuts, books and maps, the copper-hulled sailing vessels that facilitated global trade, and the red sandstone walls that once held men and women prisoner on Matakong Island.

. . .

I have endeavored to use the words of Nathaniel Isaacs and his contemporaries—African and European—wherever appropriate. The language of distant eras often sounds dated today and may offend sensitive ears. At times the documents and sources consulted—many gathered from archives in England, St. Helena, South Africa, and Sierra Leone—support one another; at times they contradict one another. In the latter cases, I have either selected the most plausible account of events or let the contradictions in the evidentiary record speak for themselves. European writers, Zulu informants, and West African witnesses all provide their perspectives. Those who prefer not to hear the voices of European colonial actors and the African men and women who abetted conquest, or those who suffered from its depredations, should turn elsewhere for more consoling visions of the past.

The primary materials I draw on include Isaacs' *Travels and Adventures in Eastern Africa*, the memoirs of others, Indigenous oral histories, government records, personal correspondence, medical and anthropological texts, grammar school primers, guidebooks, shipping registers and account ledgers, newspaper articles and long-form journalism, missionaries' reports, encyclopedia and dictionary entries, poetry and prose, maps and images, and works of political economy, as well as legal briefs and sworn testimonies. I have had to immerse myself in unexpected minutiae: the tactics of Zulu warfare and nineteenth-century elephant hunting, the economics of West African peanut farming, missionary activity in Sierra Leone, shipboard commerce, and government-backed piracy. My research has taken me to each of the locales I describe: Canterbury, England; the island of St. Helena in the South Atlantic; the lands of the Zulu Kingdom in present-day South Africa; Sierra Leone's bustling capital of Freetown; and Matakong Island. The notes typically offer citations rather than further explanation or commentary. Most readers can avoid becoming entangled in the academic rigging and instead take their bearings from the narrative. Specialists will ideally read this work as the biography of a man and a history of his times, and I hope they will benefit from my references to archival documents and published sources.

Part 1 of the book deals with Isaacs' trajectory from England to St. Helena and on to South Africa. Part 2 focuses on his time in West Africa and his return to England. The first chapter, "Jews and Other Savages," traces Isaacs' family history, his peregrinations from Canterbury to St. Helena, and to his shipwreck at Port Natal, today's city of Durban. This chapter also situates the precariousness of Isaacs' social position in the context of Georgian-era anti-Semitism. Chapter 2, "Strange Surprising Adventures," follows the castaway Isaacs as he encounters the Zulu people and King Shaka and attempts to survive amid uncertainty and upheaval. This chapter explains Isaacs' decision to bear arms for Shaka and how his bravery in battle was rewarded with a controversial land grant and honorific title. Chapter 3 centers on Isaacs' relationship with the Zulu king, his ivory-hunting forays, and the assassination of Shaka. It also examines Isaacs' influence on later Zulu and European representations of the legendary king. Chapter 4 details Isaacs' escort of a Zulu diplomatic mission to British authorities in the Cape Colony, the chaos following Shaka's assassination, and the rule of King Dingane. It also outlines the networks of African and transatlantic trade in which Isaacs participated.

Chapter 5 describes Isaacs' paternal relationship with his teenage Jewish apprentice, Ben Moss, and their voyages along the West African coast, principally to hazardous ports in today's Mauritania. In addition, it examines the reception of Isaacs' *Travels and Adventures* in England and its connection to various schemes promoting the colonization of South Africa. Chapter 6 chronicles Isaacs' activities in Sierra Leone as a merchant and a colonial commissioner dispatched to local rulers to negotiate an end to persistent slave dealing. Sierra Leone's origins in abolitionist activity are the background against which Isaacs consolidated his hold on Matakong Island. Chapter 7 details accusations against Isaacs for slaveholding and subsequent efforts to prosecute him. This chapter draws on transcripts of the recently discovered testimony of men and women held captive by Isaacs. Chapter 8 recounts Isaacs' final years on Matakong and in England and describes the indirect influence of his *Travels and Adventures* on Charles Dickens and H. Rider Haggard. A postscript explores Isaacs' surprising afterlife in early-twentieth-century Jewish nationalist discourse and offers considerations of what his checkered career may mean for readers today.

PART ONE

From England to South Africa

A few white people intended to come first and get a grant of land. . . . [T]hey would then build a fort, when more would come, and demand land, who would also build houses and subdue the Zoolas, and keep driving them further back.

—Jakot Msimbithi, a.k.a. Hlambamanzi, ca. 1828

CHAPTER I

Jews and Other Savages, 1808–1825

Nathaniel Isaacs' father was a cordwainer, a leather worker who crafted shoes. Other than that, all we know about Isaac Isaacs is that he lived long enough to start a family in Canterbury, the ancient capital of the English county of Kent. Nathaniel's mother, Lenie (Elizabeth) Solomon, was a native of Margate, a seaside town famed for its mild climate. Like most of the few thousand Jews resident in provincial England at the start of the nineteenth century, the Isaacs family would have lived a precarious existence. Still, the twisting lanes of medieval Canterbury and the surrounding countryside offered several advantages over London to struggling Jews. Isaac may have scratched out a living supplying footwear to sailors in the naval dockyard in Chatham, about thirty miles distant. Jewish merchants and craftspeople often moved into the British middle class by way of trade with the Royal Navy and its sailors. Perhaps Lenie was able to feed her young children—Benson, Nathaniel, and Hannah—by haggling over the morning's catch at the city's daily fish market. Fresh water from the River Stour streamed to public fountains, cherries grew in abundance, and the region's famous hops meant that home-brewed beer was plentiful.[1]

London was less than sixty miles away and connected to Canterbury by coach routes. Lenie's father, Natte (Nathaniel) Solomon, for whom her middle child was named, had been a silversmith and a dues-paying member of London's Great Synagogue as early as 1766. The Isaacs family likely retained close contacts in the metropolis. Canterbury Cathedral ensured a constant flow of pilgrims, news, and merchandise, and the nearby ports put the nine thousand inhabitants of Canterbury in contact with a wider world of commerce and culture. Yet the city retained its pastoral character. Gravel paths bordered by poplar trees wound along the Stour's banks. The dull gray ragstone and Roman-era bricks of the walled city framed the green hills of the surrounding country.[2]

Canterbury, the seat of the Anglican Church, was home to four hundred or so Jews at the start of the nineteenth century. At approximately 4 percent of the surrounding population, the Jews were described at the time as "very numerous." The neighborhood they had traditionally inhabited, St. Peter's, lay in close proximity to the cathedral, but they built their synagogue beyond the shadows cast by its Gothic crosses. Although Quakers, Methodists, and Catholics also circulated among Canterbury's High Church devout, the Jews were marginalized even among these Nonconformists and papists. Mostly poor and definitely alien, British Jews of the Georgian era (ca. 1714–1830) were victimized by anti-Semitic folklore. They were commonly thought to be burdened by a telltale stench; they were the devil's spawn, killers of Christian children, devoted to primitive rituals, conniving in trade, and tribal in their insularity. In Chaucer's *Canterbury Tales*, the Prioress calls the Jews a "cursed folk" for their purported murder of a Christian boy.[3]

To many of their neighbors, Canterbury's Jews would have seemed little more than savages, or perhaps even worse. At least savages could be romanticized as noble, according to writers such as Michel de Montaigne, Aphra Behn, Jean-Jacques Rousseau, and the economist Adam Smith. Jews in Great Britain, however, were imagined as ignoble and figured as "crass and venal, lacking honor and virtue, in thrall to a slave religion or unrestrained passion." They might exist on the fringes of polite society, but the trappings of civilization were only a veneer. The fourth edition (1810) of the *Encyclopaedia Britannica* made it clear that civilization—or the only one that mattered—depended on a state of grace. "The origin of civilization," the encyclopedist maintained, could not be accounted for "otherwise than by a miracle." Therefore, "Let us rejoice and be thankful . . . that we are Christians."[4] Canterbury's Jews might not have been loathed as much as Chaucer's Prioress suggested, but they were not much loved.

The idea of the savage served Georgian England much the way the idea serves us today. Savages are those we prefer to think we are not. Savages are primitive; we are modern. Savages are uncivilized; we possess culture. Savages are nomads oblivious to borders; we are settled in states. Savages practice strange and threatening rituals, speak an incomprehensible language, and live in expanded familial clans.[5] They thus present a reversal of the values that underpin Western culture: home and hearth, property and privacy, the sanctity of the conjugal family. Condemned to wander, destined by persecution and inclination to live apart, and thought to adhere to archaic rites, Jews were often considered less than fully human. The connection between Jews and savagery allowed British voyagers and missionaries to label Indigenous peoples as the lost tribes of Israel wherever they landed, whether in North America or in Africa.

These mirages persisted amid ignorance of actual Jewish life and culture. Jews had been expelled from England in 1290 and not readmitted until 1656. In the interim, anti-Semitic legends and folk mythologies flourished. When they were finally permitted to return to England, Jews became objects of crude ethnographic interest The diarist Samuel Pepys, an administrator for the Royal Navy, recorded his impressions of a visit to a synagogue less than a decade after Jews had resettled : "Confusion in all their [religious] service, more like brutes than knowing the true God, would make a man forswear ever seeing them more."[6] And yet brutish Jewry could be saved. All the stiff-necked Hebrews needed to do was convert to Christianity. But until they did, Jews were to remain second-class citizens, denied full political rights and subject to discriminatory measures.

As the Napoleonic wars raged across much of Europe, the Emperor Napoleon himself, hated and feared in equal measure, threatened British commercial and imperial interests around the world. Jews were thought by many in Great Britain to be sympathetic to Napoleon, in part because he supported their civil emancipation. Isaac Cruikshank created a number of caricatures featuring hook-nosed Jewish financiers, peddlers, and rowdies in league with a diminutive Bonaparte. In one cartoon from 1803 Cruikshank depicted a pointy-bearded Jew, stereotypically named Moses, conspiring to aid Napoleon and his allies: "I tink if I lend a little more monish . . . it will soon annihilate dem."[7] "Dem" was Great Britain, Moses's trade was usury, and his accent served as a marker of Jewish difference. The dark-caped Moses, who supports his stooped frame and crooked legs with a cane, also evokes familiar images of "the old gentleman in black"—the devil eager to cheat you of your soul.

Jews were often marked as grifters, particularly among those serving in the Royal Navy. As the navy grew, developed, and shaped the expanding

British Empire, many British-born Jews and immigrants became tailors, peddlers, or licensed agents selling clothes and other goods to petty officers and seamen. Official agent lists of the era reveal dozens of identifiably Jewish names, from Abraham Aaron to Levy Zachariah, with many Isaacses, Moseses, and Solomons in between. These agents employed fellow Jews who wandered the ports selling goods on credit, leading to the widespread charge that they exploited sailors. The ubiquity of Jews in the trade and the derision directed at them and their merchandise are reflected in naval slang of the day. *Jewing* became a term for tailoring, and a *Jewing firm* referred to a sailor who mended clothes for others.[8]

Jewish traders met naval vessels at the dockside to peddle their goods. Officers generally allowed these peddlers to lug their bundles up gangplanks or clamber out of bumboats to board ships and outfit the sailors, who were often forbidden from going ashore for fear of desertion.[9] In addition, "bumboat women" sold cheap wares, alcohol, and sometimes their attentions to men long deprived of female company. Many documents and illustrations of the era record boisterous or drunk sailors pressed together below decks and beneath hatchway grates, cajoling women and haggling with bearded Jewish merchants.

One intriguing literary sketch seemingly based on fact describes a scuffle that occurred in the 1780s involving "an individual of the Jewish persuasion, by name Isaac Isaacs, the lower extremities of whose face were distinguished by a beard of no mean pretensions." Whether this was Nathaniel Isaacs' father is uncertain. But the passage includes descriptive details not strictly necessary for a humorous anecdote, and an author seeking a name for a stock figure would more typically have chosen Moses. This Isaacs serves as a representative figure for the Jewish peddlers who frequented naval yards with their stock of "buttons, scissors, and thread,—jackets, thick and thin,—shirts of all sizes, hats, caps, and looking glasses . . . watches, gold chains, stockings, shoes, and a mass of indescribables." And "in this Jewish collection," the author recounts, "Mr Isaacs had not forgotten to provide some large strong fishermen's boots, at that time in great request for washing decks."[10] Perhaps the boots had been made by Isaac Isaacs, cordwainer of Canterbury and father of three.

A popular novel written by an officer of the Royal Navy and set during the Napoleonic Wars describes a shipboard fracas involving a similarly equipped Jewish peddler. Captain Frederick Marryat's *Peter Simple* (1834) depicts a vengeful sailor "hunting everywhere for a Jew who had cheated him," while another embittered seaman punches a

Methodist abolitionist in the face for daring to suggest that Blacks and whites are brothers. After the abolitionist preacher makes his escape and an officer ejects the Jewish merchant and various bumboat women over the side of the ship, the "intoxicated [sailors] were put to bed, and the ship was once more quiet."[11] Here the religious and racial admixture that flourished along Georgian-era wharves is expelled from a floating microcosm of a white, male, Christian Britain drunk on power—not to mention large quantities of alcohol. The message in Marryat's novel was clear: Jews, like dissenters and Blacks, existed on the sufferance of the white Anglican majority.

Marryat pioneered the serial publication of seafaring fiction and later encouraged his friend Charles Dickens to publish his novels in installments.[12] The typical Marryat novel, like many of Dickens's works, follows the fortunes of a young boy who grows to adolescence and adulthood in a surrogate family. He meets eccentric or fearsome characters, some of whom manipulate or exploit him, as well as benefactors who educate and support him. He demonstrates bravery and an often naive integrity throughout, and by chance and coincidence he comes to be rewarded with money, land, a title, or some combination of these gentlemanly riches.

Nathaniel Isaacs' own life story follows this narrative structure to a remarkable degree. And though little read today, Marryat's novels remain the Ur-texts for the later imperial romance genre, which incorporates authors and works as diverse in style and ideology as H. Rider Haggard's *King Solomon's Mines*, Rudyard Kipling's "The Man Who Would Be King," and Joseph Conrad's *Heart of Darkness*. Like all these authors, Marryat held ambivalent views of Jews, Blacks, and other domestic aliens and foreign "savages."

Across the Channel in 1807, Napoleon produced a spectacle designed to demonstrate his benevolence towards European Jewry and embarrass his British enemies. He revived the institution of the Sanhedrin, the ancient Jewish tribunal of learned men. Whether this is judged today as a cynical political ploy or a sincere gesture, the assembly caused a sensation. The philo-Semitic radical William Hamilton Reid wrote a book describing the occasion. "Englishmen will revolt at the idea," Reid observed, "but it cannot be concealed that the Jews of this Sanhedrin acknowledge the Head of the French Government as their *Deliverer*." The suggestion that Jews might view Napoleon as a messianic figure and therefore do his bidding to the detriment of British interests was clear. Isaac D'Israeli, the Jewish father of the future prime minister

Benjamin Disraeli, wrote two articles in August 1807 detailing the pageantry of the Grand Sanhedrin for curious and alarmed British readers. According to D'Israeli, the president of the Sanhedrin addressed his coreligionists with sharp words. "Nations civilized themselves," he lamented; "we [Jews] among them, alone, remain barbarous."[13] As modern, romantic nationalism took root, Jews came to see themselves as others did: as abject wanderers lacking a place of their own.

In the modern era, Jews were doomed to be thought uncivilized until they possessed territory and exerted political power, or until they achieved full civic enfranchisement, at the cost of abandoning all distinctive markers of their identity save for their forms of observance. The dream of controlling territory suggested a return to the biblical promised land, a vision that ultimately gave birth to Zionism. The aspiration to full political and civil emancipation, on the other hand, led to a raft of reforms aimed at modernizing Jewry—that is, making Jews palatable to Christians by filtering out the sediments of their tradition. Some in England at the time did believe that Napoleon might restore Jews to their ancient homeland in Israel, but the legacy of the Sanhedrin was in fact to disperse Jewry into all corners of the modern world.[14]

The story of Nathaniel Isaacs is neither one of Jewish nationalism nor one of ideologically driven assimilation. His trajectory instead charts many of the upheavals of the Napoleonic era and the long decades of Great Britain's imperial supremacy. Isaacs lived on the cusp of a recognizably modern world, and his relationship to Judaism would have been familiar to many ambitious British Jews of his period. He exchanged piety, tradition, communal ties, and limited opportunity for the burgeoning secular world of individualism and commercial networks, preferring to trust the Invisible Hand than to see the hand of God in human affairs. And he did so without apparently giving his Jewish identity all that much thought. Or perhaps he did not have the luxury of reflecting on his Judaism. Such a verdict on Isaacs may be unsatisfying for those interested in discerning the signs of supposedly unique Jewish values in his life. He was not unduly burdened by convictions, religious or moral. In fact, he swore fealty to no cause unless it served to advance his own interests.

Isaacs was no high-minded African explorer like the missionary Dr. David Livingstone, who sought to convert and succor African peoples for the glory of God and Great Britain; and no Charles Dickens, who dignified England's working poor in his popular fiction and journalism. He did not further the cause of science like the globetrotting

Charles Darwin. Nor could he aspire to power through public office, like the Jewish-born Benjamin Disraeli, who was elected to Parliament after his early conversion to Christianity. Unlike these approximate contemporaries, Isaacs was a merchant. As Adam Smith suggested in *The Wealth of Nations* (1776), to the merchant, "home is . . . the centre" around which capital circulates. Home for Isaacs was therefore mobile and transitory. Home was a process of exchange. The only imperative he obeyed was his own gain, whether through commerce or connivance. Isaacs was an extraordinary man even by the standards of his day, though in many ways his eager accommodation to the social mores that surrounded him was representative of his era—and of ours.[15]

Though Isaacs abandoned all but a vestigial relationship to Judaism, his story paradoxically remains a Jewish one, part of the narrative of the Jewish encounter with modernity. Isaacs seems to have relinquished his Judaism intentionally at times, as when he channeled the voice of an Anglican gentleman in his two-volume *Travels and Adventures* (1836). This half-hearted ventriloquism was unlikely to have persuaded many readers, however. At other times, Isaacs appears to have cast off his Jewish past without serious calculation. Here, too, his behavior is no different from that of many of his brethren, then or now. The fact that he apparently never adopted another faith is intriguing in an era when religion was a central marker of identity and baptism allowed for social advancement.

Like many Jews, Isaacs was never able—even if he had been willing—to conceal his heritage. To the patrician class, Isaacs remained a parvenu. His social betters made certain to remind one another that he was "a Jew, or of Jewish extraction."[16] Isaacs' nineteenth-century life might best be understood as a definitively modern one. He sought to forge an individual identity, to free his unique sense of self from the snares of a collective. Isaacs did not look backward to his origins but forward to what he might make of himself. Nonetheless, he could never quite outpace his Jewish "extraction," even in a remote corner of West Africa. Perhaps he even retained a sense of duty, or at least a sentimental attachment, to the religion of his birth and rites of his youth, though it seems he did so mostly when confronting death.

The year that Napoleon assembled the Sanhedrin witnessed another challenge to the British Empire, this one launched closer to home and with religious minorities again in the vanguard. Quakers and evangelicals championed abolition in the United Kingdom and forced the passage of the Slave Trade Act. The evangelical revival movement emerged from the Anglican Church under the influence of John Wesley, whose sermons inspired numerous religious and lay leaders, including the prominent abolitionist William Wilberforce. The term *evangelical* refers to those who held a constellation of convictions that inspired them to political action and often to moral crusades. Those evangelicals affiliated with Wilberforce adhered to a set of core beliefs: individual sinfulness, which required a personal conversion experience; constant spiritual conflict, which necessitated intense self-examination to prevent temptation; and the potential for salvation only through faith in Jesus Christ, which demanded proselytization of the unconverted, be they Catholics, Jews, or heathens.[17] The conversion of all three groups became a pillar of evangelical missionary work at home and abroad.

Through the efforts of Wilberforce and his circle, the Slave Trade Act (mostly) outlawed the lucrative slave trade in British territory. Slavery itself was not abolished. The act also led to the establishment of Sierra Leone in 1808 as a Crown colony for liberated Africans, with Freetown as its capital. If the first half of Isaacs' life was influenced by the political dominion and military might represented by Napoleon and King Shaka Zulu, the "Napoleon of Africa," the second half was dominated by the specter of slavery.[18] Over all his life, savagery—real and imagined, European and African—cast its shadow.

We know virtually nothing of Nathaniel Isaacs' childhood. Perhaps he did not wish to remember those years. His older brother, Benson (or Benjamin), was born in 1808. By his own account, Nathaniel arrived later that year. Their sister, Hannah, was born in 1810 or 1811.[19] Sometime after Hannah's birth Isaac Isaacs died, and his footprints, not to mention the shoes he made, were lost to history.

Nathaniel likely had some schooling, at least enough to read and write English well and to tally accounts. We cannot know for certain whether he had any religious education, though given the large Jewish community of Canterbury and its active synagogue, it is likely that he received instruction in the basic Jewish rituals and prayers. Most poor Georgian-era Jews remained within the fold and maintained religious observances to some extent.[20] Upwardly mobile and educated Jews might convert to Christianity to advance their fortunes, as Benjamin Disraeli did at the age of twelve. Nathaniel would have become an adult—a bar mitzvah—according to Jewish law in 1821, and might have been called to read from the Torah at the synagogue that once stood near St. Dunstan's Church, not far from Canterbury's turreted medieval Westgate entrance. But whatever religious training he might have had was cut short when his mother sent him to be apprenticed to her brother. That Lenie chose Nathaniel for this role, over the first-born Benson, suggests that she saw in him some spark of talent or charisma that her older son lacked. Like the biblical archetypes of Joseph, Moses, David, and Solomon—all of whom were younger sons—Nathaniel was marked for a royal destiny.

"I left England in the year 1822, at the age of fourteen," Isaacs recollected. Children of the lower and lower-middle classes were often apprenticed or put to work at a tender age: if anything, Isaacs' departure from his widowed mother was relatively late. "It was determined, several years before I quitted England," he wrote, "that my destination should be Saint Helena, in which island Mr. S. Solomons [sic], my maternal uncle, resided."[21] There were two serious obstacles to Lenie Isaacs' plan for her son's advancement. First, St. Helena lay about six thousand miles distant, in the middle of the South Atlantic Ocean. Such a voyage was hazardous and expensive, though Isaacs' uncle would likely have paid his passage. Second, St. Helena had served since 1815 as the island prison of the exiled Napoleon Bonaparte and his retinue. Although the chances of his escape were remote, few individuals were permitted to relocate to St. Helena during Napoleon's detention there.

Napoleon's fate was shaped by islands: he left his birthplace on Corsica to endure his classmates' taunts at a French military school, came to grief battling enemies from the British isles, and was bound as a captive on Italy's Elba. After his escape from Elba and his subsequent defeat at the Battle of Waterloo, the British decided to isolate him once and for all on the East India Company outpost of St. Helena. The island stood isolated in heavy seas patrolled by the Royal Navy. One of those charged with guarding Napoleon's island prison was the future author Frederick Marryat, who even drew a sketch of the emperor.[22] When Bonaparte peered through the spyhole he had gouged into the wooden shutters of Longwood House, where he was imprisoned, he could glimpse a rolling patch of grass, some stunted flowerbeds, rocky cliffs, an endless expanse of sea, and the straight-backed British sentries with loaded muskets who enforced his confinement.

Isaacs' life too would be framed by islands, and even more so by the oceans that bounded them. He recalled that his departure for St. Helena was delayed, "owing to the great difficulty which at the time existed in obtaining the East India Company's permission to proceed thither." By

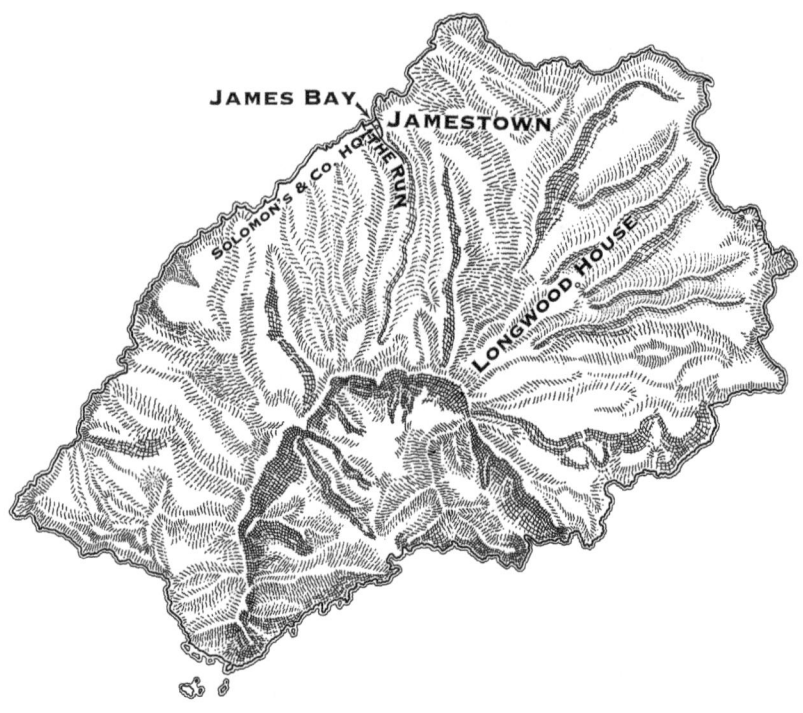

JAMES BAY
JAMESTOWN
SOLOMON'S & CO. HOUSE RUN
LONGWOOD HOUSE

MAP 2. St. Helena.

the early 1800s, the East India Company existed as a sort of extrasovereign state that ruled an archipelago of global ports "Under the Auspices of the King and Parliament of England," as its motto proclaimed. Decisions made by the company's directors had the imprimatur of the British government. Founded as a joint-stock company for the import of pepper and textiles from India and southeast Asia, the company evolved into one of the world's most formidable forces of colonization. Though it exerted its aristocratic arrogance across the globe, its stockholders back in London enjoined a more democratic mercantilism. One contemporary described noisy meetings that observed "no distinction as to citizenship—the Englishman, the Frenchman, the American [and] no difference as to religion—the Jew, the Turk, the Pagan."[23] Still, it would not be easy for the penniless nephew of St. Helena's foremost merchant, Saul Solomon, who was already suspected of being sympathetic to Napoleon, to secure an entry permit from the East India Company's powerful directors.

Ships of the day sailed along the track of the trade winds that carried them west, nearly to South America's shores, before they could head southeast toward Africa, around the Cape of Good Hope, and northwest toward India; the voyage could take up to eight months. East Indiamen returning laden with goods from Asia needed to replenish their supplies of freshwater on the return voyage. St. Helena rested along these busy trade routes. So although St. Helena appears to be an insignificant dot on the map midway between Brazil and Angola, it held outsized importance in the age of sail. Ships dropped anchor at Jamestown every other day to take on water, supplies, and cargo or to conduct repairs.[24] The sick and dying were removed to breathe their last on terra firma.

In 1796, the twenty-year-old Saul Solomon sailed on an East India Company vessel as a soldier bound for the subcontinent. He became desperately ill with an unnamed malady during the first part of the voyage and was carried off his ship, transferred by lighter to the makeshift wharf at Jamestown on St. Helena, and then likely brought to some weather-beaten barracks to recover or succumb. Once he left the cramped and fetid ship, he regained his health, but he then found himself discharged from the company's rolls. He remained on St. Helena and there built a new life.

Solomon established a boardinghouse and shop in Jamestown. He married a non-Jewish woman named Margaret, who bore him a son. They named him Nathaniel—after Saul's father—and baptized him on 25 May 1800, in accordance with Anglican custom. By the time Napoleon arrived on St. Helena in October 1815, Solomon's family included two more sons and a daughter, Phoebe. They too were all baptized, though there is no indication in the historical record that Solomon himself underwent baptism. His outward assumption of Church of England ways nodded as much to Jamestown's social conventions as to legal requirements. Proclamations from company administrators enforced the strict observance of the Sunday Sabbath on pain of imprisonment and demanded attendance at Sunday worship. To these men, Christian piety was a reformist social good.[25]

Despite his adoption of a Christian lifestyle, Solomon's Jewish descent remained an open secret. An attendant informed Napoleon that "he commonly goes by the appellation 'Saul Solomon the Jew.'" Solomon prospered by catering not only to the frequent ship traffic but also the island's most famous inhabitant. He charged Napoleon twenty-four shillings a day for Longwood House's standing order of two pounds of "maccheroni," which in England was about equal to a skilled worker's weekly wage. Other customers were charged similarly exorbitant prices. A British naval officer complained of paying Solomon £6 for four days' accommodation, which he considered excessive.[26] Even today, there are few affordable options for accommodations in Jamestown. To travel to St. Helena, you really, really have to want to go, and, you pay for the privilege.

Jamestown is nestled in a ravine. A stream known as the Run flows down from cloud-forest peaks that trap the trade winds' moisture. Ten-foot-tall tree ferns and lush thickets make these gusty slopes look like something out of a Carboniferous Era forest. At lower elevations only sparse vegetation clings to the steep, rocky ridges. Stone fortifications rise above the town; to those arriving by water, they seem to form one menacing piece with the precipitous red-brown cliffs. From the wharf, narrow lanes rise gently past a complex of whitewashed government buildings, a public garden, the steeple of St. James' Church, and Georgian buildings featuring stone facades and well-proportioned windows. The subtropical vegetation, including coffee bushes and bougainvillea, and the rugged topography make for a jarring background to the mannered symmetry of the Georgian structures that still testify to the island's prosperity under East India Company rule.

A few months after Napoleon arrived in Jamestown, his jailer, Hudson Lowe, took up his joint appointments as the island's governor and commander in chief. He officiously monitored Napoleon's movements and enforced all strictures of his exile. Lowe was the same age as his charge, forty-seven, and like Napoleon, he spoke Italian and French. Coincidentally, Lowe had also previously served on both Corsica and Elba. But these similarities only bred antagonism between warden and prisoner. Numerous statements reveal Lowe to have been ill-tempered, petty, and insecure. Even British sources claimed he "threw a sort of gloom" upon the island. Bonaparte himself described Lowe as "hideous" and possessing a "most villainous countenance."[27] Portraits of Lowe indeed depict him as an angular scarecrow of a man with a prominent nose and a hawkish scowl.

Napoleon was Lowe's physical opposite: portly, brooding, and conscious to affect a regal bearing despite diminished circumstances. Lowe rarely met with his truculent charge: he had plenty to keep him busy. In 1818, the governor and commander in chief oversaw more than 2,000 soldiers of the St. Helena Artillery, approximately 3,000 white residents, 1,000 free Blacks, 1,200 or so slaves, more than 1,500 head of cattle, 95 dogs, and precisely one *très misérable* former French emperor. In Longwood House, five winding miles from Jamestown, he dressed in full uniform, including the famous felt bicorne hat, which he always wore sideways. He insisted that his officers wear military dress at formal dinners, which flowed with wines from Bordeaux and Champagne. Once the most powerful man in the world, Napoleon now pursued and then abandoned gardening, dodged his British sentries, played billiards and chess, dictated reminiscences, groused about the rats infesting Longwood House, intrigued against Lowe, and for a brief period of time enjoyed kicking off the ground on his custom-built see-saw.[28] Bonaparte surely also contemplated his inlaid wooden globe and how his own world had shrunk. Rumor had it that he scratched the speck of St. Helena off the globe, but examination reveals that it remains intact.

Circumscribing his prisoner's movements according to his orders from London, Lowe implacably opposed Napoleon's continued pretensions to power. In turn, the aggrieved Napoleon inveighed against Lowe to his sympathizers in Europe via clandestine correspondence. The governor insisted that all Napoleon's letters be sent unsealed for his review and censor. Despite the restrictions he enforced against Napoleon, however, Lowe seems to have supported the broader cause of liberty. On Christmas Day, 1818, he took the first step toward the gradual abolition of slavery on the island. All ninety-two slave holders on St. Helena signed a proclamation dedicating themselves to a "progressive end" to the cruel institution. They included Saul Solomon. Nonetheless, when Solomon's own slaves absconded as stowaways in 1824 and 1825, he vigorously pressed for their return.[29]

As a favorite of Bonaparte's, Solomon was disliked by Hudson Lowe. Thanks to an extensive trade network, Saul and his brother Lewis Gideon Solomon, a jeweler, who arrived on the island in 1814, passed secret letters for Napoleon and his entourage that evaded Lowe's sharp eyes. "This was the way Napoleon privately received intelligence from Europe," a witness recorded. Members of Napoleon's exiled court would "visit the shops of these men [the Solomons] under pretence of purchasing articles [and] would give them such letters as they wanted sent to England or to Paris, and receive back any that had come." Such dealings did little to endear the Solomons to Lowe, who sneered at their Jewish surname, calling it "a Name which almost implies some predisposition to engage in illicit speculations." The appellation *Jew* had indeed become a synonym for criminal enterprise at the time. The 1823 edition of the *Classical Dictionary of the Vulgar Tongue* collects British underclass argot, including entries that connect Jews to specific misdeeds such as proffering worthless security ("Jew bail"), pickpocketing, and diminishing the precious-metal content of coinage ("clipping" or "sweating").[30]

Tales of Saul Solomon's instinct for profit also reveal the anti-Semitic attitudes of the day. One visitor spread word that Solomon had been idly kicking at stones on St. Helena when he uncovered a canvas bag containing a buried hoard of gold, which he then parlayed into a fortune through speculation. The merchant king Solomon, this visitor jeered, sometimes even deigned to extend a "very un-Israelitish generosity . . . towards persons ill prepared to endure [his] extortion." Napoleon, by contrast, held the Solomon brothers in esteem. Lewis Solomon reported that he often played chess with Bonaparte, "who cared little for the rank of his antagonist, if his tactics . . . were but sound and ingenious."[31] The checkmated emperor who had favored Jewish enfranchisement seemingly kept to his egalitarian principles even at the ends of the earth.

Word of Saul Solomon's sympathy for the Bonapartists led to rumors that he plotted to send Napoleon a teapot concealing a silk ladder, by means of which Bonaparte could descend the cliffs and escape in a conspirator's boat. There is probably little truth to the story, but Solomon's relationship with Napoleon had certainly grown obnoxious to the East India Company by 1821. When it came to light in March of that year that Royal Navy and company officers had dined with Solomon at his home, a lieutenant colonel threatened his subordinate with discipline were he to attend another of the notorious Solomon's entertainments. Saul Solomon caught wind of this threat and protested to Lowe, writing: "I certainly did not expect to find that my estimation in Society was progressively lower'd so that an association with my family should be considered either a degradation or a crime."[32] Lowe responded with icy silence.

Shortly after Napoleon's death in May 1821, the company's board uncovered clear evidence that the Solomon brothers had smuggled letters for him. Documents forwarded to the company proved that Lewis "had been carrying on a correspondence with persons belonging to Longwood, during General Bonaparte's life time." Copies of these illicit letters were sent to the company but mysteriously went missing. We can gather some sense of their contents from scattered references in other documents. With Napoleon's body still warm, Lowe's agents ransacked Longwood House and discovered charred papers in a fireplace grate. They were reconstructed to reveal that Lewis Solomon had provided intelligence, often faulty, regarding Lowe's movements and matters of state. Lewis Solomon was "ordered to quit the island within six months," but owing to the loss of the incriminating documents, with Napoleon deceased, and with Lowe a lame-duck governor enmeshed in preparations for his to return to London, the company ultimately allowed him to remain.[33]

Less than a year later, in May 1822, Nathaniel Isaacs received permission from the company's board "to proceed to the Island of St. Helena, for the purpose of being employed in the business of his uncles Msrs. L[ewis] and S[aul] Solomon." A sum of five hundred pounds was pledged as security should young Nathaniel prove a burden on the colony, guaranteed by a relative, Lewis Isaacs, who sold toys in Chatham, and one Michael Levy, perhaps a relative or friend of the family. Levy was a slopseller, a merchant specializing in outfitting sailors—another sign of the Isaacs family's connections to seafaring commerce.[34] Around this same time, the ten-year-old Charles Dickens, a resident of Chatham who might have purchased toys from Lewis Isaacs, faced his own expulsion from the idyllic county of Kent and found himself working long hours in a London shoe-polish factory.

Young Nathaniel was propelled from his home not only by surface currents of economic circumstance but also by deeper currents of world affairs, particularly the quickened pace of British commercial and territorial expansion after the Napoleonic Wars. As a disadvantaged and provincial Georgian Jew, Isaacs was pushed to the fringes of trade and the limits of empire—and beyond. And in these far-flung borderlands he found himself at the crossroads of imperial history.

The fourteen-year-old Nathaniel would have started on his adventures aboard a lighter that ferried him to the merchant vessel *Margaret*, a two-masted, square-rigged, wooden-hulled brig. He would have climbed twenty-five feet or so up the accommodation ladder to the deck. Although Isaacs must have seen hundreds of ships, this was likely the first he had set foot on. He would have had to step carefully among a confusion of ropes, tackle, and freight. The smell of tar and vinegar would have hung on the breeze. Men would have shouted as they clambered up the rigging and sung bawdy songs while heaving aboard the last of the outbound goods. Lines would have soared through clattering wooden blocks to the tapering masts. A hidden sailor's voice might have boomed through a speaking trumpet. Even if the boy had been excited to sail to parts unknown and become a man, the sudden riot of noise, the blur of bodies bustling around him, and the lurch of the ship must have left Isaacs longing for Canterbury's familiar lanes. Now his home was upon the deep.

The brig would have offered nightmarish confinement rather than freedom. Quarters were tight below decks, dimly lit by oil lamps, damp with seeping bilge water, noisy with the creak of timbers and the scurrying of rats, and pungent with the aromas of grease, tobacco, and unwashed bodies. Perhaps Isaacs was fortunate enough to have his own closet-sized cabin with a cot. If not, he would have slung his hammock along with the sailors, whose rough ways—drinking, cursing, gambling, feuding—would have initiated him into a masculine world of opportunism, risk, and casual violence, with little respite or opportunity to hide.[35]

Even the most private functions were public on a merchant vessel, as most lacked lavatories, except for the master's use. Common sailors made use of a toilet area at the front, the "head," of a ship. Waves regularly swept through the open-air outhouse to rinse the area clean. On deck, the roll of the ship and pitch of the horizon left passengers heaving over the rails. The constant cries of officers, the striking of watch bells, and the howling wind offered no peace. When becalmed, the crew set to tedious tasks beneath a pitiless sun and grumbled to one another about the endless inspections of rigging and the arbitrary discipline enforced by their superiors. At other times heavy seas, pounding rains, and sea swells crashing on the decks broke the monotony. Strong winds could take a grown man's breath away or sweep him from his feet. Rations of salt beef or pork, hardtack biscuits, dried peas, raisins, molasses, and tots of grog (rum diluted with brackish water from wooden casks) kept the crew working through twelve-hour shifts. Duty was hazardous and diversions few. Isaacs' passage was especially gloomy, as "calms and contrary winds" lengthened his voyage beyond expectations.[36]

Nathaniel's "affectionate mother" Lenie had entrusted her son to one Captain Johnson, the master of the *Margaret* and a man likely known to her brother from his Jamestown dealings. In *Travels and Adventures* Isaacs recalled that Johnson was in "a state of almost constant inebriety" during the voyage and repeatedly "forgot the civilities of a gentleman." He does not explicitly relate the master's actions, but his coded language provides a good idea of Nathaniel's distress: "I was too young to attempt to expostulate with a man of his propensities; and, as I was too weak to contend, I thought it discreet to succumb [and] endure the caprice and submit to the offensive behavior." Was the teenaged Isaacs assaulted or raped by the man meant to protect him? We cannot be certain. The language used to report sexual assaults at sea in the 1820s was restrained, in part because of victims' efforts to demonstrate a manly honor in the face of abuse and vulnerability.[37]

Adolescent boys like Nathaniel were targeted for sexual violence by seamen because they were easy to physically overwhelm, dependent on older crew members, and lacking in oceangoing experience.[38] He does not linger on what he endured under Johnson's all-too-watchful eyes or from his rum-rank attentions. But beneath the euphemisms with which he describes the voyage we can sense the grief of a child cast from his mother's embrace, initiated into a world of foul bodies and frightful risk, and bound for unknown shores. This may have been Nathaniel's first experience of vulnerability, one that taught him to despise weakness. Yet whatever he suffered at the hands of Captain Johnson did nothing to quench his thirst for adventure. He was intelligent, curious, ambitious, and blessed with a hardy constitution. Like many young men before and since, he found that the high seas offered him an escape to a male world where the physically strong and temperamentally bold could make their fortune. And so he turned his back on an ungrateful England and toward what he might yet make his own.

Cries of "Land ahead!" resounded aboard *Margaret* on Sunday, 6 October 1822, as the brig made St. Helena. Church bells rang out from St. James'. Isaacs' four-month ordeal was at an end. Nathaniel would still have still felt the motion of the sea in his legs as he made his way down the gangplank and along Jamestown's recently expanded Lower Wharf.[39]

By the time of Nathaniel's arrival, his uncle had restored his reputation and outlasted his bilious antagonist, Governor Lowe, who had departed the island the previous year. Solomon was a middle-aged man enjoying financial success and family life and basking in the reflected glow of his former association with Napoleon. The East India Company now considered Solomon a "respected merchant" who had taken great pains to educate his children. He would soon advertise his lodging house as the "only one on the Island which affords the convenience of Hot and Cold Baths," and he promised guests a variety of carriages for rent, as well as "superior Saddle horses." Travelers could purchase "Fancy Articles" from Europe, India, and China, while ship captains could "procure all kinds of Supplies at the shortest Notice" from Solomon's stores—all for a high price, of course.[40]

Saul Solomon's rise to prominence was due in part to his sharp instincts, in part to circumstance, and in part to a rapidly proliferating business technology: cheap credit, which enabled merchants to transact business across vast distances and over long periods of time. Financial innovations in the City of London radiated outward along global

FIGURE 1. Solomon & Company headquarters, Jamestown, St. Helena. Photo by author.

networks comprising family, coreligionists, and reliable friends who provided the necessary commercial intelligence to propel Jewish traders like Solomon, and later Isaacs himself, into a world of plenty.[41] Solomon made loads of money—literally. He and other partners minted the St. Helena halfpenny, which circulated for more than thirty years. The coins read "Payable at St. Helena by Solomon Dickson and Taylor." The East India Company countenanced this unsanctioned coinage.

Saul's first-born son would soon be granted an appointment by the company. His daughter Phoebe had recently returned from England to St. Helena and was now of marriageable age. After the death of his first wife, Saul had married Mary Chamberlain. She bore him a son in June 1822, another Nathaniel, this one baptized with the distinguishing middle name of Lee.[42] Another brother, Joseph Solomon, and his wife,

Hannah Moss, had also joined Saul and Lewis in Jamestown. And now his sister's son had arrived from Canterbury to assist him in the family business.

This aggregation of familial networks was common at the time and endures even today. In the world of business, ties of blood or marriage, shared religious conviction, or a common ethnic and national origin have long been thought of as checks against fraud. Transactions conducted over great distances and with imperfect knowledge of foreign markets required access to creditors who could bear a long wait for a return on their investment.[43] Jewish family networks filled these needs. That Jewish merchants and traders operated on the same principles that animated Quaker or Scottish concerns is thus unsurprising.

A portrait of Saul Solomon depicts him as clean-shaven and thin-lipped, with hazel eyes and an aquiline nose. His placid expression, coiffed gray hair, and white collar mark him as a gentleman of the first rank. The merchant king of St. Helena would later serve as consul for actual royalty, including the kings of the Netherlands and France. From his indulgent uncle Isaacs learned the fundamentals of the import-export business. A poor boy from a solemn corner of provincial England found himself suddenly in the subtropical sunshine of St. Helena and the luxuries of the Solomon home—the finest imported food, fashionable clothing, decorative flourishes from far-off Asia, the attendance of slaves. Isaacs' early years had been marked by deprivation; suddenly the glittering temptations of consumption were arrayed before him. Soon enough, he found his material cravings unappeasable. The "insipidity and monotony of the counting house" came to weigh on Isaacs as he sought a way to satisfy his desires.[44]

Looking back on this period, Isaacs described himself as grateful to his uncle, yet longing for a life that would suit his "fickle and inconstant" inclinations and his "strong predilection for the sea." He could hear the ocean calling him from nearly every point in Jamestown, while whispered tales of Bonaparte's gallantry circulated in the Solomon households, frustrating the ambitious young man who had arrived too late to meet him. His uncles' subterfuge on behalf of Napoleon must have offered a lesson in intrigue that would serve him well in later life. As Bonaparte's passions and thirst for power had been diminished during his confinement on St. Helena, so now Nathaniel found himself in a kind of captivity as well, galled at having to perform "irksome labors" for his uncle while dreaming of an "adventurous life" beyond the island's shores. Two of his young cousins had already sailed to England to receive a Jewish education at a well-known boarding school.[45] But Isaacs had no interest in returning to London or in pursuing a Jewish education. Nor would he remain a princeling at his uncle's table. He set his sights on other shores.

In June 1825, after nearly three years of being chained to his desk and suffering Jamestown's limited entertainments, Isaacs got his chance to escape. The *Mary*, a brig registered in Plymouth, dropped anchor in Jamestown's harbor carrying cargo for Saul Solomon. Her master was James Saunders King, a native of Halifax, Nova Scotia. King was born in 1795, enlisted as a boy in the Royal Navy, and by his own reckoning had served as a midshipman for about ten years in American and Canadian waters. Following the end of the Napoleonic Wars and a drawdown in naval strength, King worked his way into the British merchant service and cruised the eastern coast of Africa aboard the decommissioned troopship *Salisbury*.[46] His command of the *Mary* represented a step up.

King recognized in Isaacs a young man of similarly ambitious character, and the two became fast friends. Perhaps Isaacs found in King a father figure—or perhaps he was relieved to forge a relationship not built on familial obligation. King proposed that Isaacs accompany him to Cape Town, the seat of the Cape Colony and Britain's bridgehead in South Africa. Afterward they would sail up Africa's eastern seaboard to trade. Isaacs was hooked thanks to his natural "impulse of curiosity," not to mention the "attractions of commercial speculation." He resolved to sail on the *Mary* "for the purpose of gaining experience, and seeing the world."[47] Saul Solomon assented to the venture. Isaacs must have imagined that he, like his uncle, would make his fortune in trade. Like many adventurers of the day, he saw exploration abroad—especially in Africa—as a means to transcend the low and marginal status he endured at home.

On 28 June 1825, the seventeen-year-old Isaacs set sail with James King. They arrived in Cape Town's majestic Table Bay on 1 August, carrying goods from London and St. Helena. When King discovered that he had arrived too late in the season to hazard sailing up the Mozambique Channel, he and Isaacs agreed to trade whatever "cargo sundries" they could until fairer winds and weather returned. With its officials and government clerks, sailors and soldiers, artisans and shopkeepers, British and Dutch settlers, and many African residents—free and enslaved—Cape Town must have shocked Isaacs after his cramped adolescence in Jamestown. At the time, there were 8,248 white inhabitants, 1,870 free Blacks, and 7,076 slaves among Cape Town's total population of more than 18,000. The slave trade had been outlawed by Britain in 1807, but slave holding remained legal in British possessions. Records indicate there were 119,305 slaves in South Africa in 1825. Abolitionists fulminated against Britain's "slave colony."[48] Although emancipation throughout the empire finally arrived in 1838, slavery continued until 1843 under the jurisdiction of the East India Company in two locations: Ceylon (now Sri Lanka) and Saul Solomon's St. Helena.

Unlike Jamestown, with its modest, genteel facades, Cape Town boasted expansive administrative buildings, including Garden House, whose interior featured luxurious French wallpaper specially ordered by the ostentatious Governor Charles Henry Somerset.[49] Manicured formal gardens, a hospital, public houses, and shops offering goods from Asia, Africa, and Europe enabled social life and commerce to thrive. At the dockside, James King learned that an old friend of his from his time aboard the brig *Salisbury*, Francis George Farewell, had vanished during a voyage along the same route that King and Isaacs intended to follow.

Farewell, a sometime East India Company merchant, "had been absent for more than sixteen months on a very hazardous speculation . . . amongst the natives," King wrote in his journal. He feared that Farewell was being held captive as a curiosity, because "it appeared when on my former voyage on that coast, [these Africans] had never seen a white person." King was stirred to action. "I resolved on personally ascertaining [Farewell's] condition," King recorded, "and having discharged my vessel's cargo, I prepared for the voyage with my companion, Mr. Isaacs, and sailed from Cape Town on the 26th of August, with strong anticipations that I should effect the recovery of my long absent friend." This mission of mercy would have redounded to King's credit had Farewell actually gone missing. In fact, as King must have known from news dispatches, other vessels had already found and resupplied Farewell and his party.[50] King's putative search for Farewell could be cast in a far more honorable light had he chronicled the entirety of their travels together along that southeastern coast.

Born in 1791, Farewell was a few years King's senior. He had entered the Royal Navy around the time Isaacs was born and served throughout the Napoleonic Wars on various ships. He was injured fighting in the Adriatic Sea in 1811. No doubt brave, Farewell was also opportunistic and headstrong. In 1815 he left the navy with the rank of lieutenant, and like James King, he served on merchant ships during the British drawdown of naval forces. By 1822, Farewell had settled in the Cape Colony and married Elizabeth Catharina Schmidt. Retained by the navy on half pay as a merchant mariner, he represented a fair matrimonial prospect: he was able to grant his new wife the considerable sum of £3,000 on their marriage. Farewell had a high forehead, wispy hair, a bulbous nose, and a youthful countenance marked by a deeply cleft chin. He and other merchants in Cape Town formed a partnership with the aim of establishing a trading outpost somewhere along the eastern shore of South Africa. Farewell chartered James King's 120-ton *Salisbury* to undertake a reconnaissance voyage, and in June 1823 the two men and their crew set sail.[51]

While anchored in Port Elizabeth's Algoa Bay, they met Captain William Fitzwilliam Owen, whose Royal Navy squadron was engaged in surveying the coast. King and Farewell hoped Owen would be able to provide them with geographic information and maritime charts to help them locate a site for their commercial venture. The generous Captain Owen handed over a chart of the coast adjacent to the as-yet unmapped Port Natal region—today's Durban—on condition that James King reciprocate with any charts he might make in the future.[52] Possibly King knew Owen, or at least of him, from their time in Canadian waters. Owen had been commissioned to conduct a hydrographic survey of the Great Lakes from 1814 to 1816, and he and King would have known some of the same captains, vessels, and ports of call.

While aboard Owen's flagship, James King was recognized by a figure from his much more recent past. Government administrators had previously chartered King's "tight, staunch and substantial" *Salisbury* to ferry cargo, troops, and criminals to and from Cape Town. On one of these trips, King met Jakot Msimbithi, a Xhosa man who spoke some Dutch. Msimbithi had been captured by Boers while a youngster and dragooned into serving a Boer master. He later escaped and avenged himself by guiding his countrymen to Boer homesteads and stealing their cattle. During one of these raids, he was captured, tied to the back of a horse, and forced to run after the animal. When Msimbithi stumbled, he was flogged with a sjambok —a heavy whip typically made from hippopotamus or rhinoceros hide. After escaping his tormentors once more, he again rustled Boer cattle until he was rearrested and transported from the frontier to Cape Town, and from there sent to Robben Island, where Nelson Mandela would later be incarcerated for eighteen years. The windswept convict station was a place of misery and harsh labor. Inmates in the 1820s received only one blanket per year despite the cold, and, like Mandela after them, performed the grueling work of "quarrying, sawing, and polishing stone."[53]

Aboard the *Salisbury*, Msimbithi made an impression on King, who "knocked off the irons with which [Msimbithi] was manacled, gave him clothes, and . . . indulged him with an allowance of grog." Jakot, later referred to as Jacob, was subsequently freed by British authorities to act as an interpreter and was pressed into service on Owen's survey mission. Msimbithi recollected James King's generosity when he recognized him aboard Owen's ship, and after obtaining Owen's consent to transfer vessels, he eagerly joined Farewell and King as their interpreter. Owen must have released Msimbithi reluctantly: he depicted him as a favorite among the crew and a man of stature, "having been a Chief 'famed for deeds of arms.'"[54] Whether or not he had been a chief, Msimbithi was a character of great physical strength and superior intelligence.

With Msimbithi now aboard the *Salisbury*, Farewell and King set out to find a harbor from which they could establish a base for trade. They hugged the coast until sighting a promising bay, St. Lucia, nearly a thousand miles northeast of Cape Town. On 4 August 1823 the crew launched a landing craft filled with cargo: bangles and beads, a bale of cloth, and eight muskets. Heavy surf capsized the small boat, and all the goods were lost. The crew tried again eleven days later, but this second cargo of 196 bangles, 600 pounds of beads, 4,320 buttons, and two dozen knives also went down in strong seas. Farewell, who could not swim, tumbled into the swirling currents. Msimbithi, a strong swimmer, rescued him and swam with him to shore. At least three other men in the party drowned. As Farewell and the other survivors panted for breath on the rough coastline, a charter member of Farewell's trading company blamed Msimbithi for the accident and struck him. Furious at yet another instance of white perfidy, Msimbithi fled into the bush and left the Europeans to their fate. Meanwhile, the *Salisbury*'s anchor cable was cut in a sudden gale, and the ship was blown out to sea. For five weeks the *Salisbury* impotently cruised offshore while local inhabitants provided the hospitality that kept Farewell and the ill-fated landing party alive. Later the survivors claimed that the local people were "extremely inoffensive" and had "never seen a white person before."[55]

Once rescued by King and brought back aboard the *Salisbury*, Farewell and the others turned southwest, back toward Port Natal, which Owen had been unable to survey. This uncharted destination seemed a good bet for two enterprising mariners who sought to stake a claim. As the *Salisbury* approached the narrow entrance to the natural harbor, the changeable currents bore her on, and the howling winds of the Indian

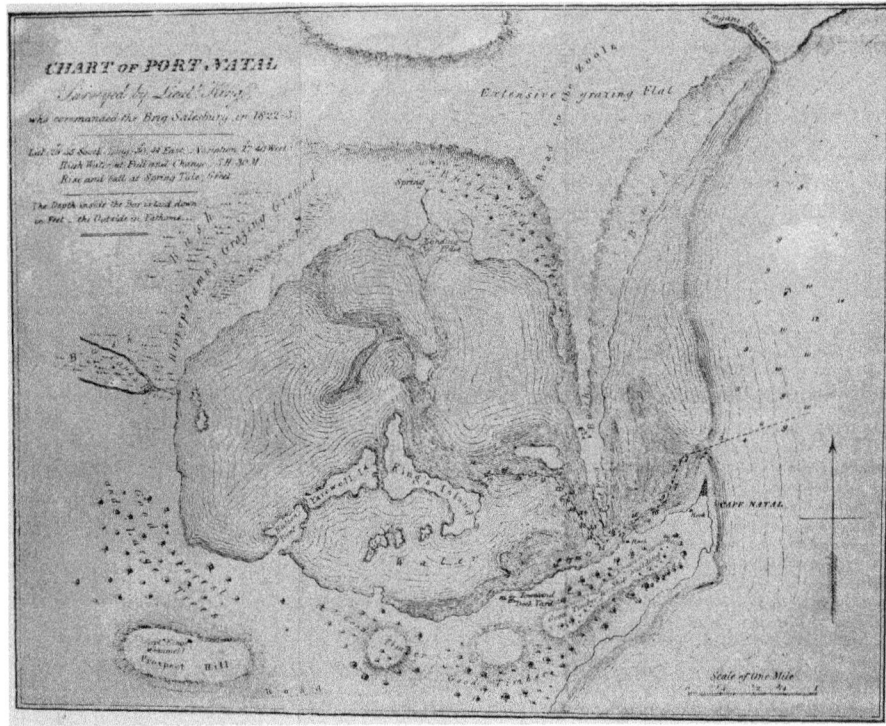

FIGURE 2. James King's chart of Port Natal, ca. 1823. Killie Campbell Collection, University of KwaZulu-Natal.

Ocean tore at the sails. By skill and chance the brig negotiated treacherous shoals and sailed into the broad bay. King set about sounding the waters and surveying the resources of Port Natal. He named a bank of fruit trees bordered by a mangrove swamp Salisbury Island, after his ship. The area was bordered by a "rich plain" that was plentifully supplied with freshwater, as well as stands of timber and "good thatching reeds." King recorded that the harbor "abound[ed] with hippopotamus" and that the country was "beautiful & blest with a salubrious air & a productive soil."[56] Armed with a sort of treasure map of Port Natal, King and Farewell returned to Cape Town, each with his own idea for colonizing the bay.

Farewell formed the Farewell Trading Company with Jan Peterssen and Josias Hoffman, Cape Town merchants. While the colonial governor encouraged Farewell and his partners to "establish a commercial intercourse and lay the grounds for civilizing the inhabitants," he sternly warned that His Majesty's Government could not "sanction the acquisition of any territorial possessions." Thanks to connections and word of mouth, Farewell and his partners persuaded twenty-five settlers to join in the venture, including fortune-hunting Englishmen, a number of hardy Boers eager for farmland outside British jurisdiction, and three "Hottentot" servants.[57]

The etymology of *Hottentot* remains uncertain. Today the term is archaic and insulting, but in the early nineteenth century, it was often employed to describe the Khoekhoe people, Indigenous nomadic peoples of southern and southwestern Africa. The Khoekhoe were thought by many Europeans to have Jewish origins. The term *Hottentot* came to connote a lack of civilization and was applied to other peoples in Africa. But as the South African writer Stephen Gray reminds us, the term was also used broadly to refer to "armed 'coloured' Christian converts . . . who played . . . an intermediary role between the whites and blacks" in the nineteenth century.[58]

Another problematic word is *Kaffir*, which is spelled in a variety of ways in English. In both Arabic and Hebrew the term refers to an unbeliever, infidel, or heretic. Muslims in East Africa used it to refer wholesale to Black Africans. Centuries later the word evolved into a slur referring to Blacks in South Africa. In nineteenth-century English, however, the term had valences ranging from descriptive to pejorative. The designation *Caffraria*, for example, was used to indicate an imprecise region beyond full British control and partially encompassing the Zulu heartland. Farewell, King, Isaacs, and other chroniclers all used the terms *Hottentot* and *Kaffir* in ways that, while not always offensive, carry connotations of racial paternalism. Though I am conscious of the potential insult of the terms, I retain their usage to represent the era's words and outlook: assumed cultural and racial superiority is part of the story of European and African contact. And bear in mind that the term *boor*, derived from the Dutch *Boer*, though not racially charged, is not exactly a compliment.

The Farewell Trading Company's pioneers set off for Natal in two ships they had chartered, the *Antelope* and the *Julia*. James King did not accompany them. Instead he sailed for London, where he hoped to distinguish himself in the eyes of the Admiralty with his prized chart of Port Natal. Once the Farewell Company's *Antelope* had landed its supplies and the colonists at Port Natal, it returned to Cape Town. The *Julia* remained behind and later transported back to Cape Town a number of disgruntled homesteaders who found "that the country and the natives were different from what was told them" by Farewell. On a second return journey from Port Natal in late 1824, the *Julia* caught fire with eleven returnees aboard. Its two hundred pounds of gunpowder exploded, and flames consumed the sloop and its cargo of dried animal skins, six elephant tusks, forty-nine hippopotamus teeth, and two dozen bull hides. All hands were lost. With the *Julia* reduced to charred staves, Farewell was stranded ashore near Port Natal with a few Englishmen, his Hottentot servants, and paltry supplies.[59]

James King, meanwhile, had sent ahead a letter to the secretary of state for war and the colonies offering to provide the government with his chart. King detailed the natural beauty and commercial benefits that Natal offered, also suggesting the potential for finding gold nearby. He reassured his superiors about the attitude of the inhabitants, who at first "appeared rather hostile . . . being armed with their assegai [spears] & shield but when we became better acquainted they were extremely well disposed & expressed a particular desire for us to remain among them." Once arrived in London, King asked a high-ranking friend to make introductions for him so that he could present his survey of Port Natal to the Admiralty.[60]

On learning of King's machinations, Owen was indignant that King had breached his promise to share his discovery and charts. In the Admiralty's offices, King offered to surrender his important chart to the Royal Navy in exchange for being made a lieutenant. In response to this impudent proposal, according to Owen, King was "dismissed . . . in the most contemptuous manner." A young sailor who recollected these events years later insisted that in fact King was a "strictly honourable man" who was "incapable of committing so base an action" as Owen claimed. This sailor, who subsequently authenticated King's charts of Port Natal, was convinced that Owen had been misled into accusing King of duplicity. Nonetheless, Owen received the credit for surveying Port Natal, even though he apparently had never entered the harbor at all. A frustrated and spurned King simply bestowed on himself the rank of lieutenant that he felt was rightly his and vowed to return to exploit his and Farewell's discovery. That journey took him almost a year, by which time "Lieutenant" King feared that Farewell had been murdered.[61]

In August 1825 James King informed the governor, Charles Henry Somerset, of his and Isaacs' intention to locate Farewell and his party of castaways. He requested two small cannons, a bevy of muskets, and a store of ammunition, which were all duly granted "for the security of his vessel." Isaacs and King arrived at Port Natal on 1 October 1825.

At the mouth of the harbor, King sent over a whaleboat to sound the sand bar. Just as the boat returned, a gale began to howl and thrum through the *Mary*'s rigging. Those in the whaleboat just managed to scramble back aboard, but they assured King that as the tide rose, the *Mary* could pass over the bar if her load was lightened. King, with "great coolness and intrepidity," ordered some of their supplies thrown overboard.[62]

The waters crashed over the ship's decks while King took the conn. Isaacs secured himself with rope, fearful of being washed overboard. The *Mary*'s prow heaved upward and crashed down with the tidal surge. Isaacs spotted rocky shallows to one side and the placid harbor ahead beyond the narrows. The ship rolled violently, but King maintained control, bellowing above the storm a refrain of "Steady!" and "Meet her, meet her!" to his helmsman. Then a powerful wave broke over the vessel, and it lurched down into the "hollow of the sea." For the first time the crew panicked, screaming, "She is going!" Forced down by the wave, *Mary* struck the rocky bar. Yet King was as relentless as the sea. He cheered his crew with cries of "Press her over, my hearties!" until the waves turned the ship broadside and overwhelmed it.[63] The rudder broke, and there was nothing more they could do to save the vessel from broaching.

A steely King ordered his men to the boats. But during the maelstrom the narrow whaleboat's planks had splintered against the *Mary*'s side. The one remaining evacuation vessel, a longboat, was lowered, and the crew climbed down a rope, timing their descent to meet the surge of the sea. One sailor misjudged the tidal retreat, which sucked the longboat with it, and he fell into the roiling water. A loyal Newfoundland dog plunged into the waves, grabbed the man by his red shirt and kept him afloat until his fellow mariners could heave both man and dog into the longboat. The raging seas battered the small craft, and the men had to bail it out while rowing against the squall to keep clear of rocks along the shore. The *Mary*, meanwhile, was pushed over the bar and beached on an inner bank. Tools, clothes, personal items, and many of the remaining stores were washed overboard.[64] Now Farewell's would-be rescuers were marooned themselves.

This was James King's first shipwreck in eighteen years of seafaring. He surveyed the shoreline with horror. "Not a living soul was to be seen on this desolate coast," he wrote." Isaacs too feared being "cast in the days of my youth amidst a people whom I imagined not humanized." In these moments of peril, his thoughts turned to Canterbury and to the "fostering care" he had received from his mother.[65] As for many travelers, it was Isaacs' thoughts of home that firmed his resolve to face whatever dangers might await.

Those dangers seemed legion. There was no sign of habitation. All Isaacs could see about him was a wilderness that convinced him that Farewell and his party "had all fallen by . . . savage hands." He scanned the horizon in despair. Later that afternoon, the crew spied a group of eight people approaching them from the opposite side of the bay, waving a tattered Union Jack. Isaacs signaled to them. But as they neared, the shipwrecked mariners noted with apprehension that some of the men were completely naked. The others—including an African woman wearing a "dungaree petticoat" and "blue handkerchief" on her head, a male dressed in European rags, and a woman only wearing a hide about her waist—all seemed hostile. King and Isaacs concluded that Farewell and his party had been massacred "and that the people we descried sought to decoy us ashore, where they would . . . make us their captives, and devote us to the gratification of their savage propensities."[66]

Isaacs found savagery everywhere he looked. The first volume of his *Travels and Adventures* features some form of the word *savage* seventy times. As he grew accustomed to his African exile, however, he warmed to the people and places he encountered. The European who expected to find a savage behind every tree was himself humanized by contact with the peoples he at first feared were less than human. Ironically, Isaacs would have been considered something of a savage in British culture of that time. Had he tarried in Cape Town, he could have attended a performance of Thomas Dibdin's popular farce *The Doctor and the Jew* (1798), in which the venal Jewish antagonist is called a savage.[67]

King and Isaacs rowed closer to get a better view of those who hailed them and discovered that one of the group was Halstead, a young member of Farewell's group of settlers. All members of the party were alive, they learned. Farewell himself was absent, having set off on a visit to the region's king. Others of the group, including Henry Francis Fynn, had gone in search of ivory. The son of a hostelry owner in the Cape Colony, Fynn was determined to make his fortune in trade. He was also dogged by suspicions that he was a fugitive from the law.[68] After ascertaining Farewell's fate, King returned to the wreck of the *Mary* with his crew to salvage whatever supplies and timber they could. Isaacs, relieved to be on land, joined Halstead and a Hottentot woman, Rachel, on the long tramp toward Farewell's camp.

Rachel warned Isaacs not to venture out at night because of attacks by African wild dogs (Isaacs calls them wolves) and leopards (Isaacs calls them panthers), which had recently "carried off two natives." Another sailor who spent that first night ashore later recalled that "about midnight one of our finest men was snatched away" by a beast. Rachel appeared to have been made Farewell's second-in-command. She enjoyed great respect from the local people and ordered them about as her servants, occasionally swiping at them with a cane when they did not jump to do her bidding. The exhausted castaway Isaacs dined on milk curds and bread and retired to sleep under "strips of bullock's hide." There he could "enjoy the sweets of repose, such as even monarchs might envy."[69] He could not possibly have dreamed that in the following days he would wake to find himself a guest of one of the most famous and feared monarchs in African history.

CHAPTER 2

Strange Surprising Adventures, 1825–1827

Nathaniel Isaacs was now marooned on unfamiliar shores. Winds roared in from the Indian Ocean, driving breakers that pounded the coast. Cloudbursts and electrical storms raged over the harbor. Isaacs huddled in Farewell's meager shelter while predatory animals roamed outside. He feared that Black Africans might devour his flesh. Describing the experience in *Travels and Adventures*, Isaacs invoked William Cowper's popular poem imagining a castaway's isolation: "Better dwell in the midst of alarms / Than reign in this horrible place."[1] The words are ascribed to Alexander Selkirk, who had survived more than four years alone on a remote island in the South Pacific in the early 1700s. His story was celebrated in Great Britain and fictionalized by Daniel Defoe in *The Life and Strange Surprising Adventures of Robinson Crusoe* (1719). In Defoe's novel, Crusoe was wrecked during a speculative voyage to Africa, where he intended to trade beads, baubles, and metal tools for ivory, gold, and slaves. Washed up alone on his island, Crusoe set about domesticating nature and battled fearsome cannibals, the last a tale that surely fed into Isaacs' panic that the natives might have him for dinner. The reference to Cowper's poem highlights a self-conscious effort to claim the mantle of literary author as well as shipwrecked adventurer.

We might indulge Isaacs these romantic flourishes. The real-life Selkirk was cast ashore at the age of twenty-eight, the fictional Crusoe at twenty-three. Nathaniel Isaacs was just seventeen. Even so, he was not

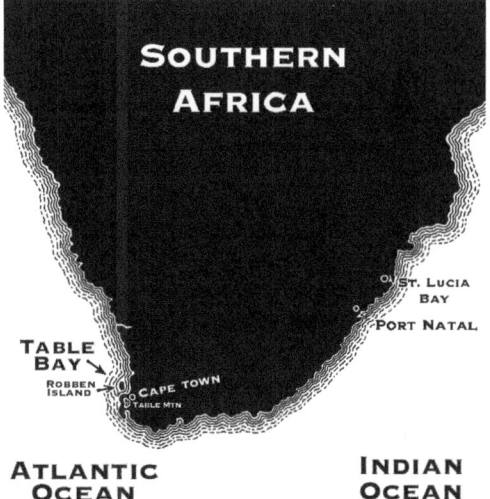

MAP 3. South Africa.

the youngest of the shipwrecked mariners. Captain King's apprentice, Charles Rawden Maclean (also known as John Ross), was a ginger-haired Scottish boy about thirteen years old.[2] Although, unlike Selkirk or Crusoe, Isaacs and Maclean had British companions, they had no way of knowing how long they would be marooned on the edge of the Zulu Kingdom.

Like Isaacs, Maclean penned an account of his time among the Zulu. Maclean's memoir, published in installments nearly twenty years after Isaacs' *Travels and Adventures*, provides valuable correctives to Isaacs' narrative. Based on his diary, Isaacs' published work remains histori-cally significant—and contentious—because he was the first European eyewitness to depict life under King Shaka.[3] He described Zulu organi-zational hierarchies, internal politics, weapons and military tactics, family dynamics, rituals and rites, clothing, dwellings, and foodways, all with an ethnologist's eye for cultural distinctions and a novelist's sense of pacing. Whether he did so accurately is a vexed question, and the debate among historians of the period of first European contact in South Africa remains acrimonious.

Popular novels and influential histories from the late nineteenth and early twentieth centuries relied on Isaacs' most sensational passages to depict a brilliant and bloodthirsty King Shaka. Such accounts served the interests of the white minority in South Africa and later emerged as justifications for colonialist and apartheid-era policies of repression. Yet as the South African historian Carolyn Hamilton has observed, colonial depictions of Shaka "were also shaped by the form and content of various African views which they encountered . . . during their stay in Natal." African peoples possessed their own ideological aims in depicting Shaka. Some of these projections of a mighty warrior-king later served African (and specifically Zulu) nationalist efforts to enshrine Shaka as a symbol of anticolonial resistance. Moreover, the Zulu recovery of Shaka's legacy has at times been reactionary, ethnically chauvinist, and elitist in ways that sometimes aligned with apartheid-era policies. Nevertheless, both European and African resistance histories base much of their depiction of the early Zulu Kingdom on Isaacs' portrayal of Shaka's rule. [4]

My own reliance on Isaacs' words is not uncritical. Published and unpublished accounts of other Europeans present at first contact illuminate what Isaacs got right and what he got wrong and help to expose the agendas he sought to further with the publication of *Travels and Adventures*. I draw on African oral traditions to amplify or challenge Isaacs' recollections. The Western notion of "oral traditions" itself suggests a hidebound discourse, obscuring the complex and dynamic origins of these narratives and the varied ends that they serve.[5] The Zulu had a rich oral history. Some of its accounts are consistent with Isaacs' portrayal of Shaka; others diverge from his reports.

Following the publication of Isaacs' work and the appearance of European-mediated versions of oral traditions whose informants (and compilers) may have been influenced by *Travels and Adventures*, insider Zulu history and memory has been inextricably entangled with outsider narrative. A "pure" history is thus irrecoverable. Though at times Isaacs was inaccurate and biased, overall his travelogue is supported by other primary documents, sometimes to a surprising degree.[6] His work offers a detailed portrait of European-Zulu interactions that, in contrast to many later nineteenth-century travelogues, grants interiority and dignity to African individuals and cultures.

Just after the wreck of the *Mary*, Isaacs felt that he was "an almost solitary European, wandering occasionally I knew not where, and in search of I knew not what." In the first shocking days of their stranding, James King provided his crew with purpose. He sent men to the foundering ship to rescue what they could of its stores. They created makeshift tents out of its sails and recovered barrels of beef, pork, wine, and spirits. They also made use of what their surroundings had to offer. Natal Bay provided plenty of food, including fish and turtle. The Indigenous inhabitants associated turtles with "everything that is nasty and abominable," and when the Europeans feasted on the creatures with gusto, the locals "were quite disgusted."[7]

Though they would not starve, the crew lamented that few of their personal effects could be saved. The only clothes Isaacs possessed—the ones he was wearing—were in such tatters that he was "almost in a state of nudity." Farewell's party had been reduced to a similar state.[8] Natal Bay's Indigenes wore very little to begin with.

Although King attempted to keep order, his authority waned on land. After crewmen ransacked the cache of spirits rescued from the ship, King poured out every last dram of grog into the harbor. Furious at the loss, three of his men commandeered a sixteen-foot longboat they had salvaged, deserted their shipmates, and aimed to sail for the Cape Colony, six hundred miles distant. They miraculously arrived unharmed. Back at Natal, King instructed his remaining men to build a more seaworthy vessel from the remaining longboat and from the stands of hardwood near the harbor. On this makeshift craft they intended to sail for the Cape. Yet they knew the task might take years. Many of the carpenter's tools were lost in the wreck, though Farewell apparently possessed a set.[9]

On Sunday, 9 October 1825, King conducted prayers for his crew and exhorted them to treat with respect the curious Africans who had gathered around them. Isaacs himself believed that setting "a good example to the natives, who sought to mingle with us, and to watch every action, might be productive of good." And he did try to help. A man suffering from a swollen throat that prevented him from swallowing begged Isaacs' assistance. Few items remained in his medicine chest after the shipwreck, but Isaacs dosed the man with an antacid, rubbed a liniment—probably a mixture of soap, alcohol, and camphor—on his neck, and bound a cloth over the affected area. He also asked the African women to prepare a folk remedy for his patient, one Isaacs would have been familiar with from his mother's kitchen: chicken soup. At first the man recovered, and Isaacs was hailed as a healer. But despite his white magic, his patient soon succumbed.[10]

Isaacs reported that the Zulu scrutinized him with a frank and disturbing curiosity. As a Jew in Georgian England, he would have been familiar with such treatment. Indeed, Jews in Great Britain were often equated in rhetoric and imagery with Blacks, and hence, according to the prejudices of the day, cast as primitive. As the scholar Sander Gilman has observed: "In the eyes of the non-Jew who defined them in Western society the Jews became the blacks." Jews and Blacks, as well as other marginalized peoples such as Irish immigrants, were represented in a way that aligned social class with a purportedly natural hierarchy of beings. An engraving by William Hogarth (*Harlot's Progress*, plate 2, 1732) offers a clear instance of Jews aping upper-class British manners.[11] Hogarth uses cross-hatching to depict a nouveau riche Jew as darker-skinned than his fair English mistress. The Jew's white powdered wig, signaling his pretensions to class and racial purity, is given the lie by his jet-black eyebrows. His mistress overturns a tea table with a kick of her mischievous leg. The Jew's pop-eyed surprise at her sudden movement duplicates the wide-eyed dismay of his Black child servant, whose expression echoes that of the pet monkey at the Jew's feet to suggest that Jew, Black, and simian are all akin. And so before the Jew could become a king, Isaacs would first have to become a white Englishman. That would not be easy, for according to the notions of that time, the truth of racial origins would out.

The British physician James Cowles Prichard, whose work constitutes an early physical anthropology, noted that depending on the climate in which Jews lived, they could become swarthy, even black. One implication of Prichard's conclusion was that environment and culture shaped race, which was thought to be a more mutable category of identity than we often consider it today. Another contrasting implication was that Jews were shapeshifters who might pass as white in cloud-covered England, but who concealed within themselves a nigrescence that would reveal itself in sunnier climates. Anti-Jewish sentiment in Europe had long considered the Jew as black metaphorically and physically, and this view became entrenched in later nineteenth-century racist pseudoscience, with Jews and Blacks occupying a lower stratum of human development and potential.[12] To Europeans, certain African peoples were imagined to be the remains of degenerate Israelite "lost tribes."

For the Zulu, the bedraggled whites who washed up on their shores were as alarming and fascinating as the half-naked Zulu were for the whites. In these early years of exploration and settlement, power lay with the Zulu. They referred to these whites as "beasts from the sea," connoting their grotesque helplessness in Zulu territory. The Zulu accompanied this description with an obvious "gesture of opprobrium," according to Isaacs. Maclean reports that the Zulu thought the white men were a mollusk-like species, with "flesh soft and pulpy like that of any oyster or shell-fish." Some Zulu pinched the Europeans to make certain that they were flesh and blood. Their hair was likened to the color of maize. Isaacs, Maclean, and the other shipwrecked whites were later known in isiZulu (the language generally known as Zulu) by a host of unflattering epithets. They were called tricksters or magicians (*abalumbi*), "the little red ones" (*ababomvana*), or "the little wild beasts" (*izilwanyana*).[13] Later they came to simply be called *abelungu* (sing. *umlungu*), "whites," a term that today may carry negative connotations in South Africa.

Isaacs viewed the first Zulu chieftain he met, Enslopee, as having a gift for mimicry when he impersonated James King.[14] Enslopee, in turn, seems to have thought that Captain King resembled a Zulu warrior, suggesting that he needed only to have black skin and wear animal hides to blend in. Less graciously, Isaacs described Enslopee as having a face reminiscent of a baboon. King's recollections of his encounters with the Zulu are free of this sort of crude racialism. He described them as "a tall athletic well-looking race, extremely clean, and very respectful." But he also found them "always in readiness for war . . . and [they] have a great thirst for the blood of their enemy." Maclean, too, believed the Zulu to be "well made, robust, muscular, and powerful [and] with the exception of the colour of the skin they might justly rank with the most perfect European."[15] These initial meetings make it clear that the physical and cultural differences between Europeans and Zulu were mutually acknowledged, at times mutually disparaged, and at other times mutually admired. What remains intriguing today is the emphasis Isaacs, King, and Maclean placed on surface distinctions—clothes, skin color, musculature, cleanliness—rather than on an imagined racial essence that might imply white superiority. In other words, both Zulu and Europeans acknowledged that they resembled one another in their fundamental features and thus that they could to some degree transform themselves through their encounters with the other.

Word soon reached King Shaka of the clumsy arrival of the *Mary*. Shaka ordered his chieftains to bring the "beasts from the sea" to him. Farewell and King made the first foray to Shaka's *umuzi*—a settlement, typically oval or horseshoe in shape and surrounded by fencing or brush, with a paddock for cattle and other livestock at its center. The site of Shaka's capital at kwaBulawayo lies in the heart of what is today often called Zululand, not far from the village of Mashishi. The journey to Shaka's royal residence—known as an *ikhanda*, a kind of sprawling *umuzi* plus military headquarters—required a journey of about 120 miles over unmapped territory, crossing rivers teeming with crocodiles and hippopotamus.[16] The trip to kwaBulawayo and back took more than two weeks.

While fording a river during their return trip, Farewell lost his balance and was swept downstream. Farewell's African escorts dove in to rescue him from the crocodile-infested waters and helped him struggle to safety from a reed-filled eddy. He and King later returned to Port Natal driving a gift of cattle from Shaka and bearing promises of the ruler's protection as well as stories of his fierce mien. The reports excited Isaacs.[17] This time he was determined not to miss his chance, as he had with Napoleon, to meet a powerful sovereign face to face.

Armed with a fowling rifle, Isaacs set off for kwaBulawayo at Farewell's request on 29 November 1825. His first duty was to collect a store of perhaps four tons of ivory that Farewell had accumulated in the interior through trade and hunting. The value of ivory often outstripped even that of gold. Elephant tusks arrived in Europe and North America by the tens of thousands to be shaped into dominoes, shaving brushes, buttons, doorknobs, toothpicks, pistol grips, combs, piano keys, billiard balls, and decorative figurines fit for the parlors of a growing urban bourgeoisie. The trade in ivory, like the trade in slaves and arms, was a global phenomenon made possible by the emergence of an industrial economy. To procure the ivory, African peoples were enslaved or coerced by European, African, or Arab taskmasters into hunting elephants with imported firearms. Their rapaciousness was abetted by the support in England for free-trade policies.[18] Free trade, which demanded consistent access to emerging markets, was believed to benefit not only the English but also the supposedly benighted peoples the world over who provided the raw materials and labor. Farewell and Isaacs would have been convinced that they were extending British prestige and advancing Black South African interests while enriching themselves.

Isaacs' task was to supervise the ivory's transport to Port Natal by sixty Africans recruited by Farewell. Each of the principal Europeans in the party had collected a number of followers who attached themselves to the shipwrecked survivors out of curiosity, necessity, or the hope of rewards for their labor. Some of these retainers were remnants of peoples defeated and amalgamated by the Zulu. These outcasts were marginalized among their conquerors. Others were coastal peoples denigrated by the Zulu as *amalala*, those who sleep with their fingers in their anus. The whites at Port Natal therefore encamped amid a heterogeneous and despised population whom they recruited for labor. The whites "exercised the most absolute control" over this population; in this they acted just like Zulu headmen. Along their route to the cache of elephant tusks, Isaacs and his men found hospitality at various *imizi* (the plural of *umuzi*). Isaacs noted that the Zulu women and girls he met were attractive.[19] Some were so intrigued by the white teenager that they made nocturnal visits to his hut. He recorded being annoyed by these interruptions, though possibly he was secretly pleased by their feminine attentions.

Once Isaacs located Farewell's ivory and sent it back with the porters, he continued to King Shaka's *ikhanda*, stopping to bathe in a stream in preparation "for entering the capital of this barbarous sovereign." He arrived in kwaBulawayo on the night of 3 December and was taken to meet Shaka. Indistinct figures emerged now and again from the smoke and shadows as Isaacs was led along by firelight. Isaacs found Shaka seated on a mat near a half circle of headmen. Isaacs took his seat on the ground about twenty yards from the king, who soon ordered him to approach. An overawed Isaacs had to be thrust forward. The king made meaningless marks on a scrap of paper, waved them in Isaacs' face, and demanded that his guest interpret them. When he could not do so, Shaka laughed and dismissed Isaacs contemptuously.[20] Zulu existed only as a spoken language at the time, and King Shaka ridiculed the act of writing and the *abelungu* who attached such importance to the scratchings.

The crestfallen Isaacs did not realize, or neglected to mention in his published account, that Francis Farewell had introduced Shaka to the written word more than a year earlier. Shortly after his abandonment at Port Natal in 1824, Farewell visited Shaka at his capital. The Zulu king, Farewell reported, was happy that he and the other whites intended to settle in the area. A Zulu elder later confirmed that Shaka "deliberately made friends of the first settlers at the Bay [and] not only permitted them to settle at Port Natal and trade with his people [but also] made them a grant of a large piece of land in the neighborhood of the Bay." In August 1824, Farewell prepared a written grant for Shaka to sign. That document elevated Farewell to a chieftaincy and put him in possession of a large swath of territory. Although the governor had explicitly warned Farewell before he sailed against acquiring "any territorial possessions," Farewell claimed the territory and cheekily informed the governor that he had "hoisted the English colours" in the name of His Majesty.[21]

He then supplied the colonial government with a copy of the document signed by Shaka:

> In the presence of my Chiefs, and of my own free will, and in consideration of divers goods received, [I, Shaka] grant, make over, and sell unto F.G. Farewell and Company, the entire and full possession in perpetuity to themselves, heirs, and executors, of the Port or Harbour of Natal . . . together

with the Islands therein and surrounding country, as herein described, viz.: The whole of the neck of land or peninsula in the south-west entrance, and all the country ten miles to the southern side of Port Natal, as pointed out and extending along the sea coast to the northward and eastward as far as the river . . . now called "Farewell's River," being about twenty-five miles of sea coast to the north-east of Port Natal, together with all the country inland . . . extending about one hundred miles backward from the sea shore with all rights to the rivers, woods, mines, and articles of all denominations contained therein. . . . I, of my own free will and consent, do hereby . . . acknowledge to have fully consented and agreed to . . . and perfectly understand all the purport of this document, the same having been carefully explained to me by my interpreter . . . before the said F. G. Farewell, whom I hereby acknowledge as the Chief of the said country, with full power and authority over such natives that like to remain there.

The long-winded document was signed by Shaka, who was identified as "King of the Zulus," using a mark. This encounter with handwriting explains why Shaka had jeered at Isaacs' inability to read his meaningless fireside scribbles. The king was mocking both Isaacs and the technology of literacy, and perhaps thereby also signaling his indifference to his agreement with Farewell. Shaka's mark was as much a sign of contempt as of consent.[22]

Until the postapartheid era, Farewell's land grant was treated as proof that the Zulu had invited the colonizers into their land. More recently, the document has been dismissed as a swindle perpetrated by Farewell at the expense of Zulu sovereignty. Evidence exists to support both positions, but the dispute obscures a third possibility: that Shaka shrewdly exploited the whites' reliance on facts on paper to suit his own efforts to create facts on the ground. Shaka possessed a nimble strategic mind, and there is evidence that he used the presence of the settlers to consolidate his rule and extend his dominion and trade. Europeans were not the only ones interested in expanding their political, cultural, and commercial horizons.[23]

Shaka was an empire builder who, Zulu elders later insisted, "established colonies like the Europeans." As he conquered neighboring chieftaincies, he assimilated the young men into his swelling ranks of *amabutho*—well-drilled bands of warriors—and incorporated their lands and cattle into his realm through political confederation and cultural assimilation. Certainly a Zulu understanding of territorial rights would not have coincided with Farewell's idea of private property as recorded in the land grant. Farewell may have sincerely thought he would own the territory and maintain the right to settle future colonists in the region. However, Shaka would have understood his grant as simply allowing Farewell to establish what would become a Zulu vassal settlement. Like other peoples incorporated into Shaka's growing polity, Farewell would have had to supply soldiers and tribute (*khonza*) to the royal house. While Farewell viewed himself as chieftain of an independent territory, Shaka would have viewed Farewell as the headman for a ragtag clan of "beasts from the sea."[24]

The morning after his arrival at kwaBulawayo, Isaacs presented Shaka with "twelve brass bangles and a bottle of sweet oil," the latter a liniment for soothing bruises. Shaka insisted that Isaacs rub his legs with the oil, an intimate honor. Then, through an interpreter, he and Shaka discussed their respective kingdoms. When Shaka learned that Great Britain possessed the mightiest empire among all the *abelungu*, he proclaimed: "King George and I are brothers; he has conquered all the whites, and I have subdued all the blacks." One aspect of the comparison nagged at Shaka, however. He asked Isaacs whether "King George was as handsome as he was."[25] Isaacs prudently flattered Shaka.

Isaacs does not describe Shaka's physical appearance in his travelogue, though an earlier article that he likely authored comments that "it is reported, that [Shaka] is . . . of white extraction," and descended from people described as having "a Hebrew expression, but *uncircumcised*." No evidence is provided for the bizarre claim that Shaka had white ancestry. However, the reference to his "Hebrew expression" suggests that Isaacs found in the Zulu king a reflection of his own countenance, if not of his excised foreskin. In *Travels and Adventures*, Isaacs reproduces James King's description of Shaka as "upwards of six feet in height, and well proportioned." King and Farewell both record Shaka as being about forty years old.[26]

No contemporaneous likenesses of Shaka exist. The closest we possess may well be the illustration by the artist William Bagg in Isaacs' first volume. It shows Shaka in his battle dress of "monkeys' skins, in three folds from his waist to the knee [and] round his head . . . a neat band of fur stuffed, in front of which is placed a tall feather, and on each side a variegated plume." He has a muscular frame and wields a traditional shield and long assegai. His brow is broad and his jaw square, and his eyes gaze fiercely off the page. Bagg's respectful treatment of his subject reveals no sign of racialist scorn.[27]

In oral histories, Zulu elders described their king as being "light" or "light brown (like a lizard)" in complexion and relatively tall. Others challenge this image, recalling Shaka's skin as being "dark but not

FIGURE 3. William Bagg, "Portrait of
Chaka" (King Shaka Zulu), from
Nathaniel Isaacs, *Travels and
Adventures in Eastern Africa* (London:
Edward Churton, 1836).

black" or "dark-brown," and stating that he was of medium height.
What these informants do agree on is that Shaka's nose was peculiar.
One witness recalls him having a "prominent and long and rather narrow nose." Another remembers that his "nose used to perspire" so much that Shaka would often have to wipe the sweat off it.[28] Perhaps the reported size of Shaka's nose accounted for Isaacs' depiction of the king as having a "Hebrew expression."

Sadly, no portrait or photograph of Isaacs himself exists. We possess only scattered descriptions: his prominent facial features and high forehead, his height (about five feet, five inches), his hair color (light brown), his complexion (sun-bronzed), his full beard, and his substantial physical presence despite his lack of stature.[29] I imagine Isaacs as looking something like Edward G. Robinson playing Dathan in *The Ten Commandments* (1956), perhaps because the film depicts the Israelite Dathan as a vainglorious orphan who insinuates himself into power thanks to his charisma.

According to both Zulu and European recollections, Shaka was concerned with his appearance, even vain. Maclean records the king's wonder at first beholding his reflection in a mirror. He regarded himself intently for a few moments, finally spotting some white hairs in his beard. Shaka then became anxious at this sign of aging and asked whether King George's people possessed medicine to preserve one's youth. Another Zulu informant relayed that Shaka "looked into the water (our former looking-glass) [and] found himself ugly," with a "nose so large that it filled much of his face—[it] was as big as a toad." Here Shaka emerges as a sort of anti-Narcissus, turning away from himself in disgust rather than drowning in self-regard. If the Europeans and Shaka's own subjects were attentive to his complexion and his nose, the king for his part was "astonished" at the color of the Europeans and their physical characteristics. He "walked round them, looking at and surveying them." Shaka became enamored of the young Maclean because of his "brilliant red" hair—perhaps the first ginger-haired individual the king had ever seen.[30]

By daylight Isaacs could view the full scope of kwaBulawayo. He estimated the *ikhanda* to be about three miles in circumference and to contain around 1,400 huts. At the head of the royal *ikhanda* stood Shaka's palace, which consisted of roughly one hundred huts for his wives. Traditional Zulu huts are built by placing saplings or poles in a circle and then bending the tops inward toward a center support pole. The frame, about nine feet in diameter, is then woven with sticks, reeds, and thatch to form a beehive shape. These huts remained remarkably cool in the dry season and dry in the rainy season. A cooking fire was set in the center of the hut, but lacking a chimney, the interior was smoky and dark.[31] Low doorways forced everyone but small children to stoop or crawl through.

To reach kwaBulawayo, Isaacs would have traveled over green hills and red earth, beneath the orange-red canopies of flame trees, and over watercourses lined with reeds and palms. Today spindly trees grow in a horseshoe pattern on the grounds of Shaka's former *ikhanda*. Thorny bushes grow from the still-discernible outlines of concentric oval humps, perhaps the remains of kwaBulawayo's earthworks or middens. The limbs of euphorbia trees, which look like a cross between a cactus and a grasping hand thrust upward, pierce the cloudless blue sky. Few cars pass on the road that cuts through the site, and one can hear birdsong, cows lowing, and muffled sounds from a nearby village—a much quieter scene now than in Shaka's time, when his headquarters served as home to thousands of men, women, children, and the king's warriors.

A wide-eyed Isaacs reported on Shaka's royal court at its zenith. The king gestured, a cow was conjured, and the beast was swiftly slaughtered for a welcoming repast. Servants appeared from nowhere balancing gourds of water on their heads for Shaka to wash. He bathed

FIGURE 4. "Emigration to Natal," showing Natal Bay, from *Illustrated London News*, 16 March 1850.

himself from head to foot while his chiefs approached to report on affairs of state, and then daubed himself with a red paste of animal fat and ochre that gave his muscular frame a glossy sheen. Warriors dropped to one knee to salute their ruler. They had good reason to treat Shaka with deference, for he could summon executioners with a nod. On this first visit to the "sanguinary chief," as Isaacs called Shaka, he witnessed the king order the seizure of three unfortunates. Their necks were broken, and they were dragged away to the bush. There they had sharpened sticks forced up their anuses to puncture their bowels and were left to die, or to be consumed by beasts before they had breathed their last. The terrified Isaacs could not understand the cause of these executions. Possibly the condemned had been accused of cowardice or witchcraft.[32]

Although Isaacs was shocked by this episode, he must have been aware of the grisly business of execution in England. Condemned men and women could be hanged—at the time a painful process of slow strangulation—or hanged and then dissected, or hanged and then beheaded, or hanged and then placed in irons (gibbets), or drawn and quartered, or even burned alive. During the Napoleonic Wars, the British military punished cowardice by hanging or death by firing squad.

Insubordinate mariners in the Royal Navy were subject to a more inventive assortment of cruelties: they could be flogged nearly to death with a cat-o'-nine-tails, made to run a gantlet and beaten by their fellow sailors with thick knotted cords, gagged with a bit like a horse and tied up at the captain's pleasure, or hanged from a ship's timbers. When the young Maclean explained British modes of execution to the Zulu, he reported that "these savages . . . shuddered at the cruelty of our practice of hanging a man up by the neck."[33] Sanguinary punishment, like savagery itself, is all in the eye of the beholder.

Following the executions at kwaBulawayo, Isaacs returned to Port Natal, now dubbed Fort Farewell. He traveled on horseback and on foot, all the while lamenting the poor state of his shoes. Perhaps he wished he had learned his father's shoemaking craft to make use of the animal hides at his disposal. He eventually abandoned the shoes to walk barefoot, but was then tormented by thorns lacerating his feet. Isaacs was exhausted by the time he returned, but Fort Farewell could offer little comfort to him. Its main dwelling consisted of a structure about twenty-four feet long by fifteen feet wide, partitioned to function as a kind of dormitory. African men and women lived in the beehive huts nearby. Maclean referred to their encampment as looking at once "semi-civilized and semi-barbarous."[34]

After Isaacs had recovered from his painful trek, the boredom of routine at Fort Farewell and his desire to learn isiZulu drove him to make another journey to Shaka. By mid-March 1826, King Shaka had again summoned representatives of his white people. Isaacs set off for kwaBulawayo with a company of Europeans carrying gifts for Shaka, including the carved and painted figurehead scavenged from the *Mary*—the bust of a woman—as well as various medicines and salves.[35] The leader of the party was one of Farewell's original would-be settlers, Henry Francis Fynn.

Fynn had arrived in the Cape in 1818 when he was about fifteen years old to work with his father, a trader and innkeeper. He later wandered in search of a job to the edges of British-controlled territory. Rumors circulated that he took advantage of a period of violence and unrest to steal from a shop in Bathurst, about six hundred miles beyond the reach of Cape Town's authorities. By the time Isaacs met him, Fynn was familiar with the peoples and customs of the region. Isaacs described him as "tall [and] with a prepossessing countenance." A surviving photograph of a much older Fynn captures him as a pug-nosed, round-faced man with jowls, a double chin, deep-set eyes peering from beneath

bushy brows, and a high, balding pate; he looks something like a bull-
dog. Fynn wandered into Isaacs' life at Fort Farewell from the bush,
where he had spent weeks in pursuit of ivory. At their first meeting,
Isaacs recalled, Fynn wore "a crownless straw hat, and a tattered blanket, fastened round his neck by means of stripes of hide, [that] served to
cover his body, while his hands performed the office of keeping it round
his 'nether man.'" Other observers also recorded Fynn as having
adopted African dress—"a skin [worn] round the waist." The sight of
the half-naked Fynn impressed Isaacs, who also admired Fynn's fluency
in isiZulu and his fortitude.[36]

The rainy season had now settled on the Zulu Kingdom, bringing
squalls, wild lightning storms, and swollen rivers. But Fynn, Isaacs, and
their armed party forged onward to kwaBulawayo. At one point they
crossed a shallow stretch of river known to be free of crocodiles. The
Europeans stripped and waded into the rushing water, supported on
each side by retainers. Isaacs noted that the women who watched from
the shore were "more than ordinarily diligent in scanning" their nearly
nude white bodies. And he returned their gaze, staring at "the young
females [who] were rather handsome, displaying much symmetry of figure." Yet his tone shifted at once from admiration to a "disgust at their
general habits."[37] What those habits might have been goes unremarked.

Isaacs' disgust indicates more about his ambivalent response to the
attractive Black women than to any behaviors the Zulu themselves
exhibited. Among his fellow castaways in Fort Farewell, and the British
reading public who consumed *Travels and Adventures*, the expression
of European superiority was encouraged. No doubt the teenaged Isaacs
in Port Natal, anxious to be a man among men, and the adult Isaacs, as
an author eager to shed his Jewish marginality, would have conformed
to such cultural prejudices. That his description of the Zulu women's
fascination with his body and his own appreciation of their bodies is
followed so quickly by a declaration of unexplained disgust points to a
contradiction between Isaacs' emotional state and the norms of the
prevailing racialist system.[38]

Throughout his memoir, Isaacs pays a great deal of attention to
clothing, or the lack thereof. Among men of his era, clothes—including
naval uniforms—offered clear indications of class, rank, and social
identity. At the time he sailed from England for St. Helena, fashionable
women would often have worn veils over their faces, along with several
layers of underclothing, including a corset. Their dresses would have
revealed little save for the skin of their neck and wrists. To Isaacs, the

bare skin of Zulu women and men, not to mention Fynn's peekaboo "nether man," might have suggested sexual availability or potency. For a provincial seventeen-year-old, the attractions of this exposed flesh and the internal conflicts aroused by those attractions must have been powerful. His many mentions of clothing and nakedness also suggest that Isaacs suffered a kind of identity panic when faced with the disintegration of external markers of belonging, such as his literally unraveling wardrobe. Who was civilized now? Who was savage?

The European party straggled into kwaBulawayo during daylight. This time, Isaacs' admiration for Shaka's *ikhanda* was clear: "When the abode of his Zooloo majesty opened upon us, its appearance was singularly magnificent, and the scenery imposing and attractive." But he and Fynn were soon confronted with a chaotic and bloody scene. A chief had died, and a frenzy of mourning followed. Those Zulu who did not express their grief through extravagant weeping—a weeping surreptitiously aided by quantities of snuff that brought tears to their eyes—were dragged away to be beaten and killed. Isaacs was confused and horrified by these and further executions at Shaka's command, and he urged the "dissemination of more civilized notions ... among the natives of this part of Africa, as may eventually root out these savage and brutal propensities." Here Isaacs invokes the common colonialist perspective that Europeans had a responsibility to intervene to save so-called primitive peoples from themselves. In 1899 Rudyard Kipling, in a much-quoted verse, would enjoin his readers to take up the "White Man's burden." Kipling's phrase distills the notion of a Eurocentric and paternalistic mission to civilize "sullen peoples, / Half devil and half child."[39] The obvious racism of this notion did not prevent it from proving tragically seductive to generations of humanitarians, missionaries, soldiers, free-trade acolytes, and other emissaries of European culture.

When the bloodletting at kwaBulawayo abated, Isaacs and Fynn bestowed on Shaka the gifts they had brought. The king disdained their presents in front of his people but privately expressed gratitude and fascination with the items, especially the medicines and muskets. These muskets were probably a version of the famed Brown Bess muzzle-loading long guns used by the British for about a century. Though not particularly accurate, the Brown Bess was lethal enough, and was bought, sold, bartered, and fired wherever Great Britain maintained a colonial presence. Shaka and his council of warriors discussed the efficacy of the weapons and asked the *abelungu* to fire at a kettle of vultures soaring overhead. Before they could do so, however, a musket-bearing African stepped from Shaka's circle, raised his weapon, and fired at the scavengers. "The effect which this produced was astounding," Isaacs recorded. The Black marksman was none other than Jakot (Jacob) Msimbithi, the one-time cattle rustler and prisoner freed from irons by James Saunders King, the former interpreter for Captain Owen, and the man who turned his back on Farewell's party after rescuing the accident-prone lieutenant from drowning in the shipwreck. After being walloped for his troubles, Msimbithi had abandoned the whites and turned inland from St. Lucia Bay, later finding his way to Shaka's imperial residence. There he regaled the king with tales of European barbarism. Shaka elevated Msimbithi to his inner circle because of his familiarity with the *abelungu* and their language and bestowed on him a praise name (*izibongo*), Hlambamanzi, meaning "Swim the Seas" or "Great Swimmer."[40]

Shaka then requested that the whites exhibit the power of their muskets and display their battle formations. Afterward Shaka noted that during the time required to reload their weapons, his *amabutho* foot soldiers could attack and destroy the helpless whites. He asked the Englishmen whether King George "possessed as many cattle and if he had as many girls" as he himself did. When Isaacs explained that it was customary in England to have only one wife, Shaka laughed and suggested that he and King George were like-minded because they avoided "promiscuous intercourse with women."[41] In *Travels and Adventures*, Isaacs highlights the contrasts between British and Zulu life, with European ways almost always presented as superior to African ways. Yet he depicts King Shaka as a perceptive man who sought to identify affinities between his kingdom and that of King George IV. In a sense, Shaka even adopted the British King as Zulu, often referring to him as "umGeorge," *um* being an isiZulu noun prefix. For Shaka, of course, Zulu ways triumphed over European ways. Both Isaacs and Shaka retained their curiosity about the alien, even when they judged their own customs as superior.

If Isaacs records Shaka's character accurately, the Zulu king was an ambitious, inquisitive, and clever leader able to extrapolate from and understand a great deal about a foreign culture based on his encounters with just a few disheveled white men. However, if this portrait is instead the product of a young man's fancies and fears, as some critics insist, then something else, perhaps even more remarkable, emerges: Isaacs' ability to imagine English manners and norms as refracted through Zulu eyes.[42] Though he rejects the Zulu perspective on the *abelungu*, he is able to temporarily adopt it. In other words, if Isaacs' King Shaka is indeed more fiction than fact, we can conclude that Isaacs possessed the empathic creativity of a talented novelist. And if Isaacs' two volumes are generally accurate, though at times embellished, as I strongly believe, then readers today should be grateful to him for incorporating differing perspectives on the events he records. Such an approach allows for a single image to come into focus: the mutual fascination between Shaka, a mighty kingdom builder, and Isaacs, a marginalized Jew and headstrong teenager.

Isaacs' willingness to seriously consider an alien other's point of view may derive from his youth spent on the fringes of the Anglican Church in Canterbury as part of a despised minority. Given the importance of religious belief as a marker of identity at the time, Isaacs was unusually circumspect about religion. In the more than 750 pages of *Travels and Adventures*, nowhere does Isaacs invoke Christ or Christian duty, except to offer a withering rebuke to a missionary. Isaacs quotes the New Testament exactly once, and only in the context of his ironic reproof to that overzealous Christian missionary: "Go thou and do likewise" (Luke 10:37). Isaacs does cite the Hebrew Bible on several occasions. To be fair, he also mentions conducting Sunday services for "his" Africans.[43] Isaacs, of course, would have learned the fundamentals of Christian worship during his time with his uncle on St. Helena, attending St. James' Church as required by law.

By contrast, Maclean's memoir of life among the Zulu piously invokes the saving grace of God, quotes from the New Testament, and emphasizes his Christian faith nearly twenty times over the course of eleven installments published serially between January 1853 and March 1855 in the *Nautical Magazine and Naval Chronicle*, which was obviously not a theological publication. One could ascribe this difference merely to varying degrees of personal piety, but we should also bear in mind that Isaacs would not have felt comfortable invoking his Jewish heritage (even if his surname made it obvious to most readers).

Isaacs' references to religion are few, and to the extent they reveal anything about his beliefs, they suggest a deist mindset. During his second visit to kwaBulawayo, Isaacs had a discussion with Shaka about faith. "The religion of our nation taught us to believe in a Supreme Being, a First Cause, named God," Isaacs explained to Shaka. He then clarified that God "created all things, and was the giver of light and life," a statement that greatly interested Shaka. The king "seemed as if struck with profound astonishment" when Isaacs regaled him with the biblical account of creation.[44] It must have been an extraordinary scene: we can imagine the Zulu leader dressed in his finery, seated on a straw mat before flickering fires, ringed by servants, elders, and warriors all listening to the teenaged Isaacs, who was clad in tatters, his voice shaking and struggling to form basic isiZulu phrases, his hands tracing shapes in the air while he freely translated the first verses of the Hebrew Bible, which any Jewish lad in England at the time would have known by heart.

The king's undoubted charisma and a fascination with the Zulu way of life spurred Isaacs to make several trips to kwaBulawayo. Yet his audiences with King Shaka on his third and fourth visits in 1826 left him shaken. In Isaacs' presence, Shaka executed suspected cowards or those who mourned for the condemned. Isaacs recorded numerous brutal killings, including those of women and children. At one point, he hid in his hut, dreading future massacres and sickened by the howls of animals drawn by the mutilated corpses dragged outside the village. Yet his fear of this violence did not keep him away from Shaka's magnetic presence. Isaacs suspected that the king orchestrated these brutal displays to awe him and the other *abelungu*, yet he seemed to feel that by virtue of being one of "the king's white people," he would be spared Shaka's wrath so long as he indulged the king's caprices.[45]

These journeys to kwaBulawayo challenged Isaacs physically and emotionally. He suffered from hunger, exhaustion, and extreme heat. Walking barefoot like the Zulu, he rubbed his swollen feet with animal fat to soothe them at the end of each day. Rats gnawed the toenails from his feet as he slept in huts along the way.[46]

While Isaacs made these wearying treks, Farewell and James King quarreled at Port Natal. Both were resourceful, ambitious, and resolute. Like many men with these traits, they were also stubborn, reckless, and arrogant. Circumstance and character drew the two seamen into an adversarial friendship. But King was even more driven than Farewell. King was a self-made man who had worked his way from the hinterlands of empire to a stint in the Royal Navy, and then to the command of a merchant vessel. Yet he could not help but be aware of the precariousness of his station. He was sensitive about his status, as his self-conferred rank of lieutenant suggests. When Farewell argued that he should be the one to act as the "principal" in all dealings with Shaka, and further demanded that all gifts be sent to the Zulu monarch in his name, King at first assented. But as time wore on, he chafed at Farewell's high-handed manner. The clash of wills intensified when it became clear that Shaka preferred King to Farewell, whom Shaka dismissed as being "too much like an old woman." Farewell had set himself up as the white settlers' chief, a position he believed that Shaka had ratified in the land grant of August 1824. But now Isaacs referred to Farewell and King as the "two chiefs" of their shabby European settlement.[47] The politically astute Shaka may have inflamed their rivalry to advance his own interests.

The disagreement between Farewell and King dated back to the *Mary*'s wreck. Isaacs and another sailor later swore before a notary public that Farewell had stolen some of the brig's salvaged stores. They also claimed that Farewell had refused to share cattle given to him by Shaka when food supplies had run low. More lurid still, Isaacs testified that Farewell intended to "destroy his (Mr King's) . . . [property] and people." Later Farewell and King squabbled over who should present the gift of a brass crown and a bundle of peacock feathers to Shaka. When they could not agree, Isaacs volunteered to take the feathers with him when he visited kwaBulawayo in November 1826. Once Isaacs arrived, Shaka asked after James King and expressed his disappointment that his friend had not come to visit. Shaka was unimpressed by the peacock feathers; the king sported the brilliant green-blue iridescent feathers of the lourie (touraco) in his headdress.[48]

By the time Isaacs returned from this visit, the petty jealousies and bickering had escalated. Farewell accused King of undermining his position with Shaka and of secretly "obtaining the ivory Chaka was able to collect." Isaacs tried to reconcile his friends, but he finally sided with James King. As King's junior partner, Isaacs assented to receiving one-quarter of their ivory haul. He was an enthusiastic hunter, and his quest for ivory led him deep into the interior. He was lured by rumors of elephants and a tribe of expert hunters who practiced cannibalism. With his servants, he crossed miles of undulating hills, rocky precipices, and hazardous rivers barefoot, carrying—or more likely having his servants carry— his musket, a reed mat, and some trinkets for gifts and barter. Some of the peoples he encountered had never seen a white man before.[49] Their first sight of a European was not likely to instill awe: a trouserless, tender-footed teenager traipsing through their territory.

Nonetheless they seemed impressed by Isaacs' singular appearance. And he reveled in their attention, enjoying the songs, dances, and curiosity that greeted him as he wandered the interior. "All ages and sexes," he wrote, "gathered round me; some afraid to approach me, others placing their children at my feet, entreating me to touch them, conceiving that my doing so would be a good omen." On another occasion, he was accosted by women who prodded at his arms and ruffled his hair, all the while saying to one another, "Look at his hands and feet, how pretty they are, just like ours." He indulged the inquisitive men, women, and children who surrounded him, even as he dismissed them as superstitious and credulous. Of course, Isaacs himself was no less naive, chasing after reports of elephant-hunting anthropophagi for two months. He returned fatigued to Fort Farewell in December without having collected any ivory or encountered any cannibals.[50]

Isaacs was a restless young man willing to risk danger at the ends of the known world for the chance to prosper. All he possessed was a remarkable constitution, a penchant for exploration, and a loaded musket. His entrée into this global commerce was abetted by complex international affairs that involved the leading powers of Europe as well as their colonies and peoples, the rulers of African kingdoms and their fractious populations, the dissemination of emergent technologies, and the militarization and financialization of corporate trade the world over. As a result of the Jewish diaspora, some Jews possessed international kinship ties that facilitated trade and the extension of credit across global ports.[51] In one sense, Nathaniel Isaacs' relationship with his prosperous uncle on St. Helena, and Saul Solomon's network of contacts in Great Britain, the Cape Colony, and beyond, granted him privilege. But this privilege endured only so long as his Jewish heritage could be concealed or proved inoffensive to those who held money and power. And so in another sense, like Farewell and King, Isaacs was just another hotheaded roving Englishman with few prospects at home who was willing to risk his life at the frontier of a new market. Seen thus, his Jewish identity was mostly incidental. What mattered was his ability to leverage his position as part of an established trade diaspora, but that in turn was made possible precisely because of the Jewish people's history of sojourns and expulsions. Just as frontiers became sites for trade, territorial expansion, and self-definition for African peoples—the Zulu among them— they also opened these possibilities to ambitious Jews like Isaacs.[52]

Weeks of inactivity in Port Natal followed Isaacs' failed ivory expedition. He had grown "accustomed to roam in search of new objects" and was presumably impatient to find some ivory and relaunch his stalled trading scheme.[53] He and King decided to follow Fynn's route into the bush along the crocodile-infested uMlalazi River, where Fynn had managed to kill fifty or so hippopotamuses and collect more than six hundred pounds of their teeth. These teeth provided a harder and whiter ivory than elephant tusks and were also much in demand. At the start of their journey, they visited Shaka to gauge his approval. The Zulu

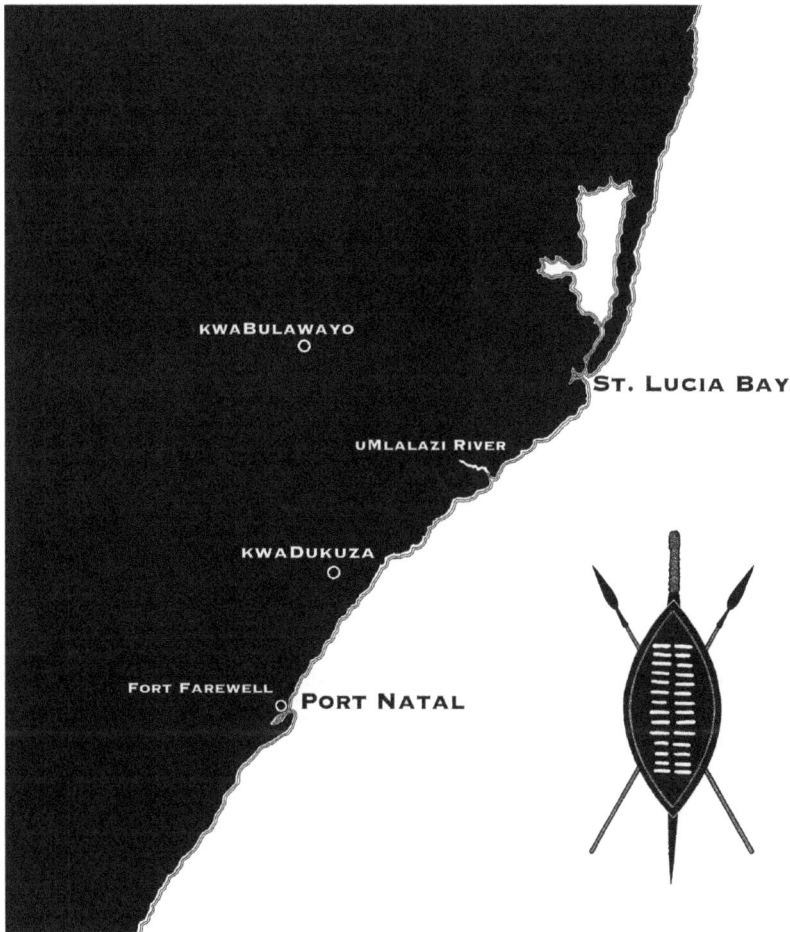

MAP 4. The Zulu Kingdom.

ruler had by this time abandoned kwaBulawayo to construct a new capital at kwaDukuza, approximately forty-five miles northeast of Fort Farewell. He did so in part to keep an eye on his newest vassals, the motley group of whites and their retainers encamped at Port Natal. In addition to prospering through trade with the *abelungu*, he also sought to distance himself from elements in the Zulu heartland who resented his rule.[54] Shaka sanctioned the ivory-hunting expedition and promised the hunters a grant of land to establish a settlement with trading rights near the mouth of the uMlalazi, about eighty-five miles northeast of Port Natal.

Isaacs and King explored the uMlalazi region until they located a favorable span of rich soil and thick forest, where they planted the Union Jack, "taking possession of it as a grant to us from Chaka to inherit, for the purposes of trading under the auspices of the Zoola monarch." No documents survive to attest to this grant. However, one of the *Mary*'s shipwrecked crew, John Cane, testified that he knew from Farewell—who would have been keen to sabotage King's commercial venture—that in fact "there was a grant made by Chaka to Capt King of certain lands," and his description accords with Isaacs' own report.[55] Based on Cane's testimony, it seems that Shaka had indeed extended to Isaacs and King the right to establish a homestead and trading post.

The day after they staked their claim, Isaacs and King awoke to learn that two servants employed by Farewell, whom they referred to as Michael and John, had raped a favored wife of a local chieftain. The men "successively violated the unwilling girl" at gunpoint. Their crime had already been reported to Shaka by fleet-footed messengers. Isaacs and King knew they must hurry to kwaDukuza to seek Shaka's forgiveness. Like the abandoned kwaBulawayo, Shaka's new capital was circular in layout. The *ikhanda* was estimated to be slightly less than 1,000 yards in diameter and contained approximately 1,500 beehive huts, with the King's inner palace complex covering an area about 360 yards by 50 yards. This interior zone boasted a dense collection of huts and contained another fifty or so larger beehive dwellings. Shaka's compound was protected by an eight-foot-high fence "composed of two and sometimes three thicknesses of compact wickerwork, carefully worked with young twigs of saplings" that created a formidable bulwark against attack.[56]

Isaacs and King entered through a gate and were received "very coolly" by Shaka, who directed "a good deal of wrath" at the *abelungu*. As retribution for the rape, he threated to "kill all the white people in Natal." Even had they handed over the rapists, the Europeans would still have been subject to punishment, as they were held responsible for their retainers' actions. Isaacs had witnessed enough of Shaka's rages to fear for his life. While he decided how to exact punishment, Shaka detained his prisoners "under a guard of thirty warriors, with instructions to prevent [their] escape." King's African assistant, Nasopongo, interceded with Shaka, explaining that the whites "thought it hard the innocent should suffer for the guilty."[57] Nasopongo may have saved their lives.

Later that evening, Isaacs and King were brought before Shaka, who had just received a message from Farewell apologizing for the crimes of his servants. Farewell assured Shaka that he was "anxious to make his majesty every satisfaction for the outrage."[58] This expression of remorse served Shaka's plans. He had sought a way to appease the chiefs bent on vengeance for the rape without actually killing any of his white allies. Now Shaka arrived at a shrewd solution.

The King ordered the *abelungu* to muster a force to go to battle for him, though he exempted Farewell and King from fighting. In other words, Shaka treated the Europeans as he might any other vassal clan and conscripted soldiers from them. The two chiefs—Farewell and King—were to supply him with mercenaries on pain of death. In a missive that King penned and hid in a calabash to be sent back to the Cape for posterity, he recorded that Shaka told them: "You shall live, if you go to war for me." Shaka wanted the white conscripts to attack an unconquered chieftaincy. "We knew the crimes of the Hottentots entitled us all to a severe punishment [so we] at length agreed on the attack," James King wrote that night.[59] The enemies Shaka sought to destroy had found refuge on a rocky outcrop and hurled rocks down at their attackers, whose assegai could inflict little damage from below. But the *abelungu* had gunpowder, shot, and muskets.

Isaacs and King returned to Fort Farewell with news of Shaka's demands. They recruited a detachment of ten men, including the settler John Cane, several Africans (including the guilty Michael), various dependents, and Isaacs himself. King remonstrated with his friend, urging him to let another man go in his place. The headstrong Isaacs, however, "resolved to be one of the party." They equipped themselves with muskets and twenty rounds of ammunition each and then hastened back to kwaDukuza to muster prior to the attack. King Shaka supplied the European squad with an accompanying force and instructed them to kill every last man, woman, and child of his enemy. Isaacs claimed to have protested that the women and children were innocents who could not be slaughtered. Yet Shaka insisted there should be no survivors: "They can propagate and bring children, who may become my enemies. . . . I command you to kill all."[60] Isaacs' possibly self-serving recollections of Shaka make him out to be reminiscent of Pharaoh in the Book of Exodus (1:10), who orders the death of newborn Hebrew males "lest they multiply, and . . . when there falleth out any war, they join also unto our enemies." Reluctantly now, Isaacs set out on the bloody mission.

The force once again had to march over rough terrain. This time Isaacs participated in a ceremony designed to protect them from crocodiles at a river crossing. Local diviners collected dried crocodile dung, or possibly the scales of dead reptiles, chewed them until soft, and then spread the resulting paste over the soldiers' bodies "as a charm to keep off those voracious animals."[61] Thus anointed, Isaacs and the rest of the detachment successfully crossed the three-hundred-yard span of river.

Encamped near the enemy's stronghold, Isaacs fashioned a leather bandolier for his ammunition and prepared for battle. "I did not feel the want of courage," he wrote; nonetheless, "I reflected also that on the cast of the die all our hopes depended." Should the company triumph, their lives would be saved; if they failed but survived the battle, Shaka would condemn the *abelungu* to death. Isaacs had no choice but to trust in his musket. The Europeans and Zulu escorts advanced on the stronghold and were met by enraged defenders. "I rushed forward and got on the top of a rock; one of the enemy came out to meet me, and at a short distance threw his spear at me with astonishing force, which I evaded by stooping," Isaacs recorded. Then he added coldly, "I levelled at him and shot him dead."[62] At eighteen, he had killed a man for the first time.

Musket fire reverberated and scattered some defenders, but rocks continued to rain down on Isaacs and the other warriors. They advanced slowly. Isaacs was hit in the shoulder. As they continued their scramble toward the summit, they burned any huts they came across and killed the enemy's dogs, either shooting them dead or spearing them through. Still the defenders would not relent. Isaacs fired into a crowd of attackers and saw a second man fall. Then, while readying his weapon to take another shot, he felt something strike him from behind. "I took no notice," he wrote, "thinking it was a stone, but loaded my musket." The pain continued and he soon saw a bloody "stream running down [his] leg."[63] A barbed spear blade had lodged in his back.

Six of Isaacs' comrades tried in turn to remove the spear. Finally, an unnamed servant forced his finger into the wound and deftly worked the barbs from Isaacs' flesh. He staggered to a stream, washed his wound, and now "weak from loss of blood," retreated downhill until he fell to the ground and had to be carried back to camp. That night he suffered "excruciating pain" and in the morning was visited by a diviner. The Zulu healer slaughtered a heifer, removed the excrement from its entrails along with a portion of its bile, boiled these foul substances together with some roots, and then presented the concoction to his patient to drink. Isaacs dryly noted that he was asked "not to drink too much." He was then left to sleep off his pain and to hope the aftertaste of the "infernal mixture" would fade.[64]

The next day the last defiant enemy warriors surrendered, overwhelmed by the "king's white people" who were able to "spit fire" from their muskets. Contrary to Shaka's original orders, the conquered were not slaughtered wholesale. Instead, the defeated men handed over their cattle and goats to signal their submission. They then bestowed on the victors "ten young maidens by way of cementing their friendship by nuptial ties."[65] In this way, the rape of a Zulu woman by Michael and John was compensated for by the death and subjugation of Shaka's enemies, who in turn paid their own blood money by delivering up ten female subjects to whatever demands, sexual or otherwise, the victors might make.

Readers learn nothing more of these "young maidens." Perhaps some were indeed married off to Zulu warriors and became part of that nation. Isaacs describes Zulu women as subservient to their husbands and other male kin. Married Zulu men, he recorded, "rule over their wives with a despotic sway, and constantly put the poor creatures' lives in jeopardy." King noted that "every command of the husband must be obeyed even in the most trifling cases, or their life is the inevitable sacrifice." And Maclean too wrote that Zulu women were seen by their men "as beings of an inferior order." All three men described cases of women who, condemned by their husbands to violence or death, fled to Fort Farewell to seek the Europeans' intercession. When young women were eligible for marriage, which might occur at the age of fourteen, male relations bargained over their dowries "with as much composure and indifference as they would for one of their cattle."[66] Of course, most women in Great Britain also lacked property rights and control over their fates and bodies. The age of consent was thirteen in England, and a woman without a dowry was unlikely to make a good match. Whether in the United Kingdom or the Zulu Kingdom, women were treated as little better than chattel.

Isaacs does not disclose what sort of relations he himself had with local African women, but the evidence suggests that he and his fellow settlers cohabited with them. As chieftains, Farewell, King, and Isaacs were granted the right to establish homesteads; these *imizi* functioned as the fundamental sociopolitical units in the Zulu Kingdom. Women, and the dynamics of polygamy, formed the central nodes of an *umuzi* and hence the backbone of Zulu social organization. More than one Zulu informant noted that the whites "intermarried . . . and had children" by African women.[67] European writers also claimed that Isaacs and his fellow settlers had multiple wives, in the Zulu custom.

In 1833, an outraged Methodist missionary, Stephen Kay, maintained that the Europeans in Port Natal possessed "domestic circles [that] embrace eight to ten black wives or concubines!"[68] It is difficult to know whether Kay was more disturbed by polygamy or by interracial relations. Moral scruples aside, sensationalizing African customs was a way for him to rally political and financial support for converting the Zulu to Christianity. Missionaries at the time, like some well-intentioned activists today, felt a need to represent the people they assisted as victims of degrading cultural practices. Of these, only slavery disturbed African missionary societies more than polygamy.

The construction of missionary stations in South Africa was designed to banish polygamous relations. Arrayed around the church and school, square homes housed the nuclear families of a missionary's flock. Such structures replaced the sprawl of beehive huts or circular dwellings that housed extended kin and subsidiary wives in Zulu *imizi*. The missionary stations' European-style homes typically contained two rooms to enable privacy for a husband and (sole) wife. Missionaries further regulated their converts with calendars, clocks, and church bells that signaled the proper times for prayer, study, and laboring in the workshops, gardens, and agricultural fields. These routines encouraged dependence on the missionaries and their networks of external contacts at the Cape and beyond.

Economic salvation and spiritual redemption were intertwined goals for London's missionary societies. David Livingstone's later assertion that he had gone to Africa "to open up the country to commerce and Christianity" made that connection explicit.[69] In nineteenth-century evangelical theology, individuals existed in an "economy of redemption" in which conversion entailed recognizing the "gift" of Christ's sacrifice. The true believer was thereby redeemed and further credited for his efforts to save others. The nonbeliever remained spiritually insolvent. Those who rejected Christ or sinned against Him would find their souls in default, ultimately to be repossessed by the devil. Those who accepted the good news would inherit eternity, which promised an everlasting spiritual wealth likened to "mines flashing with gems of richest lustre."[70] Such a clearly defined commercial relationship between the individual evangelical believer and the Christian Godhead raises the question of why it is Judaism that has so often been accused of being a materialist religion. Perhaps the slur against Judaism is merely projection.

The devout Maclean dismissed Kay's hearsay accusations as the work of a "religious fanatic." Isaacs also denied Kay's charge of polygamy and accused him of an "intemperate and malignant desire for reproving individuals upon the questionable testimony of natives, or on the more suspicious evidence of low and discarded Europeans." Here Isaacs places the Indigenous peoples of South Africa *higher* on the scale of trustworthiness than many whites. Isaacs concluded: "I challenge the pious author [Kay] to prove his allegations of . . . concubinage." A later and more objective author who had lived in Natal for fifteen years wrote in the 1850s that "whatever [Isaacs] and others may say to the contrary, . . . with scarcely an exception, they all had Kafir wives and concubines, and as many of them as suited their wishes or convenience, varying from one or two to ten."[71] There is little doubt that Kay and the later scandalized informant were correct.

Whatever public protestations Isaacs made to the contrary, he must have cohabited with at least one woman while in the Zulu Kingdom. Two documents reveal that Isaacs took African lovers. In a letter to Fynn dated 10 December 1832, Isaacs asks his friend, still roaming the Zulu Kingdom, to

> get my little fellow Porter and place him under the superintendence of . . . the Missionary Station. I trust, dear Fynn, that you will not neglect this promise, and have him baptized as Henry Porter Isaacs; any reasonable expence on his account that you may deem expedient you may always rely on getting disbursed from me. This will empower and authorize you to act on my behalf towards my dear little boy, and I beg of you to attend particularly to his education, as when he gets old enough I intend either to give him a trade or take him with me. I cannot forget this duty altho' an illegitimate child; and had he been old enough I should have snatched him from his mother's arms when I left Natal.[72]

Though he expresses affection for his (abandoned) son, Isaacs here betrays a lack of concern for the child's mother, whose name he does not mention.

The *South African Commercial Advertiser* published excerpts from Isaacs' journal that mention a woman named Macowzala, who worked for him at Port Natal. Her facial features and "skin several shades lighter" than that of others, he noted, "indicated [her] European extraction." Could Macowzala have been the mother of Henry Porter Isaacs? Perhaps he reckoned that this woman, whose skin was "several shades lighter" than others', was appealing enough by British standards of beauty to bear his child.[73] Isaacs' eagerness to enroll his son in a Christian missionary school may signal his indifference toward Judaism, a recognition that a child born of a non-Jewish mother is not Jewish according to rabbinic law, or, more likely, that Isaacs saw no other options for the boy's future. Nonetheless, he makes explicit the responsibility he feels for his son. Perhaps in penning this letter Isaacs recollected his own childhood, which came to an abrupt end when he was, in a sense, snatched from his mother's arms and sent to his uncle's distant home.

An indication that Isaacs fathered additional children is preserved in the diary of a British soldier who traveled through Natal. He records meeting a grown woman,

> the child of a Mr. Isaacs formerly living in Natal at the time of Chaka. . . . She is married to a Caffir, a native teacher at one of the missionary establishments at Natal. It is singular that with such strong evidence of white blood in her veins she should have allied herself to a Caffir instead of one of her own caste. But the reason assigned is that a brother and a sister by the same father having died from neglect and ill usage whilst in the service of some white people, she renounced all connection with the civilized part and allied herself to the savage part of the community.[74]

Looking beyond the racist presumptions evident in these words, we can infer that Nathaniel Isaacs had at least three (and possibly four) children by at least one African woman, and that the woman described by the soldier might have been the sister of Henry Porter Isaacs. There is no record that Nathaniel Isaacs ever sent for Henry Porter, his other illegitimate children, or his concubines. Nor does it seem as if he prevailed on Fynn to aid any of them except Henry Porter.

Affection and devotion might, however, sincerely exist between European men and African women. Maclean recorded the story of the *Mary's* shipwright, Hutton, who rescued a young woman, Dommana, from a brutal beating by her own father. Grateful, she insisted on remaining with Hutton, who presented her father with a gift of beads to placate him for the loss of her domestic labor. Dommana came to live with Hutton and nursed him with tender dedication during a later illness.[75]

Isaacs, too, was faithfully attended during his recovery from his battlefield injuries. His men constructed a kind of stretcher from a bull hide to transport him back to kwaDukuza after a three-day convalescence. News of the victory traveled ahead of Isaacs' party. Along the route, warriors offered their tributes to him, "surprised at the bravery of the white men." Isaacs was granted the praise name Dambuza, which Isaacs rendered as meaning "the brave warrior who was wounded." A Zulu informant records a fuller version of the name as uDambuza mtabate, literally meaning "the one who waddles; off he goes at speed."[76] The Zulu used the sobriquet Dambuza to refer to Isaacs ever after.

Dambuza-Isaacs soon recovered enough to abandon his makeshift stretcher, and he waddled at speed back to kwaDukuza, where he found James King awaiting him. Shaka too was eager to see him and asked Isaacs to display his spear wound. When Isaacs complied, Shaka rebuked him: "That is a cowardly sign; you must have had your back turned to your enemy." In fact, Isaacs had been outflanked by an assegai-wielding warrior during the battle. But then Shaka's mood shifted, and he insisted "that he was only jesting, and that he had heard of [Isaacs'] gallantry."

Both Isaacs and Maclean record examples of Shaka's malevolent humor, which frequently discomfited the Europeans.[77] The references to Shaka jesting at others, especially at "his whites," provide more insight into the Zulu king's aggressive personality than do the probably sensationalized claims of his bloodthirstiness. Perhaps Shaka's confrontation with unfamiliar European technologies—writing, mirrors, muskets, sailing vessels, industrially produced glass beads, medicines, and cosmetics—spurred him to mock the whites in order to enhance his sense of self and his standing among his people.

Nonetheless, Dambuza-Isaacs grew in Shaka's estimation after the skirmish. When fully recovered from his wound, Isaacs once again set out collecting elephant tusks and hippopotamus teeth. He bartered and hunted in various locations with his clan of Black African men, each trained in the use of weapons. Between these forays, Isaacs and King met at kwaDukuza, sometimes also convening with Fynn, who continued to seek ivory on solo raids into the wilderness. During one such rendezvous, Fynn penned a letter to Farewell. Shaka entered Flynn's hut unannounced and watched attentively as he wrote. Shaka then asked Fynn "if the ink would wash off the paper [and on] my replying in the negative, he asked it if would stain a shield. I told him that it would not, but the hair and the skin might appear slightly tinged." This information intrigued Shaka. The king pressed Fynn about the ways Europeans altered their hair color. Fynn explained "that there was a substance, called Macassar oil, which was said to darken the hair. . . . On being assured the oil was easily procurable, [Shaka] expressed the greatest satisfaction. . . . To defray all expenses [for purchase of the Macassar oil] he would commence hunting elephants, whose ivory should belong to Mr. King [who] would . . . supply the oil." Fynn, King, and Isaacs all attempted to discourage this demand, but obtaining the hair tonic was Shaka's "constant theme and induced him to take the whole of his force elephant hunting," Fynn later wrote.[78] Shaka hoped the oil would be forthcoming after James King set sail aboard the craft Hutton was then constructing.

Isaacs recorded a similar scene. Summoned to Shaka's residence, he, King, and Fynn found him reclining on a mat awaiting their arrival. He dismissed his attendant and then motioned the white men into his inner sanctum for a private conversation—an unprecedented invitation. "I could not help perceiving that there was a reserve, or a peculiarly grave manner in him," Isaacs recalled, "to which we had not been accustomed." He served them boiled beef and sour beer made from fermented maize, and after a period of weighty silence, Shaka proclaimed that he would like "to cross the water to see King George," but in the meantime he would send the British monarch a gift of elephant teeth as a message of goodwill in the charge of a senior and trusted chief. This Zulu diplomat would sail with James King aboard the vessel under construction to forge an alliance with King George, whom Shaka "esteemed as a brother." Shaka urged King to return from this proposed diplomatic mission with more medicine and "particularly some stuff for turning white hairs black"—Macassar oil. Isaacs noted that Shaka was "more than ordinarily anxious to obtain this . . . preparation, and promised . . . an abundance of ivory and droves of cattle" in exchange. Further, Isaacs described Shaka as begging them "in the most entreating manner, to keep this request a profound secret."[79]

What would have induced Shaka to fixate on the Macassar oil? Why would he instruct Isaacs and the others to keep his request a secret? And what is Macassar oil anyway? According to Isaacs, Shaka's white hair jeopardized his rule. The king had first noticed this sign of aging when he encountered the looking glass. White hair offered his rivals "proofs of [Shaka] having become unfit and incompetent to reign," Isaacs wrote. "It is therefore important that [Shaka] should conceal these indications so long as [he] possibly can." From the moment Shaka "heard that a medicine could be obtained which would prevent his hair from becoming white," Isaacs recalled, "nothing could appease him." Fynn reported that all Shaka wanted from the Europeans "was Macassar oil and medicine." In anticipation of these gifts, he presented King and Isaacs with a down payment of eighty-six elephant tusks.[80]

Macassar oil was a popular men's hair tonic and pomade used from the late eighteenth through the nineteenth century. It was marketed to invigorate growth, cure dandruff, and give men's hair a dark sheen. The most famous brand, Rowland's Macassar Oil, was mentioned by Isaacs, though he thought the claims for its efficacy preposterous. The product was named for a port, today Makassar in Indonesia. According to Alexander Rowland Jr., who popularized and sold his barber father's formula, the peoples of the region extracted oil from the seeds of a local tree and used the product to fortify their lustrous hair. At one time, Macassar oil may have been extracted from the seeds of what is commonly called the Ceylon oak. The resultant yellowish-white oil smelled of bitter almonds, which would have been masked by other scents in Rowland's patented mixture, including the fragrant ylang-ylang flower.[81] Because Macassar oil was viscous, and its residue discolored clothing and upholstery, a fashion developed of having washable fabric coverings spread over the arms and headrests of upholstered furniture. These slips of cloth, which one can still find in quaint bed-and-breakfasts and on the backs of some airline seats, are known to this day as antimacassars.

This redolent episode induces a sort of capitalist vertigo. The invisible hand materializes into stark black-and-white thanks to the twinned pursuits of Macassar oil and ivory. King Shaka sought to blacken his white whiskers through use of a British hair colorant, one derived from trade with Dutch Indonesia and supplied by a white colonial vanguard often in the employ of quasi-governmental enterprises like the East India Company. The mass-marketed pomade would then be bartered for "white gold," which was to be supplied by roaming bands of elephant-hunting Africans in the nominal employ of white Englishmen. The Englishmen would sell the tusks to merchants, who purveyed the ivory to a middle class in British industrial cities eager to accessorize their antimacassar-festooned salons with exotic emblems of luxury. Isaacs and his compatriots appeared not to have noticed this intersection of commerce and colonialism. What they did sense was that Shaka's demands—even for a hair-coloring agent—could not easily be dismissed.

CHAPTER 3

Black Napoleon, 1827–1828

Isaacs roamed the Zulu Kingdom throughout the summer of 1827, shooting elephants with his band of trained hunters. The men wandered barefoot through high brush, in and out of shadow, beneath blistering sun, along pathways the beasts tramped through vegetation, following the huge prints left in baked mud until they caught up with the herd. As the hunters approached, they selected adults with large tusks, careful to stay downwind and silent as the elephants flapped fleshy ears against gray bodies. Then one or two men torched the brush to panic the animals, and as the blaze caught, those with muskets fired their weapons. Wounded bulls heaved and groaned. Dying cows sprayed their orphaned calves with blood. More fortunate animals trumpeted and thundered their way to safety. Hunters then approached a downed elephant and shot it at close range. Once certain it was dead, they climbed onto the carcass and used assegai or hatchets to hack at the thick flesh above the elephant's trunk and along the nasal bone up to the eye sockets. Then, slicing back the skin of the head to reveal the pink and spongy tusk pulp, and careful not to damage the ivory, they would plunge elbow-deep in gore, carving the six- to eight-foot-long tusks from the skull. The corpses were left in the brush, where they attracted vultures, clouds of flies, and other scavengers that soon reduced the body to a heap of bones and maggoty scraps.[1]

Shaka too urged his warriors to fan out and collect ivory, with the goal of amassing enough to trade with his whites for supplies of cloth, blankets, medicine, knives, and the coveted Macassar oil. Fynn, who preferred to hunt with his own men, fell in with Shaka's hunters during one of these excursions in early August 1827. He accompanied them to the region of the King's former *ikhanda*, kwaBulawayo. Before dawn, Shaka asked Fynn to visit his mother, Queen Nandi, who lived nearby. She was reported to be ill, and the king believed that European medicine might cure her. Fynn entered her stifling hut to find it packed with nervous Zulu healers and the atmosphere thick with smoke. He pushed his way into the throng to examine the patient. The queen lay dying from dysentery, he thought. There was no hope. Fynn reported back to Shaka, who received the news in stony-faced silence.[2]

No full portrait of Shaka's relationship with his mother can be reconstructed, though from the remaining fragments of oral tradition and European reports we can assemble a sense of a bond shaded by resentment and overlaid with a gloss of perverse devotion. Shaka's father, Senzangakhona kaJama, reigned as chief of the Zulu at the end of the eighteenth century. Nandi was one of his many young wives and Shaka just one of his numerous children. Isaacs records that Shaka's birth was considered miraculous because Senzangakhona had been uncircumcised at the time of the child's conception, and the Zulu believed circumcision to be necessary for reproduction. According to some legends, this circumstance gave Shaka his name, for Nandi could not believe she was pregnant, and her swelling belly was ascribed to an intestinal complaint called *itshaka*. A Zulu source suggests that Shaka's name instead derives from a word related to martial prowess. European and Zulu informants agree that Nandi and Shaka were exiled after the young, headstrong mother fell from Senzangakhona's favor.[3]

At some point, possibly in his teens, Shaka was sent from his mother's side to Ngomane kaMqomboli, a neighboring chieftain. Ngomane became a father figure to Shaka and instructed his young charge in the art of warfare and the politics of succession. Much like Isaacs, Shaka had been raised fatherless, had suffered displacement, and was entrusted to an older man whose mentorship might help him advance. And advance Shaka did, assassinating his father's handpicked successor and wresting control of the Zulu chieftaincy sometime around 1815. Nandi achieved power through her son, guided his ambitions, and gained a measure of revenge for having been banished by her late husband. Isaacs observes that King Shaka afterward granted her "strong filial affection," except on those occasions when he struck her.[4]

Whatever the precise contours of Shaka's relationship with his mother, Fynn describes an extraordinary response to her death. He watched as Zulu men and women stripped themselves of ornament and "commenced the most dismal and horrid lamentations." Shaka hung his head over his shield, wept thick tears, and sighed, "Maye ngo Mama!" (Alas, my mother!) As a sign of her singular importance, Nandi's name (meaning "pleasant" or "sweet") was no longer to be used. The deceased queen would from now on be referred to as Ndlovukazi ("Mother of Elephants"), in acknowledgment of her son's supremacy.

Isaacs soon received word that he was expected to pay his respects to the grieving king. He tarried, but hearing that Fynn still remained at Shaka's side, decided to set off three days later and meet his friend there. On the way, Isaacs encountered a grisly scene at a neighboring *umuzi*: the smoldering bodies of three women and two children who had been burned alive in their hut, reportedly because their headman had not hastened to Shaka to offer condolences. Isaacs blamed the "savage monarch" for the slaughter.[5]

The repeated references throughout Isaacs' *Travels and Adventures* and other period documents to "savage" Zulu customs deserve to be set in context. In British discourse of the 1830s, savagery was considered a stage of human development. No matter what their race, cultures and individuals could progress from a state of savagery to a state of civilization. Charles Darwin observed that "savages" could be made to quickly "change . . . into . . . complete & voluntary Europeans." The signs of civilization were reason, science, private property, and commerce; those of savagery were irrationality, superstition, communal living, and barter. But these traits were seen as mutable, or what we might today call socially constructed. Isaacs' construal of Shaka and Zulu life as savage certainly indicates his presumed cultural superiority, but not that he considered the Zulu in essence inferior to Europeans. Until about the mid-nineteenth century, English portrayals of Africans in print were often positive, at times even laudatory, even if tinged with paternalism.[6] Such chauvinism, of course, can easily transmute into coarse biological racism, and expressions of cultural superiority may be a metonym for racism. The pallid ethnocentrism of Isaacs and others of his day contained germs of racism, apartheid, and genocidal policies. It is no coincidence that in Joseph Conrad's *Heart of Darkness* (1899), Kurtz scrawls his murderous testament—"Exterminate all the brutes!"—in the margins of his own pamphlet written for the fictional International Society for the Suppression of Savage Customs.[7] That Kurtz was once in the vanguard of Western culture remains one of Conrad's most potent ironies.

Even in the Cape Colony in 1828, when 119,918 slaves languished in captivity, *savagery* did not carry the obvious racist valence that it does today. Cape Town's *South African Commercial Advertiser*, helmed by the humanitarian reformist John Fairbairn, proclaimed that "men, however savage, still resemble our civilized selves in a far greater number of points—and these the most important—than those in which they differ from us." Another editorial likely authored by Fairbairn described the Port Natal settlers' "friendly relation with the people under Chaka," which might lead to the Zulus' "improvement," and he concluded that "the Human Race is ONE, and . . . in every variety of the species those mental powers and moral qualities which are essential to the social state, will be developed and acquired under favorable circumstances." Isaacs was friendly with Fairbairn, and both he and James King published articles in his paper. Nor was Fairbairn, a former theology student, alone in promoting Christian humanitarian views regarding the South African peoples.[8]

Fairbairn's influential ally in the cause, the Reverend John Philips, was also his father-in-law. Philips served as superintendent of the London Missionary Society and published an exposé on the state of affairs at the Cape. His *Researches in South Africa* (1828) leveled blistering attacks on British exploitation and violence. The abolitionist crusader Thomas Fowell Buxton, Philips's powerful friend in London, later aired these grievances in the House of Commons. A short-lived outpouring of humanitarian sentiment in South Africa was tied to an emerging middle-class value system that combined evangelical faith and abolitionist agitation with a belief in the British duty to aid the less fortunate.[9] That duty was premised on a notion at once ethnocentric and universalist: that all peoples, regardless of race, possessed the potential for achieving civilization—and that it was the mission of the British to help them along.

Isaacs reminds readers of *Travels and Adventures* that "the Zoola men are, without exception, the finest race of people which Southern or Eastern Africa can furnish, or that I have ever seen." He praises Zulu men and women for their athleticism, musculature, attractiveness, agility, bravery, cleanliness, endurance, discipline, generosity, hospitality, and friendliness. We can discern in his catalogs of positive Zulu characteristics those traits that Isaacs himself valued—precisely those masculine British virtues that he, as a Jew, was presumed to lack. The literature and popular culture of the day routinely depicted Jews as passive, weak, and feminized.[10] Isaacs identified the Zulu with vigor and power, and he refined in himself those qualities he valued in the Zulu, strengthening his body, toughening his mind, and adopting a fierce camaraderie with his Black African dependents and fellow Europeans. The privileged classes of Britain sent their sons to public schools where tutors, prefects, and peers enforced an emotional regime of stoicism through quasi-military organization, sport, and corporal punishment. Lacking such an education, Isaacs sought to forge his mettle by deprivation, maritime perils, and the hazards of life among strangers. Paradoxically, the Zulu represented idealized Anglo-Saxon values to Isaacs, and it was life among the Zulu that taught him to become British.

Pseudoscientific racism had not yet taken hold in early-nineteenth-century England, though Eurocentrism (as we term it today) held sure sway. Adam Smith, for example, had once characterized pastoral nations as existing in a "rude state." To Smith's disciples in the nascent global marketplace—and Isaacs must count among them—savage societies illustrated the dim origins of their modern civil state. Fairbairn's *Commercial Advertiser* reminded its European readers that though African peoples were subject to the despotism of chiefs like Shaka, "till very lately, we could not boast of being greatly in advance of them." Reverend Philips compared the peoples of South Africa to ancient Britons at the time of the Roman invasion: "It is here we see, as in a mirror, the features of our progenitors and ... our own history."[11] Readers of Isaacs' travelogue were primed to see the author not only as an adventure capitalist tramping through exotic places, but also as a traveler going back in time to illuminate the wellsprings of European civilization.

Despite the paternalism of such an attitude, readers should beware of condescending to such writers from the past. We often think we know better than they did. Indeed, some of us believe that our ideologies allow us to know even before we endeavor to understand. But for an intelligent nineteenth-century man like Isaacs, what might be known had first to be understood, even if imperfectly, and such understanding demanded firsthand experience.

The gruesome sight of the charred bodies of women and children forced Isaacs to reckon with a dilemma. Should he proceed at once to the king and apologize for offering his belated sympathies? Or should he return to safety at Port Natal but thereby risk incurring Shaka's displeasure for a further delay? Isaacs decided to forge ahead, trusting that his privileged status would protect him from the king's wrath. But while making his way there, he received a message from Fynn, who warned him to turn back because "the chiefs and people were being killed in every direction by command of Chaka for not presenting themselves to mourn." Isaacs hastily changed course for Fort Farewell. Along the way, he witnessed further signs of the violence touched off by Nandi's death. "We saw many dead bodies, that seemed to have been recently killed," Isaacs wrote, and these corpses had been left "for wild animals to devour in the night." Fynn, too, recorded the punishment visited on those who did not express their grief according to Shaka's exacting standards. "Those who could not force more tears from their eyes— those who were found near the river panting for water—were beaten to death by others who were mad in excitement," he reported. Zulu informants reported similar tales of wanton violence and ritual killing.[12] Although both Isaacs and Fynn likely exaggerated the horrors, it is clear that the king was plunged into manic despair by his mother's death. Those loyal to Shaka may have used the chaos as a cover to settle scores and compete for his favor.

The King's adoptive father, Ngomane, offered a eulogy for Nandi that demonstrated the Zulu elite's deference to Shaka, as well as the solemn affection felt for the queen. Ngomane referred to Nandi as "the great female elephant [whom] the heavens and earth would unite in bewailing."[13] Today the countryside no longer seems to mourn her. Nandi's grave rests off a lonely dirt track that winds up a rise in the heart of KwaZulu-Natal. When I went to see it, a handful of women carrying umbrellas to ward off the sun, and children dressed in their Sunday best, lingered on the edge of the rough trail to escape the clouds of dust kicked up by my passing vehicle. Cows and goats wandered the fields beyond. Traditional round houses, some with thatched roofs, dot the region. Winds whip over the hills, at times silencing the birds that shriek from the branches of stunted acacias. Today no fleet-footed warriors race their way across the landscape. No elephants lumber toward the watercourses. Only a twisted aluminum fence surrounds the rough stone cairn that marks Ndlovukazi's final resting place. A battered, hand-lettered wooden sign identifying the grave lies on the ground.

Isaacs settled in at Fort Farewell to wait out the violence. He proceeded to kwaDukuza to express his sorrow only after he had equipped himself with gifts of medicine from his stores. Perhaps he wanted the material means to barter for his life should the king resent Isaacs' long delay in paying his condolence call. He finally set out on his journey at the end of August 1827. His trail wound through an unsettled countryside. Fires still smoldered at scattered *imizi*. Dispossessed and fearful Zulu wandered from ruin to ruin, mourning those slaughtered.

At sunset on September 3, more than three weeks after learning of Nandi's death, Isaacs arrived at Shaka's headquarters. The king greeted him warmly and honored him with an invitation to the interior of his palace. Later that night Shaka suffered a bout of insomnia and summoned Isaacs from a nearby hut: "a common occurrence," Isaacs noted, "for when [Shaka] cannot sleep he sends for the white people to amuse him." Isaacs left no record of what they spoke about, but like David soothing the disturbed King Saul, he "talked [Shaka] to sleep" and then retired. The scene repeated itself a few nights later, as the official mourning period for Nandi was drawing to a close. The king was tormented by thoughts of his own mortality: he mentioned that he desired British medicines to "make him live until he was very old." Shaka slept for much of the next day. When he regained his composure, he set about performing the ceremonies required to mark the conclusion of the month-long period of mourning.[14]

Shaka's insomnia improved, but his restlessness persisted. On 8 September, Isaacs found the king still in a contemplative mood. He listened to Shaka's concerns for the future as they circled the inner domains of the royal *ikhanda*. Shaka felt embattled and alone, bereft of his mother and without advisers he could trust. He likened himself to a solitary beast roaming the plains, with nowhere to hide from packs of hunters who waited in ambush.[15] Long fatherless, with his mother thousands of miles away and his father figures distant or squabbling, Isaacs must have been sympathetic to Shaka's plight.

Perhaps it was Isaacs who suggested to the king that he seek support from Great Britain, though Shaka would have needed little encouragement. He thought the whites who lived among his people "strange [and] wonderful," and increasingly captivating. "If I understood writing, I would write to King George," he mused, "and tell him all that I feel, and what I think of the Moolongas [*abelungu*]." But lacking the means to correspond with the representatives of umGeorge, Shaka determined that he would "go to the other side of the water and see King George." First though, Shaka urged James King and Isaacs to escort two hand-picked Zulu emissaries to Cape Town to negotiate a formal alliance with the representatives of George IV. Shaka knew that the ship under construction at Port Townsend was nearly complete. Soon his *abelungu* could sail for the Cape to trade ivory for those commodities he most desired, especially Macassar oil. But Shaka also craved British recognition of his stature as leader of the emergent Zulu nation.[16]

The establishment of long-distance economic and cultural ties with other peoples was a priority for Shaka, much as it was for Great Britain and its formal and informal empire. The Zulu king's overtures to His Majesty's servants in Cape Town were no crude pantomime of British diplomacy. Rather, they demonstrated his ability to distill a sense of European cultural norms and strategic motivations from the lawless *abelungu* who straggled into his *ikhanda* between rounds of wanton elephant slaughter. Shaka expressly wanted to acquire European technology—mass-produced goods and medicine—as "tangible proof from the British government of its friendly alliance."[17] He knew how to make use of his whites no less, and perhaps even more, than they knew how to leverage their intimacy with him.

In the meantime, as shipbuilding progressed under Hutton's supervision at Port Townsend, Shaka convened a war council to plan an attack on neighboring peoples to the west. He explained to Isaacs that his goal was to expand Zulu territory until he "should reach the possessions of the white people" at the frontiers of the Cape Colony. Shaka cornered Isaacs to seek intelligence from him about these westward polities, especially whether they possessed muskets or operated under umGeorge's protection. Isaacs provided what information he could and did not trouble himself overmuch with Shaka's expansionist aims. Although the Zulu monarch's designs would bring his *amabutho* into close contact—and perhaps confrontation—with British civilians and military detachments, Isaacs single-mindedly pursued the ivory Shaka had promised him, "particularly as Mr. Farewell was about to proceed" to the Zulu king and might appropriate the ivory stores for himself. Isaacs had little to worry about. James King had already succeeded in "obtaining nearly all the ivory that Chaka was able to collect," at least according to a sulking Farewell. Shaka consistently preferred Isaacs and King to Farewell, lavishing on the two partners gifts of cattle as signs of his favor.[18]

For Isaacs, amassing a personal fortune was more important than alerting Cape administrators to the threat of Shaka's attacks. British authorities later suspected that Isaacs and the other settlers at Port Natal had even encouraged Shaka's aggression.[19] King and Isaacs would have surely welcomed a Zulu-dominated overland route to the colony's frontier to achieve their ivory-colored commercial dreams. Isaacs' loyalty—to the extent he had any loyalty to a cause greater than his own financial speculations—appeared to reside in this respect more with the Zulu than with the British.

This laissez-faire approach to matters of national interest may not be surprising. Adam Smith proclaimed in *The Wealth of Nations* that "a merchant . . . is not necessarily the citizen of any particular country." Isaacs had spent nearly three years as a mercenary and adviser to Shaka, and he recognized that his security depended on his fealty to the Zulu monarch. Although Isaacs was legally a British citizen, he had ample reason—as a member of an outcast minority, as a provincial castoff sent to a remote island, as a castaway in uncharted lands, and as a speculator casting his eyes on an ivory monopoly—to feel that he belonged everywhere and nowhere. Whatever his fears of Shaka's despotic character—Isaacs referred to him as a hypocrite, a monster, and a coward—he also felt indebted to him for his protection. And Isaacs admired Zulu bravery and stamina more than he did the pomp and class snobbery of the colonial authorities. Likewise, Isaacs had learned from his uncle Saul Solomon that fidelity to a powerful individual, whether the Emperor Napoleon or the "Napoleon of the Zulus," could be more lucrative than loyalty to the state.[20]

But was Shaka really the Napoleonic figure that histories, popular writing, and cinema have made him out to be?[21] Legend enshrines him as a brilliant commander who revolutionized Zulu warfare and whose tactics later contributed to the humiliating defeat of the British at Isandlwana during the Anglo-Zulu War of 1879. H. Rider Haggard, who traveled widely in South Africa as a colonial servant for nearly a decade and witnessed the humbling of British military might, regarded Shaka as a "Napoleon"—a "colossal genius and most evil man." Haggard had learned of Shaka's exploits from African storytellers who recalled the dread king's era and from accounts compiled by historians, who in turn drew on Isaacs' *Travels and Adventures*.[22] Thus a dotted line can be drawn from Isaacs' travelogue to works consulted by Haggard, who can be credited with disseminating tales of Zulu military prowess to a worldwide audience, principally in his novels *King Solomon's Mines* (1885) and *Nada the Lily* (1892).

Now often reviled, *King Solomon's Mines* stands as the template for later adventure tales such as Arthur Conan Doyle's *The Lost World* (1912), Edgar Rice Burroughs's *The Land That Time Forgot* (1918), the cinematic romps of Indiana Jones (1981–2008), and more recently the 2018 Marvel film based on the comic-book superhero Black Panther, a title first launched in 1966. Black Panther's father, not coincidentally named T'Chaka, was king of the imaginary lost world of Wakanda. There is little doubt that the two (Jewish) creators of Black Panther,

Stan Lee and Jack Kirby, were influenced by Haggard's novel, which has never gone out of print. In *King Solomon's Mines*, Haggard depicted the breakaway Zulu kingdom of Kukuanaland, which had once been exploited by the biblical Solomon for its mineral riches. One of the heroes of the novel, the noble Ignosi, speaks a dialect of isiZulu and presents himself as a Zulu warrior. He returns to his ancestral home to claim his rightful throne from a cruel tyrant who had instituted a system of militarization even more ruthless than that of "Chaka in Zululand."[23]

A few years later, in the preface to *Nada the Lily*, Haggard asserted that the "Zulu military organisation" had been one of "the most wonderful that the world has seen," and that its warrior spirit had been rendered irrelevant in what he disdained as his own more "polite age of melanite and torpedoes." Modern wars fought with explosives and artillery sapped British men of heroism. The Zulu, by contrast, fought manly battles at close quarters with shields, spears, and daggers. *Nada the Lily* establishes an unexpected inversion in which the Zulu perform an idealized fantasy of British bravery.[24]

European accounts supply stories of Shaka's pioneering use of the lethal, stabbing assegai (*iklwa*) in battle, instead of the longer throwing spear. Assegais were strong, lightweight wooden staffs about two feet long, affixed with double-sided iron blades that could be a foot or more long. Other tales circulate that Shaka directed his warriors to discard their fiber and animal-hide sandals and inure their bare feet to thorns and stones so as to increase their speed and mobility in warfare. Most famously, Shaka was said to have masterminded a battle formation based on the posture of charging water buffalo, the so-called horns-and-chest attack, during which warriors on the flanks gradually envelop and gore the centrally massed enemy force. Whether these innovations can be attributed to Shaka, or whether he even used these tactics, remains murky.[25]

Rather than boldly charge against his enemies, Isaacs claims, Shaka preferred to use spies in order "to find out positions in which the enemy might be attacked with impunity." Farewell similarly indicated that Shaka followed in the rear of his forces. The Zulu did not "meet their enemy openly, if they can avoid it," Isaacs recalled, preferring "to conquer by stratagem, and not by fighting; and to gain by ruse what might be difficult for them to achieve by the spear." Isaacs attests to the swiftness, endurance, bravery, and ferocity of Shaka's standing army of fifteen thousand, but he does not describe Shaka himself as a battlefield tactician.[26] In fact, Isaacs does not place Shaka at any of the skirmishes he recollects. Nor do Farewell, Fynn, or James King.

Shaka's true genius may have been in refining a system of rule—Isaacs called it "Zoolacratical"—that allowed him to exercise and consolidate his power. Unusually among European observers, Isaacs recognized the sociopolitical sophistication of the Zulu Kingdom. Since Shaka's troops lived "entirely on plunder," they were "always elated with the thoughts of war." Shaka could thus easily direct their passions. The king demanded that his warriors remain unmarried and celibate, thereby manipulating his warriors' ardor while also precluding the formation of domestic or romantic halters. The Zulu monarch himself did not abide by this prohibition: Isaacs reports Shaka flashing him a knowing smile as fifty women appeared, saluted the king, and entered his palace chambers. Presumably these women had been gifted to Shaka by his tributaries. Shaka followed them inside and shut the gates.[27] Shaka can be credited with an unprecedented militarization of Zulu society. He organized legions of warriors into cohorts by age—amabutho—and subsumed conquered chieftaincies into his polity. He may have been less a god of war, the "African Mars," as Isaacs termed him, than a shrewd empire builder.[28]

In *Travels and Adventures*, Isaacs presents an epic account of the Zulu at war and peace. Isaacs was no Tolstoy, of course. (In 1836 even Tolstoy wasn't Tolstoy; he was only a boy of eight.) Unlike *War and Peace* (1869), Isaacs' work utterly lacks a metaphysics, despite offering some finely observed and richly detailed passages. But anachronistic comparison between the unknown Isaacs and the towering figure of Tolstoy is not what I am after. Instead, I gesture towards the novelist Saul Bellow's dismissive and much-criticized question (ca. 1988), "Who is the Tolstoy of the Zulus?" Bellow's remark, which is understood to express a dismissive attitude toward non-Western cultures, remains a minor tempest in American literary studies. The Nobel laureate's shrill tone implies a hierarchy of cultural achievement. Yet Bellow, who studied anthropology, claimed that he was merely drawing a "distinction between literate and preliterate societies."[29] Whether or not we take him at his word, the Zulu were indeed a preliterate people at the time of Isaacs' sojourn among them. Thus one might answer Bellow's challenge with the observation that anonymous storytellers were the original Zulu Tolstoys, or at least the Zulu Homers.

More troublingly, one might suggest that it was Isaacs who became the Tolstoy of the Zulus with the publication of *Travels and Adventures*. Almost all subsequent accounts of Shaka, including those penned by Black African writers, draw on Isaacs. The acclaimed Zulu playwright

Welcome Msomi compared Shaka to Napoleon and surely consulted Isaacs' work in crafting *UMabatha*, an adaptation of *Macbeth* that revisits Shakan history.[30] The Zulu scholar, author, and exiled African National Congress activist Mazisi Kunene's epic poem *Emperor Shaka the Great* (*UNodumehlezi KaMenzi* [1979]) presents a clear example of the intermingling of Zulu and European mythologies of the Shakan era. The influence of Isaacs' version of events on Kunene's work is unmistakable, as is the impact of Isaacs' often self-serving perspective.

The plotline of Kunene's several-hundred-page epic follows the chronology of *Travels and Adventures*. Isaacs appears in *Emperor Shaka the Great* in a walk-on role, twice introduced with the epithet "kindly Isaacs." Elsewhere in Kunene's epic, the character of King Shaka states that among the whites, "only [James] King and Isaacs possessed true humanity." Kunene has one of Shaka's closest advisers maintain that Isaacs' "patience and kindness command respect everywhere. / I often forget he is a foreigner or white man."[31] Isaacs, always the hero in his own mind, thus becomes one of only two honorable Europeans in Kunene's poem, which advances Zulu nationalism in the face of apartheid-era oppression.

Even if Kunene emerges as a more authentic Zulu Tolstoy than Isaacs, his work nonetheless owes a debt to *Travels and Adventures*. There is a great deal of irony in suggesting that a British Jew became the Zulu Tolstoy *avant la lettre*. And there remains a great deal of discomfort, not to say an echo of cultural chauvinism, in suggesting that the Zulu people awaited the arrival of a (nominally) white Englishman to narrate their history and preserve their folkways. Yet at least for most non-Zulu readers, Isaacs' *Travels and Adventures* remains the touchstone for memories of the Shakan era.

The beginning of 1828 found Isaacs recovering from a painful inflammation of his ribs and a severe case of dysentery. He was incapacitated for weeks with the fevers, stomach cramps, diarrhea, and bloody stool that accompany the disease. According to King, Isaacs was well enough to visit Shaka in February, though Isaacs himself does not describe this audience. At this meeting, which was to be the final conclave before the embarkation of the Zulu diplomatic mission, Shaka reportedly elevated King to the position of *induna*, an officer, headman or chieftain. Shaka then bound under King's "charge and protection Sotobe one of my principal chiefs, [and] Jacob [Hlambamanzi] my interpreter and . . . I desire him to convey them to His Majesty King George's dominions to negotiate with His Britannic Majesty on my behalf . . . a treaty of friendly alliance."[32] The purpose of this treaty would be not only economic but also strategic.

Isaacs notes that the high-ranking Sotobe kaMpangalala and another trusted *induna*, Mbozamboza, had been instructed by Shaka to travel under King's protection in order to gauge whether the British would oppose his war of western expansion. He was determined to extend the Zulu Kingdom's borders to those of the Cape Colony. The Zulu, ruled by Shaka, and the British, under umGeorge, were to be equal powers who would divide and rule southern Africa between them. This territorial division likely would have made organizational and even cosmological sense to the Zulu, whose oval-shaped *imizi* displayed a bilateral symmetry and provided a spatially legible hierarchy of forces.[33] In other words, King Shaka's intention to expand west to meet the eastward-advancing British reflected to some degree the symbolic array of power expressed in the most fundamental of Zulu cultural patterns.

The Cape Colony's acting governor had gotten wind of Shaka's military ambitions as early as October 1827. A notation in the governor's correspondence ledger registers concern over Shaka's plan: "Progress made by the Caffres in Civilization are threatened with an attack by the Chief Chaca."[34] As is made clear from subsequent entries, the reference to "Caffres" denotes the frontier peoples with whom the colonists traded and sometimes skirmished.

By January 1828, citizens of the colony were warned of Shaka's aggressive posture. The *Colonist* newspaper printed the missive James King had composed the previous May at Shaka's *ikhanda* and had smuggled to the Cape inside a calabash. "I feel it is my duty, to acquaint the Government with the determination of the King (Chaka), that they be on the alert," King wrote. "[Shaka] is positively planning an attack upon the frontier tribes. His force is so very powerful, that I fear success evidently awaits him . . . though his taking up his abode so near the Colony would be by no means desirable." In the same breath, and contrarily, King urged the British not to oppose Shaka's military venture, as the Zulu ruler was "now well disposed towards our nation."[35]

Here King opposes two related concepts: tribe and nation. To an early-nineteenth-century Englishman, the terms *tribe* and *tribal* referred to any non-European, and presumptively uncivilized, ethnocultural group inhabiting a vaguely demarcated territory. (I use these terms in quoted materials from the era but avoid them elsewhere in recognition of their colonialist assumptions.) By contrast, *nation* would have registered to an Englishman as the preferable model of his own political modernity. The South African scholar Mbongeseni Buthelezi has observed that "corralling colonized peoples into administrative units termed 'tribes'" necessarily invokes an opposing concept—the supposedly superior European nation-state—and thus implies a right to colonial domination. This point is essential to understanding the history of the Eastern Cape and Natal, which were "the first settings where the idea of tribe was systematically developed and imposed" under colonial rule. Nineteenth-century British discourse also employed *tribe* and *tribal* to define Jews and Jewish identity—sometimes innocently, sometimes to impugn Jews as clannish, backward, or stagnant. Interestingly, *tribe* first appears in Middle English to translate the term *shevet*, referring to the kinship clans in the Hebrew Bible —the tribes of Israel—who had been awarded particular territories. The fact that African peoples were so often mistaken for, or believed to be, Jewish lost tribes serves to connect British attitudes toward the domestic Other and the colonized Other.[36]

King's special pleading on behalf of the "well-disposed" Shaka was a means to preserve his own commercial interests among the Zulu. Nonetheless, King expressed more concern to authorities about Shaka's aims than did his junior partner, Isaacs. King and Isaacs' audience with Shaka in February 1828 had resulted in Shaka's making them a further land grant, which provided King with yet another compelling rationale to urge the colony to come to terms with the Zulu's westward march. This grant apparently superseded Shaka's previous declaration to Farewell, because the territory indicated was overlapping: "I hereby grant him, my said friend, J S King, in consideration of the confidence I repose in him . . . the free and full possession of my country near the sea coast and Port Natal . . . including the extended grazing flats and forests . . . together with the free and extensive trade of all my dominions containing all my former grants." The document was signed by Isaacs as a witness. King Shaka and the ubiquitous cattle rustler, lifeguard, and interpreter Jacob, alias Jakot, alias Msimbithi, alias Hlambamanzi, also scrawled their marks across the paper. It is doubtful whether Hlambamanzi's English was subtle enough to express the full implications of this document to the Zulu monarch. Indeed, "the former part of this document might have been explained to Chaka but not the latter part," John Cane later testified.[37] But perhaps the specific wording mattered little to Shaka, who was accustomed to reward or punish his *induna* as he saw fit, and who in any case did not fully comprehend the importance the *abelungu* attached to written documents, borders, or private property.

In March 1828, Hutton completed the ship he had crafted from the salvaged timbers of the wrecked *Mary* and the stands of hardwood in the vicinity of Port Townsend. The Europeans of Port Natal christened the schooner twice. They initially named it *King Chaka*, writing the name on its makeshift stern. Late, it was rechristened *Elizabeth and Susan* in honor of Farewell's long-suffering wife, Elizabeth, and King's mother, Susannah. Once the *Elizabeth and Susan* had been jury-rigged with what remained of the *Mary*'s sails and found seaworthy, it crossed the treacherous bar at the entrance to Port Natal bound for Port Elizabeth (now Gqeberha), about six hundred miles to the southwest. From there it was to head to Cape Town, a further five hundred miles. Its crew and passengers included King, Isaacs, Hutton, the two representatives of Shaka (Mbozamboza and Sotobe) and their two wives, an additional six-man Zulu retinue, and Hlambamanzi, who was to serve as interpreter.[38] The man Shaka called "Swims the Seas" for his rescue of Farewell now crossed the waters by ship. The vessel anchored off Port Elizabeth on 4 May, after a mere five days at sea.

After several years of hard living on the Natal coast, the Europeans were anxious to enjoy the comforts of the coastal town. It is easy to imagine that the Zulu approached the bustle of Port Elizabeth, with its ships and cargo, its clattering carriages, raucous taverns, and multistory dwellings with trepidation. Isaacs waited aboard overnight for his friends to return with clothes in which he could decently appear in public. All he wore were trousers and the ragged shirt he had crudely stitched together out of duck canvas, probably taken from the *Mary*'s stores. He had long since abandoned any hopes of fashioning shoes. On his head he wore a cap cut from the brindled fur of a civet cat. When no clothes arrived by lighter the next day, Isaacs impatiently made his way to shore and was met by locals who gawked at his outlandish appearance.[39]

James King sent word to colonial authorities in Cape Town that he had arrived with a royal Zulu delegation dispatched by King Shaka. His friend D. P. Francis, an attorney and the newly appointed customs officer and port captain, hailed King for "his industry, workmanship, and perseverance" in overseeing construction of the *Elizabeth and Susan*. The townspeople treated Shaka's ambassadors as curiosities, amusing themselves by encouraging the Zulu to drink to excess, which "occasioned serious quarrels between Sotobe and the interpreter, Jacob."[40] Meanwhile, Isaacs and King were anxious to weigh anchor and proceed to Cape Town.

When they had first visited that city in 1825, the aristocratic Charles Henry Somerset had been governor. Now his lieutenant governor, Major

General Richard Bourke, was in command. A career military man who had served with distinction in the Napoleonic Wars, Bourke was determined to avoid the scandals of his predecessor. Somerset had been forced back to London after facing accusations of abuse of power: he had shut down Fairbairn's *Commercial Advertiser* after it criticized his policies. Adversaries claimed he had also spent extravagantly—recall his purchase of French wallpaper. Most damagingly, Somerset had been mired in a sex scandal. The claim that Somerset had engaged in "unnatural practices" with his physician had been posted on placards for all to see along a central bridge leading to Cape Town's Grand Parade one spring morning. A contemporary recorded that the charge against Somerset had "thrown the whole Cape in consternation." Homosexuality was still a capital crime, at least in theory. Branding the accusations as a "Defamatory Libel, tending to disgrace [his] Character and Honor,"[41] Somerset endured the infamy and whispered gossip until his recall. Yet had the naked truth been known, Somerset might have had less to worry about. His reputed lover was Dr. James Barry, born Margaret Anne Bulkley, a woman who passed as a man and served an entire career as an officer and physician in the military.

As if financial irregularities and hints of moral turpitude—not to mention the perpetual friction with Xhosa peoples on its borders—were not enough to agitate the colony, Governor Bourke now had to contend with an upstart, half-pay, self-styled lieutenant turned Zulu chieftain petitioning him on behalf of King Shaka, whom previous reports had described as bloodthirsty and warmongering. Nonetheless, the initial reply James King received to his request to escort the Zulu diplomats to Cape Town was promising. Governor Bourke instructed King and the Zulu emissaries to proceed "with as little delay as possible." The government even agreed to pay for the Zulu delegation's stay in Port Elizabeth. But King was warned not to allow the visiting dignitaries to travel to the colony's frontiers. Bourke reasonably feared that the Zulu entourage would use their forays to the border to gather intelligence for Shaka. King, however, insisted that he be allowed to escort Sotobe and Mbozamboza to the frontiers precisely in order to frighten them with a show of British military superiority. He further irked administrators by noting that the *Elizabeth and Susan* was not a large enough vessel to hazard a journey to Cape Town as the winter gales in the Southern Hemisphere approached. And so King demanded that he and the Zulu entourage remain in Port Elizabeth to await passage on another, more substantial ship.[42]

Governor Bourke was losing patience with King's impertinent delay in bringing the Zulu diplomats to Cape Town. He appointed Major A. J. Cloete, a young man of wealthy Dutch parentage, to investigate just what the Zulu chiefs and their half-feral handlers, King and Isaacs, were up to. Cloete's blonde hair, smooth cheeks, and delicate features gave him an almost feminine appearance. Oddly, Cloete had once fought a duel with Dr. Barry over an insulting remark. Cloete was wounded in the exchange of gunfire, but he and Barry later became close friends, and both moved in Somerset's inner circle until the governor's ignominious recall. Cloete had been born to a slave-holding family at the Cape and served as an officer in the Napoleonic Wars. Earlier in his military career, he had met King George III, who reportedly asked him for his Christian name. On receiving the answer—Abraham Josias—the flustered king spat out, "Then, damme, sir, you must be a Jew!"[43] He was most definitely not: Cloete was instead a member of the Cape gentry.

As such, Cloete lorded over those he saw as his social inferiors, including King, Isaacs, and the Zulu delegation. Aristocrats and the governing class in colonial Africa generally treated traders with contempt. In South Africa, such men were considered, in the words of the Reverend Philips, "the refuse of English society." We can only imagine what Cloete thought when he got wind of the outrageous sights lately seen in Port Elizabeth: Zulu nobility dressed in skins and carrying assegai wandering the streets, attended by a white English "chief" and a shoeless Jewish castaway in a civet-fur cap. Cloete wasted little time in reaching Port Elizabeth to make sense of the situation and to question the Zulu without King present.[44]

First, Cloete met with Sotobe, Shaka's designated representative. Zulu informants described the elder Sotobe as "very dark, with a sloping-back forehead, rather bald, [and with a] protuberant chest." A "very big man," Sotobe cut an impressive figure when he appeared in Port Elizabeth crowned with a cluster of red feathers. The Zulu ambassadors were described as "very dignified, and [of] great intelligence and talent," though the locals continued to enjoy sporting with them in the town's *winkels*—taverns. But Cloete, a martinet, refused to concede the Zulu their dignity. On one occasion he insisted on treating them "as I would the lunatic"; on another he referred to them as "mere children."[45]

Cloete began a hostile cross-examination of Sotobe, asking him, "Did you come of your own free will and consent?" At first Isaacs functioned as interpreter, perhaps because Hlambamanzi was holding a grudge against Sotobe following a drinking bout and a subsequent quarrel. The chieftain responded, "We were sent by our king to show his friendly disposition towards the governor and the white people; also to ask for medicines." Sotobe's answers did not satisfy Cloete, who interrogated him and Mbozamboza on multiple further occasions, typically slipping in to question them when King was absent. Yet Sotobe maintained his loyalty to King, as Shaka had ordered. The fealty the Zulu ambassadors displayed for their white *induna* angered Cloete to the point of distraction.[46] He posed increasingly suspicious questions to Sotobe, designed to undermine King's standing.

Rumors had likely reached Cloete of King's dubious reputation. Farewell took the opportunity of the voyage to Port Elizabeth to send a letter back to England accusing King of scheming to establish his own independent settlement at Port Natal on land previously granted to Farewell by Shaka. The swath of territory ceded to Farewell in August 1824 may have later been granted to King in February 1828, in addition to whatever grant King and Isaacs had already received for lands stretching along the uMlalazi River. Farewell begged his contacts at the Cape to inform the secretary for the colonies in London "that Chaka, King of the Zoolas, four years since granted me thirty five miles of Coast and One Hundred miles of Inland Country for some remuneration of Merchandise," and that therefore "neither Mr. King or any other person or persons have a right to form an establishment without my sanction."[47] Farewell was not entirely honest in staking his claim. In this letter, he overstated the span of coastline Shaka had (probably) granted him, from twenty-five to thirty-five miles. He closed by asking the British government for their protection against King's ploy to usurp his own dubious title to Port Natal.

By the time King had decided to disclose to Cloete his February 1828 land grant from Shaka—on 19 July, nearly three months after first arriving in Port Elizabeth—the authorities had lost all faith in him and his claims. King protested that he had intended to present the document to Governor Bourke when they met in Cape Town, but Cloete had prevented that audience from ever taking place by delaying their voyage with his interrogations. Incensed that King had concealed the grant, Cloete dismissed its significance, and the governor himself indicated that colonial authorities "would pay no attention to the document."[48] Personal and commercial enmity between Farewell and King had earlier been exploited by Shaka to his advantage, but now their discord echoed up the colonial hierarchy and hampered Shaka's diplomatic overtures.

The situation on the colony's eastern border also contributed to the authorities' unreceptiveness to the Zulu ambassadors. Strife between Xhosa groups, reprisal raids by British-led commandos against Xhosa cattle rustlers, and the ever-expanding range of land-hungry Boers created an atmosphere rife with bloodshed and rumors of bloodshed. Authorities feared a destabilizing flood of refugees into the colony. News arrived in mid-June that Shaka was advancing westward "with a very numerous body of men divided into 8 companies, each of which is supposed to be 2 or 3,000 strong."—that is, with sixteen thousand to twenty-four thousand troops. Fearful military bulletins described "the impetuosity of [Shaka's] character" and the "great number of armed savages" who planned to "assail the less warlike Kaffres"—that is, the frontier Xhosa. Dire predictions arrived from field officers that Shaka "will hardly await the return of his messengers [Sotobe and Mbozamboza] to begin his attacks." According to Fynn, Shaka did indeed launch a military operation only a few days after the *Elizabeth and Susan* set sail for Port Elizabeth in April 1828.[49] Reports of this military incursion were to vex the British, the Zulu delegation in Port Elizabeth, and Shaka himself.

Shaka arrived at Fynn's kraal (an Afrikaans word referring to a homestead) during the Zulu's westward push. Fynn recorded that Shaka's warriors had marched west that June to "exterminate the whole of the [Xhosa] tribes between him and the Colony," but he maintained that he and Shaka had innocently remained safe in Fynn's own kraal.[50] Fynn claimed that he passed the time behind the front lines teaching the Zulu king how to fry pancakes. This claim is either one of the most absurd alibis in colonial history or a great moment in intercultural culinary history. Sadly, Fynn almost certainly lied.

A British missionary on the borderlands reported at the end of June that "Chaka . . . is coming down the coast . . . with immense Hordes of people . . . One of his principal commanders bears the name Umbulawe; that is *The Murderer*, or the Killer of Men. The troops under him are compared to locusts for number." A postscript to another report by the same missionary adds that "there is a white man . . . who assists them with fire-arms." This was Fynn. Shaka had granted Fynn the praise name Umbulazi (also Mbuyazwe or Mbuyazi), an epithet that can mean "murderer," and of which Umbulawe is the missionary's own transliteration. Fynn's band of African gunmen were known to the Zulu and later to Cape authorities as the "locusts"—*iziNkumbi*. Fynn alludes in his diary to the reports being circulated "of an intended invasion by the Zulus, with me at their head," without dismissing them.[51]

Colonial forces patrolling the region likewise reported that "a party of armed Englishmen" accompanied Shaka's forces, and that "Fynn was present with the invading army" in early July. By August, Governor Bourke was convinced that Fynn had been at the head of "the late plundering expedition." These reports of Shaka's movements and Fynn's involvement were more than just rumor. Fynn had either joined Shaka's *amabutho* on their westward advance in order to increase his standing with the Zulu king, or else, under cover of waging war, had spearheaded a commando of his own to obtain ivory from wherever—and from whomever—he might.[52]

As reports trickled in to Port Elizabeth of Shaka's reputed aggressions, Cloete pressed on with his interrogations of the Zulu emissaries. Did Sotobe consider James King to be a chief? Did he believe him to be an authorized agent of His Majesty's government? Sotobe answered plainly, "We look upon Lieutenant King as a subject of King George's, and a Chief" under Shaka's authority. Cloete tried to separate Sotobe from James King. Would he agree to proceed to Cape Town and abandon King as his escort? No, Sotobe explained: "I have no objection to go with you, but I cannot leave Lieutenant King, he is sent with us on this mission; our king has put every confidence in him, and we consider ourselves under his particular care." Accused by Cloete of refusing to accompany him, Sotobe clarified his position: "We do not refuse to go with you to the governor; we say that we cannot go without Lieutenant King, as our king has made him [James King] a chief, and he is our principal on this mission, he knows the road, we do not . . . and cannot proceed without him." Sotobe complained that Cloete and King were competing for his loyalty. "You white people you must talk together," he lamented. Cloete, perhaps distrusting of Isaacs, now demanded that Hlambamanzi interpret and repeat his offer to Sotobe. Again Sotobe endeavored to make Cloete understand his position: "Your path is from the governor here [Port Elizabeth], and our path with Lieutenant King is, from Chaka to the governor [in Cape Town]."[53] But neither Cloete nor Sotobe was getting anywhere at all.

Though a military man himself, Cloete seemed incapable of comprehending that Sotobe was bound by Shaka's orders to maintain his allegiance to King. The failure was not one of linguistic interpretation but one of imagination. By this time, Isaacs possessed a fair command of isiZulu, and Hlambamanzi could have made himself understood in English or Afrikaans. Cloete simply could not conceive that the Zulu chieftains were constrained by a hierarchy of command and a code of honor every bit as rigid as those of the British military. And so he continued his effort to undermine King. He insisted that Governor Bourke "knows nothing of Lieutenant King, he is not a chief, neither is he a person authorized by the governor to act for him; if you like to go to the governor with me alone, you can." Yet again Sotobe refused. "Lieutenant King is a chief in our country," he insisted, "and sent by Chaka to communicate with the governor, and we cannot go with any other but him."[54]

A frustrated Cloete changed tactics. He now tried bribing Sotobe with "a large present" for Shaka, but only if he abandoned King. An insulted Sotobe clammed up: "We have told you all that we have to say, and that we wish to see the governor. You make us quite unhappy talking to us so repeatedly about one thing; and I now begin to think that you suspect us to be spies . . . and will not allow us to go back again." He was right. Cloete informed the governor that he believed Mbozamboza had already been through the frontier zone "spying the countries [the Zulu] were to conquer." Cloete accordingly urged the governor to allow him to delay the diplomats' return. Bourke agreed and ordered Cloete to "keep [the Zulu] at Port Elizabeth."[55]

Cloete continued to browbeat the captive ambassadors. Asked what Shaka intended, they sensibly responded, "How could [we] know, [we] being here and Chaka there"? Published reports circulated that the Zulu diplomats "complained of their detention and treatment." Cloete himself acknowledged that they "objected much to being . . . questioned." Sotobe and Mbozamboza feared they would remain Cloete's prisoners. When rumors reached them that Shaka's forces were nearby, they tried to escape. On one occasion, Mbozamboza ran away, and Sotobe was subjected to yet another of Cloete's angry interrogations, but "the old chief . . . pretended total ignorance on the subject" and frustrated the major. Isaacs and King eventually located Mbozamboza and returned him to colonial supervision. Cloete then threatened the two chiefs and "forcibly gave them to understand that if they attempted any thing of the sort [escape], they would indubitably run into certain destruction." Isaacs, once again demonstrating his sympathy for the Zulu and his antipathy to colonial bureaucracy, applauded the chiefs' attempted flights to freedom.[56]

At times, Isaacs' reckless disregard for authority eclipsed the prudent pursuit of his own ambitions. Like the Zulu, whom he reported as "hav[ing] no thought for the morrow; all [they] think of is the present," Isaacs' responses to events are of the moment. He demonstrated an innate talent for living in the present as much as any fictional picaro, and his resulting actions render him at times fearless, at times fool-hardy. Isaacs' behaviors varied as much as the seas off Port Natal: he could be placid or ferocious, generous or pitiless, all with a consistent undertow of self-interest. Tempestuousness in a shipwrecked teenaged ivory hunter in thrall to the charismatic King Shaka is understandable, of course. Fynn described Shaka as possessing equal parts "delicate feeling and extreme brutality"; the same might be said of Isaacs.[57] These contradictory traits—compassion and cruelty—in Isaacs partially account for his horrified fascination with the Zulu king.

By the end of June 1828, Governor Bourke had received written updates on Cloete's interrogations. These reports, coupled with spotty intelligence from the eastern borderlands, led Bourke to conclude incorrectly that Shaka's diplomatic mission was a ruse. He was convinced that Sotobe and Mbozamboza aimed "to return overland and thus to reconnoiter the Caffre country" to gather intelligence for Shaka. Another attempted escape by the Zulu ambassadors—this one foiled by Isaacs, who worried about the chieftains' safety—only increased the concerns of the colonial authorities. Cloete added fuel to the suspicions by casting doubt on King's integrity, certain that placing confidence in him "would mislead the government into serious errors." King was well aware of Cloete's disdain for him. The two carried on a correspondence full of mutual recriminations. King was stung by the lack of respect granted him and the Zulu ambassadors. He lashed out at Cloete with angry expressions of wounded pride; Cloete responded with the sneering punctiliousness that had been elevated into an art form by those of his caste. Had King presented his land grant from Shaka "on the first day of your interview with me, the recognized agent of the colonial government," Cloete informed him, then "this might have determined the relation in which my duty would have led me to consider you." Cloete continued, "I can now only regard you in a private capacity," thus denying King any formal recognition by the colonial government.[58]

Cloete and Bourke simply could not accept that Shaka's ambassadors were expressing their sovereign's honest desire to establish diplomatic relations with Britain and trade "ivory for medicines," as Sotobe had maintained all along. For British functionaries attuned to political subterfuge, Shaka's very sincerity was suspect. Farewell's damning portrayals of King's commercial intrigues wound their way to Whitehall and presumably reached the ears of Governor Bourke while in transit. The pair were not judged "very respectable characters," according to official letters dispatched to London.[59]

Meanwhile, eager to demonstrate his command, Bourke forwarded details about "the proposed attack by Chaca on the Caffre . . . tribes residing on the frontier of the colony" to the secretary for the colonies in London. The Governor painted a stark picture: by threatening the uneasy accord between the whites and Xhosa along the marchlands of British control, Shaka and his army presented a clear and present danger. The governor insisted that "the restless ambition of Chaca" left little hope that he could convince the Zulu king to "lay aside his project against the Caffres; and as the latter are unable to resist him alone, I apprehend that the colonial forces must shortly be put in motion."[60] Bourke was bound and determined to protect the fragile peace on the colony's borders, even if that meant going to war.

Unbeknownst to the Zulu ambassadors, King, or Isaacs, Bourke had in fact ordered troops to engage the Zulu in the field. Records of the governor's correspondence indicate that as early as 29 June 1828, Bourke "expected attack on the Caffres by Chaca" and intended "supporting the Caffres." He sent a phalanx eastward to present British demands to Shaka, whose forces they anticipated encountering as the Zulu moved westward. The man at the head of this armed mission was Lieutenant Colonel Henry Somerset, son of the recalled former governor. He rode out to the border region along with Major William Bolden Dundas, scion of a prominent Royal Navy family. The younger Somerset had seen combat at Waterloo, and Dundas had lost his left arm while under the Duke of Wellington's command. As merchant adventurers of no social standing, King and Isaacs were outmaneuvered, caught in a web of military, diplomatic, and administrative connections that stretched from the Cape to London.[61] Those who ruled in the colonies, like those who commanded in England, were part of an aristocratic coterie that used hierarchy and heredity to continually reinforce and sustain their status.

Panicked reports from the frontier, King's sensational newspaper article, and rumors from Xhosa lands all made Shaka and his warriors out to be a brutal, almost supernatural force. These monstrous projections undermined King and Isaacs' attempts to represent Zulu interests as well as their own commercial concerns. The Zulu ambassadorship failed in large part because of prior lurid accounts, some penned by King and Isaacs themselves. Reports of Shaka—the "black Napoleon"—alarmed the colony's leadership, for whom the terrors of Napoleon—the white Shaka—were a recent memory.[62] Many of the dramatis personae in the Cape Colony's farcical negotiations with Shaka's delegation were themselves veterans of the Napoleonic Wars, including Bourke, Henry Somerset, Cloete, and the one-armed Dundas.

The Zulu diplomats meanwhile remained in Port Elizabeth, embittered by their treatment and despairing of ever returning home. King asked that the authorities extend a gesture of goodwill and supply Sotobe and Mbozamboza with gifts in anticipation of their homeward voyage. These presents were to include "beads of every description . . . buttons, rings & looking glasses, tinder boxes, [and] blankets." Several of these items were indeed paid for by the government, along with a European "suit of clothes" requested by Hlambamanzi. At the same time, dispatches reporting the "hostile movements of Chaka King of the Zoolas" mounted, and these frequent updates alarmed colonial authorities as winter in the Southern Hemisphere wore on.[63]

Somerset and Dundas patrolled the eastern marchlands via different routes and met up at an encampment on the Kei River on 1 August 1828. Dundas was charged with making contact with Shaka in the field to present Bourke's demands for an immediate Zulu withdrawal. At their rendezvous, among the gray rocks and scattered vegetation, Dundas surprised Somerset with news that his small force had already skirmished with "Chaka's people" the previous week (25–26 July). Reporting that he had retaken cattle stolen by the Zulu from the British-allied abaThembu people, a triumphant Dundas informed his superiors that the depredations of the "ambitious and troublesome" Shaka had been repelled. Further dispatches noted that together with abaThembu fighters, Dundas's men had "attacked the forces of Chaka [and] completely routed them." Bourke in turn crowed to London of the "defeat of the Zoolas" and the resulting "60 or 70 Zoolas killed." Instructions from Bourke tasked Somerset and Dundas with supporting the "friendly tribes" against any future incursions by marauding Zulu. And now Bourke ordered HMS *Helicon* to ship the disgruntled Zulu emissaries and their nettlesome white chaperones back to Port Natal, along with large boxes containing gifts for Shaka.[64]

Bourke was happy to be rid of them. He believed that "Lieut. Farewell, Mr. King & Fynn . . . have encouraged Chaca to Invade the Caffres." This assertion was in flagrant contradiction to Cloete's own interrogation of Sotobe, which concluded that "no inducement . . . was held out by any of the Europeans at Natal to Chaka's advance upon the colony." Nonetheless, Bourke remained convinced that King, Isaacs, and the other whites were behind the Zulu incursion. He alerted the secretary for the colonies in London that Shaka's "designs were formed and encouraged by King and those of his party at Port Natal, for their own interested purposes." And so Bourke believed that when the Zulu ambassadors and the scheming white rabble returned to Port Natal, they would find a chastened King Shaka licking his wounds after his soldiers, supported by "Englishmen [with] their fire-arms," had been routed by Dundas—ahem—single-handedly. Bourke's only disappointment was that Dundas had "failed in . . . making a direct communication to Chaka of the object and intentions of the Colonial Government . . . to resist the attempts he may make to subjugate the neighboring tribes."[65] There was a very good reason Dundas had not completed this part of his mission: he had never encountered, much less defeated, the Zulu, as he had so proudly reported.

Shaka had in fact ordered his troops to "sit on their shields" or retreat should they encounter British forces. Fynn-Mbuyazwe, who had been active in the region with his predatory band of locusts, also records that Shaka had ordered his warriors to avoid clashing with any whites, though Fynn's armed mercenaries were not bound by the same scruples. Missionary witnesses, who offer a more disinterested perspective than either Isaacs or Fynn, also reported that Shaka's warriors stopped about 60 miles from the remote outpost of Butterworth, which stood approximately 120 miles distant from the frontier village of Grahamstown (now Makhanda). At least one missionary realized early on that Dundas's forces had skirmished with people who were not part of the Zulu polity: "Though the language of this people, their dress, their mode of wearing their hair, their weapons of war . . . agree exactly with . . . the tribe of Chaka; though they have come the very route which Chaka informed the British Government he should advance, and precisely at that period . . . still . . . all declare that they are not Chaka's tribe, but a distinct people." The invaders with whom Dundas and his abaThembu allies fought were almost certainly the amaNgwane.[66]

Somerset must have had some suspicion that Dundas had not actually encountered Shaka's *amabutho*. A field report stated unequivocally: "They are not Chaka's people. . . . They act, however, in the same manner as Chaka." Another subsequent dispatch from a junior officer who had chased the cattle raiders made no mention of the Zulu whatsoever. Somerset soon stopped referring to Shaka and the Zulu in his briefs to Bourke, opting instead for vague updates on the "movements of the enemy." Later that month, a disgruntled Oxford graduate, former colonial appointee, and lawyer bearing the Dickensian name of Saxe Bannister petitioned Governor Bourke, charging that the abaThembu forces

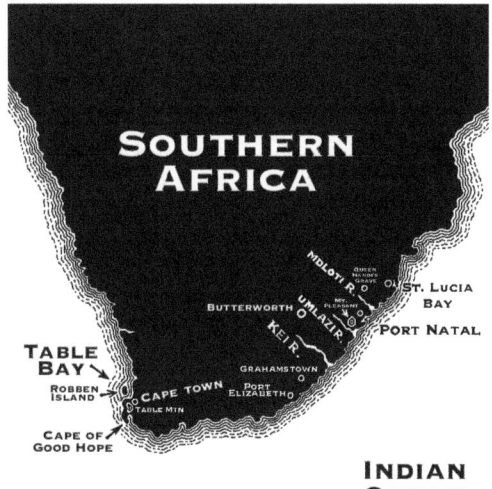

MAP 5. Detail of South Africa.

fighting alongside Dundas's troops had committed atrocities in their rout of the amaNgwane (whom Dundas had claimed were Zulu). They were guilty of "butchering the wounded, taking infants by the heels, and dashing their brains out, and were actually detected in ripping up women with child, and cutting off their breasts," Bannister wrote. He charged that Dundas and his fellow officers were "utterly ignorant" of facts on the ground and had failed to stop the massacre. He further argued that snobbish authorities—like Cloete—had squandered whatever goodwill had been built up in the eastern Cape by the first Natal settlers. "The enterprise of a few individuals, Mr. Farewell and others, at Natal had proved how well one able native chief, Chaca, appreciated some of the fruits of civilization," Bannister argued, "yet we threw away the advantages which his sentiments and ambition afforded for its extensions."[67] Bannister's brief to the governor demonstrates how some Europeans living at the Cape not only sympathized with Indigenous peoples but also hoped that British commercial contact with them would serve humanitarian ends.

Aboard ship, Isaacs and King were nearing Port Natal with Sotobe, Mbozamboza, Hlambamanzi, and the rest of the delegation. They would soon learn that Shaka had not fought against the colonial military and had in fact withdrawn from the western areas by mid-June, opting to send his regiments on cattle-raiding forays to other regions entirely. If any Zulu had attacked peoples settled along the rim of the

Cape Colony in July 1828, they would have been members of Fynn-Mbuyazwe's band of locusts.[68] Isaacs likely knew of Fynn's raids, but he shielded his friend from suspicion.

King had sickened during the *Helicon*'s voyage. Sotobe and Hlambamanzi remained at his side. Isaacs watched from aboard the *Elizabeth and Susan*, which made a smooth passage back to Natal. The *Helicon* carried a bale of red cloth and two large boxes, one containing medicines and another sundry gifts for Shaka. Michael, the rapist who had nearly gotten them all killed, swam out to meet the small flotilla as it entered the harbor. A sudden riptide carried him out to sea; no one, it seems, much mourned his drowning.[69]

King's condition worsened after making landfall, and Isaacs had to help him up the bluff to the home he had built and named Mount Pleasant. Since King was too ill to make the journey to kwaDukuza to parley with Shaka, the task fell to Isaacs. On disembarking at Natal, Isaacs and Fynn opened the two large boxes of gifts for Shaka, which contained "sheets of copper . . . a piece of scarlet broad-cloth . . . some medicines . . . a few knives and trinkets, or gew-gaws," and then repacked the contents into three smaller, more portable cases. Isaacs added bottles of medicine, and King himself supplied a "valuable looking-glass."[70] The repacking was to prove a mistake.

Isaacs led a column of men toward kwaDukuza, where he found Shaka seated amid two hundred of his warriors. Messengers had brought news of their imminent arrival with gifts from umGeorge. Shaka all but ignored Isaacs and looked with "ineffable contempt" on the arrayed gifts. Sotobe then addressed his sovereign in a manner calculated to curry favor: "You mountain, you lion. . . . Has any other black king sent people to cross the waters as you have done? We have been to a small town, and seen an officer from government, who annoyed us by asking us numerous questions, and we know not whether he looked upon us as friends or foes. Our long absence has been . . . misery to us." He then presented his king with a white box containing two cats, one male and one female. Shaka was disturbed by their meowing at first, but soon welcomed the animals when Isaacs explained that the cats would prey on the field mice that gnawed at his warriors' shields and "nibbled at one's feet and ears" at night.[71]

Shaka next asked Isaacs what he had done with the two boxes of gifts the governor had sent. When Isaacs explained that he and Fynn had opened them and repacked them for transport, Shaka berated the absent Fynn, dismissing him as "a monkey [who] wants to peep into everything." Hlambamanzi chimed in to diminish King, his one-time savior. He maintained that "King was an outcast . . . not an Englishman, and not known by George's people; that the presents sent by the Government had been appropriated by . . . King to his own purposes." Shaka, rather than expressing anger at this intelligence, evinced sorrow when he learned his friend was too ill to come to kwaDukuza. "I think they have been giving him poison on the other side of the water," he concluded.[72] Of the gifts, Shaka and his people were impressed only by King's looking glass.

Shaka's warriors gazed warily at themselves in the mirror. They waved their hands behind it to see if they could catch their images and hid their eyes with their hands, "glancing from the corner, to take a peep to see if it were imitating them." Isaacs explained to the uneasy men that the mirror held no enchantment and "was simply a production of art, and used in the white man's country for the purposes of his dressing-room." This may sound fanciful, yet perplexed responses to mirrors by previously uncontacted groups were recorded even into the twentieth century.[73] We take the mirror for granted, but it remains an uncanny object that reproduces a depthless world on a surface smoother and shinier than any found in nature. For a people whose experience of their environment was primarily lived through their senses—touch, sight, hearing, smell, taste—the slick and dimensionless, silent, odorless, and coolly bland surface of a mirror must have been disconcerting.

Shaka left his people to puzzle out the mirror and invited Isaacs to his inner sanctum. There Shaka asked him whether he had brought the Macassar oil "to make white hair black, and to prolong age, so that he might live as long as King George the Third," whom the Englishmen had spoken of.[74] George III did in fact live until the age of eighty-one, but he had spent his last decade blind, in chronic pain, and stark raving mad. Either Isaacs did not know this or he neglected to relay this information to Shaka.

Isaacs opened his medicine chest and proceeded to catalogue the contents for the anxious Zulu king. First he produced a tincture of bark for fever and expounded on its efficacy to Shaka, who sulked: "I am strong enough: do you think we are such weak things as you are?" He then proffered an ointment used to heal wounds, which Shaka dismissed: "Do you think we are such scabby fellows as you are?" Next, Isaacs opened a fragrant bottle of lavender tonic and explained that it restored one's energy. Shaka scoffed and demanded to know whether Isaacs thought his spirits "were ever dull." All of Shaka's "hopes seemed to hang" on the missing hair colorant, Isaacs recorded. "The medicine I want is the stuff for the hair," a frustrated Shaka repeated. When Isaacs failed to pull Macassar oil from the medicine chest, the king turned his back on him, stretched out, and promptly fell asleep. In the morning, Shaka again sent for Isaacs and reiterated his demands for the Macassar oil.[75]

This tale of the king's desperate desire for the hair tonic sounds like a fairytale. Yet multiple sources attest to Shaka's obsession with Macassar oil, which promised a cure for his fears of mortality. One document carried by a loyal foot soldier inside a bullock horn to HM Government reads: "King Chaka of the Zoolas sends to King George a present of Elephants teeth and want[s] in return cows tails medicine large dogs Macassar oil." On the reverse, beneath a note reading "Chaka's mark made by him," an elaborate scribble appears, in which the shaky forms of English letters half emerge, as if in cypher: this is Shaka's autograph. Another document records one of Shaka's soldiers, Monagali, stating: "Chaka has understood that Macassar Oil will prevent his hair turning grey & he requests a few bottles may be sent to him. This is a private request to the government which Chaka is desirous should not be revealed to his people." And a later deposition of John Cane records that Shaka "expected a present . . . such as Macassar Oil [and] Chaka requested . . . to keep it secret that he desired Macassar oil, and not to let any of his people know it."[76] Shaka feared that the few white hairs on his head and beard—now more apparent to him than ever, thanks to James King's looking glass—would lead to his overthrow.

Enraged by the lack of Macassar oil among the gifts, Shaka threatened Isaacs' life. Hlambamanzi fueled the king's ire by insisting that the settlers had deceived him, that the English were a puny people who might easily be conquered, and that James King was not from Great Britain. Weary of Shaka's threats and Hlambamanzi's corrosive influence on the sullen King, Isaacs made the unprecedented decision to leave kwaDukuza without Shaka's permission. He hurried back to Port Natal and found King bedridden. Isaacs consulted with Fynn about treatment, and the two nursed their patient using a copy of William Buchan's *Domestic Medicine*, a medical reference for laypeople first published in the eighteenth century. They initially diagnosed King as suffering from a liver complaint but later thought he might have dysentery. Isaacs and Fynn administered calomel and jalap, both purgatives. Then they dosed King with various concoctions of rhubarb, cinnamon, rice water, and cream of tartar.[77] While these ministrations had the makings of a passable pudding, they did little to improve King's condition.

Isaacs knew that his friend would not long survive, and so he made his way down from Mount Pleasant to Fort Farewell. He urged Francis Farewell to visit King on his deathbed, but Farewell refused to be reconciled. King was aggrieved that his former friend retained such bitterness, and he lamented to Isaacs: "Oh, Nat, what a pity it is. . . . Oh! What will my friends think of me?"

As illness ravaged his body, King spat thick phlegm, vomited, and suffered convulsions. Killing men had been easy for Isaacs; watching his friend die was more difficult. King begged Isaacs to give him sixty drops of laudanum—a tincture of opium—to put him out of his misery. Isaacs could not bring himself to do so and instead prepared fifteen drops diluted with a small amount of water. King took the glass and looked askance. "You have been my companion a long time," he said to Isaacs, "and can you now with confidence tell me, here are sixty drops of laudanum?" King attempted to speak further, gave one last rattling breath, and lay back dead. He was interred on Mount Pleasant, his body protected against scavenging hyenas by flagstones.[78] His body lies there still, surrounded by prosperous-looking homes that ring a small park set in a traffic roundabout. Grass encroaches on the flagstones. Teenagers sit on a cement bench playing cards near a brass plaque that proclaims King's specious rank evermore as "Lieut., R.N."

A week after James King's death, Isaacs, Fynn, and Hutton led a delegation to kwaDukuza to relay the news to Shaka. He forced the *abelungu* to wait while he oversaw a round of executions and then asked them to rehearse the circumstances of King's demise. Nothing could shake Shaka's conviction that King had been poisoned. Isaacs reported that Shaka "expressed great sorrow and, although a savage, shed tears." The king ordered the performance of a purification ritual during which a calf was eviscerated, the membrane surrounding its liver removed, and the fat sprinkled around Shaka, who then intoned: "I look upon the deceased as one of my family, and had he been a brother of my own mother I could not have felt the loss more."[79]

Shaka considered Isaacs to be the rightful inheritor of King's title of *induna* and of any land previously granted to him. And so Isaacs-Dambuza reigned as chief over a tract of land stretching along approximately twenty-five miles of coastline from the Mdloti River north of present-day Durban to the uMlalazi River south of the bluff, and about one hundred miles inland. A document dated 17 September 1828 reveals that Isaacs' land grant superseded Farewell's original allotment as well as the apportionment of much of this same territory to James King in

February 1828. The language of this grant resembles that of previous documents, though a certain warmth of phrasing—surely supplied by Isaacs—is evident:

> I, Chaka, King and Protector of the Zooloos, do hereby create, in presence of my principal chiefs and strangers assembled, my friend, Mr. Nathaniel Isaacs, Induna Incoola, or Principal Chief of Natal, and do grant and make over to him, his heirs or executors, a free and full possession of my territory . . . with one hundred miles inland from the sea, including the Bay of Natal, the islands in the Bay, the forests and rivers between the boundaries here enumerated. . . . I also grant him a free and exclusive right of traffic with my nation and all peoples tributary to the Zooloos. . . . So does the powerful King Chaka of the Zooloos recompense Mr. Nathaniel Isaacs for the services rendered to him. . . . All this and my former gifts I do now confirm, and wishing peace and friendship I sign myself.

Isaacs described Shaka and Hlambamanzi setting their marks to the paper in a solemn manner. But Shaka was annoyed that his interpreter's mark was so large and insisted on taking hold of the pen and paper again in order to dwarf Hlambamanzi's scrawl with his own grandiose autograph.[80]

Although the provenance of this particular document is doubtful, I believe that Isaacs was indeed invested as a "chief of Natal"—an *induna*—by Shaka, at least by verbal decree. Despite Isaacs' penchant for self-promotion, the published version of his journals accords well with other extant documentary sources. The intrusion of self-serving explanations, simple errors, and exaggerations in *Travels and Adventures* is no greater, and often much less egregious, than in other works on Africa of the period. And Isaacs' understanding of the Zulu Kingdom's political dynamics and Shaka's personal aims far surpassed that of the colonial authorities at the Cape. His biases—a sense of European superiority, an impatience with colonial gentility, an aversion to Farewell's self-righteousness, and an enthusiasm for settling Natal for financial gain—do little to dull his sharp observations of the landscape he traversed and the peoples he encountered.

Shaka permitted the European entourage to leave kwaDukuza and return to Port Natal the next day. On the morning of 18 September they began trudging back along the coast through torrential rains. The sodden ground made their march of fifty miles or so a slippery, squelching misery. They arrived in Fort Farewell two nights later in a subdued state. Isaacs found an unrepentant Farewell tending the colony's garden. Hutton was busy recaulking the *Elizabeth and Susan*.[81] The next day, pleasant sunshine and a hearty breakfast refreshed Isaacs. But the idyll was shattered by the arrival of a breathless messenger who had raced all the way from kwaDukuza bearing shocking news: Shaka was dead.

Contrasting versions of Shaka's assassination have accumulated over the years. Isaacs and Fynn, who are among the most proximate witnesses we have, reported that on the night of 22 September 1828, the King's personal attendant, Mbopha kaSithayi, raised a clamor. Mbopha, described as dark-skinned, about six feet tall, and stout, was the only man trusted to carry a spear in Shaka's presence. While Mbopha distracted Shaka, two of the king's brothers, Mhlangana and Dingane, approached from behind and stabbed Shaka in the back with assegai. The wounded king threw off the animal skin he was wearing and leapt away from his assailants. They raced after him, along with Mbopha, who caught up to the weakened Shaka and jabbed him with his long cattle spear. The king begged the plotters not to kill him and prophesied that his death would end Zulu sovereignty. "*You are killing me,*" Shaka cried as he fell to the ground, "*but the land will see locusts and white people come.*"[82] The conspirators butchered him where he lay.

Shaka had been assassinated, the messenger relayed to Isaacs, in order "to put an end to the long and endless wars, and mourning for that old woman Umnante [Nandi], for whom so many have been put to a cruel death." A later Zulu informant also claimed that Shaka had been murdered because he had become known as "*the wrong-doer who knows no law.*" Other oral sources noted Shaka's cruelty, with people being "killed off indiscriminately . . . without any trial being held." One Zulu man highlighted another contributing factor: King Shaka was gray-haired when he was murdered, a sign of vulnerability.[83] Perhaps Macassar oil would have saved Shaka's kingdom after all.

CHAPTER 4

Appetite for Consumption, 1828–1832

Shaka's corpse lay unburied overnight in kwaDukuza, twisted in horror where he had fallen. Sotobe, once Shaka's loyal ambassador, advised the regicidal princes that the king's body must be interred to placate his spirit. Solemn men dug a large grave, stuffed a piece of Shaka's clothing in his mouth as if to stop his screams in the afterlife, and removed the assegai still clutched in his lifeless hand. He was buried with his woven mat, a blanket, and the beads that signified his wealth. A detachment of warriors stood guard over the burial place.[1] Today, a white marble monument in kwaDukuza commemorates Shaka, "the Founder, King, and Ruler of the Zulu Nation." But a knowledgeable local told me that Shaka's true final resting place lies across the potholed street, beneath a butcher's shop bearing his name.

After the murder, Dingane and Mhlangana waited to press their competing claims to the throne until the bulk of the Zulu army returned from raids to the east. Meanwhile, the Europeans worried that the country would descend into civil war.[2] Isaacs surely feared that his land grant from Shaka and his investiture as *induna* would be null and void. And now that James King and Shaka were dead, Isaacs had lost two father figures and protectors. In this tense atmosphere, King's former footman, Nasopongo, returned from a march with John Cane and Mbozamboza to Port Elizabeth. Shaka had sent them on a follow-up Zulu mission to British government representatives shortly before his assassination. With Shaka in his grave, Mbozamboza now aligned himself with Dingane. Nasopongo, who attached himself to Isaacs, brought news from the Cape.

Shaka's second delegation had been treated with greater dignity. The carpenter John Cane, and not Mbozamboza, had been the one subjected to interrogation. Cane was judged to be "shrewd and intelligent" by officials, perhaps because he verified the colonial government's suspicions of King, of whose demise they were still unaware. He reported that Shaka intended to grant territory to the British to establish a settlement that stretched from ten miles west of Port Natal to the present-day site of Port Edward, a distance of about eighty-five miles. Shaka was "anxious to maintain friendly relations with the white people," Cane insisted, but was concerned because "Mr. King tells him one thing and Mr. Farewell another, and consequently . . . he cannot believe either of them." Therefore, Shaka requested that the British send him an "accredited agent . . . in order that he may fully and clearly understand the desires of the government to which he is willing to conform."[3]

Before his assassination, Shaka had realized that his communications with the British were compromised by the use of messengers, whether the avaricious whites, the canny Hlambamanzi, or the loyal but insufficiently shrewd chieftains Sotobe and Mbozamboza. Instead of using these intermediaries, Shaka hoped to imitate the *abelungu* and send letters to the government. That correspondence would be sealed with a signet ring that Shaka had requested from the governor. Cane informed Major Cloete of Shaka's wishes: "A seal with [his] name engraved and some sealing wax" for the communiqués, as well as "medicines . . . emetic and salts, blankets and thick warm clothing, oxtails, dogs, and some bottles of Macasar [*sic*] oil to prevent his hair turning gray." Nowhere in the extant record of these missions does Shaka request European weaponry or threaten British interests. Even the wary Cloete noted that Shaka sought to "send ivory for medicines."[4] Nor does Cane's later testimony depict Shaka as the bloodthirsty monster of colonial fever dreams. Rather, Shaka emerges as a rational statesman and eager trading partner frustrated by the Englishmen's games of pass-the-parcel.

Through Cane, Shaka had communicated to the authorities a "strong suspicion" that King had appropriated a gift of "fifty Elephants' teeth" he had sent as a goodwill gesture to King George IV. All but two of the tusks entrusted to King were indeed sold off as his personal property. Isaacs, as King's junior partner, may have profited from the sale. Shaka's plan to establish an alliance with umGeorge was apparently frustrated by King's greed and duplicity. On his deathbed, King realized that Cane would discover the truth when he arrived at Port Elizabeth and that his double dealing would be reported to the volatile Shaka. "Cane has gone away to injure me," he lamented to Isaacs in his last hours. Indeed, Shaka had upbraided Isaacs and his partner: "All of you . . . would tell me anything for the sake of elephants' teeth."[5] His rebuke of the Europeans was right on the mark. As Shaka was aware, Farewell and King had severed their friendship in bitterness over ivory, and Fynn had slashed a bloody trail through the western marchlands in quest of "white gold."

Goods and correspondence could take weeks or months to cross the eight thousand miles of ocean that separated Cape Town from London. So although the secretary of state for war and the colonies, Sir George Murray, did not yet know of Shaka's assassination or of James King's death, he had been apprised of their respective schemes for territorial expansion—the one through conquest, the other through commerce. Murray dispatched a letter signaling his belated approval to deploy forces to protect the frontier groups "whom Chaka has determined to exterminate." The claim that Shaka aimed to destroy British-allied peoples was of course based on faulty intelligence from Cloete and Bourke. All Shaka had intended was to expand west to the colony's frontier and then formalize an alliance with the British.[6]

Murray knew that armed Englishmen had been seen among the Zulu regiments: "The names of all these Englishmen have not been reported . . . but it is known that two British subjects, to wit, Lieut. Frank Farewell of the Navy, and Mr. James Saunders King, who has served . . . in the Navy, are settled at Port Natal, where they enjoy considerable grants of land from Chaka." Murray's letter almost, but not quite, accuses King and Farewell of being soldiers in Shaka's service. (He knew nothing of Isaacs, who had in fact fought for Shaka, nor was he aware of the lawless Fynn and his roving band of "locusts.") The secretary also indicated that King and Farewell were "placed in positions, in which their personal interest is at variance with their duty to their country" and suggested that the pair be warned "of the consequences to which they will subject themselves, if they should assist Chaka in his hostile designs."[7] But by the time the admonition arrived, both Shaka and King were in their graves.

Of those named by Secretary Murray, only Farewell remained among the living. He attended James King's funeral and then made amends, sincere or not, to Isaacs, who was his only ticket back to Port Elizabeth. In the turmoil following Shaka's assassination, Farewell determined to sail to the Cape aboard the jury-rigged *Elizabeth and Susan,* which was now in Isaacs' possession. Once back in the colony, Farewell would press his suit with the upper echelons of government.[8] He had endured shipwreck, twice survived drowning, avoided the open hostility of Shaka, and outlasted his detested adversary, King. Now Farewell sought whatever advantage rank and circumstance might provide him as he navigated his dual loyalties to the Crown and to the princes aspiring to the Zulu throne.

His streak of bad luck—King's machinations against him, his loss of influence over the other Europeans, and Shaka's dismissal of him as an "old woman"—appeared to be at an end. Farewell gambled that the governor's mandate to maintain stability on the colony's borders and his willingness to employ the military to that end would guarantee his own freehold in Port Natal, even if the murderous princes would not uphold Shaka's grant of land to him. And so Farewell launched a private diplomatic mission to convince the colonial government to establish a beachhead in the Zulu Kingdom before it collapsed in internecine war. He would be vindicated, and Whitehall would hail his foresight. But first Farewell needed Isaacs' cooperation.

No longer a callow youth, Isaacs led the preparations of the *Elizabeth and Susan* after the vessel's principal designer, John Hutton, fell ill. Isaacs prescribed what medication he could. Farewell, too, was solicitous of their sick comrade. Fynn, meanwhile, had gone to pay a visit to Dingane and Mhlangana in order to test the atmosphere at the Zulu capital. He returned with a gift of two elephant tusks and good news: the princes were anxious for the whites to remain in Port Natal. Fynn reported that Dingane had already overturned some of Shaka's unpopular decrees, including the prohibition against warriors marrying, and he had reduced the number of his tribute wives, permitting them to wed others.[9] As a result of these and other measures, Dingane was proving popular, and his half-brother, Mhlangana, was being marginalized. Fearing that the sidelined prince was plotting to ascend the throne, Dingane then had his henchmen kill both Mhlangana and Mbopha, the third conspirator and Shaka's intimate. Sotobe, a former ambassador to the British, was put to death as well.[10]

As the unchallenged Zulu king, Dingane, unlike Shaka, did not seek to rival umGeorge; he merely sought increased trade with the *abelungu* and the enjoyment of his power. Farewell and Isaacs prepared to set sail on the *Elizabeth and Susan* to deliver Dingane's message of peace to the colonial government. Naturally, they would also tout the opportunities that awaited the British at Port Natal with the warlike Shaka in his grave and his supposedly peaceable half-brother on the throne.

But death struck again before the *Elizabeth and Susan* could sail for the Cape. Hutton succumbed to dysentery despite Isaacs' attentiveness. Dommana, the young woman whom Hutton had rescued from an abusive father, had kept vigil by his sickbed. Her efforts to nurse him back to health were a "labour of love," and he returned her affections. After Hutton's death, Dommana's "amiable disposition" altered. She sank into impenetrable grief and refused to eat. Although the red-headed Charles Maclean—a favorite of hers—did his best to cheer her, "the heart-broken creature . . . lay extended on a mat, her eyes presenting a painful vacant stare," and only occasionally would she emit "a deep drawn heavy sigh" as a sign that she yet lived.[11] Dommana survived Hutton by just three days and was laid to rest beside him.

Isaacs gathered Hutton's and King's personal effects and loaded them aboard the *Elizabeth and Susan*. As King's executor, Isaacs assumed command and set sail with Farewell, Maclean, and a skeleton crew. They anchored in Algoa Bay on 15 December 1828. Isaacs' first acts in Port Elizabeth were to inform King's friends of his death and to arrange to send his possessions to his parents in far-off Canada. A friend of King's, the customs officer D. P. Francis, offered to help with the shipment of these paltry items. The only valuable item in King's estate was the *Elizabeth and Susan*, which Francis had praised on her maiden voyage to Port Elizabeth. As soon as Isaacs received Francis's promises of assistance, he began drafting a report on the recent events at Natal to update the governor. And as soon as Isaacs had turned his attention to that important task, Francis impounded the *Elizabeth and Susan* on the pretext that she was an unregistered foreign ship.[12]

Official schemes to seize the vessel had been hatching for months behind Isaacs' back. While King was in Port Elizabeth in early June, Governor Bourke had prevented the *Elizabeth and Susan* from being registered under the British flag. The decision rested on a legal technicality: Port Natal was "a foreign settlement and no foreign built vessel is intitled [*sic*] to a British register." King had protested that much of the rebuilt schooner—including "her sails, her rigging and her iron work"— had been salvaged from the *Mary*, a British-owned and British-registered vessel. Furthermore, the replacement of timbers "felled in the wilds of Africa" had been accomplished by British labor, and the whole cost of repairs had been borne by a British subject. But Bourke's interpretation of the British Registry Act trumped common sense. King had been furious. As an unregistered ship, the *Elizabeth and Susan* could be "lawfully seized and confiscated by every flag" on the high seas. King had in fact delayed the ambassadors' return to Natal on the grounds that "having black people on board"—that is, Shaka's emissaries and their retinue—"she might be taken as a slave trader," and that lacking proper registration, he, his crew, and his Zulu passengers would be subject to capture.[13] It was partially for this reason that Bourke had ordered the *Helicon* to escort the *Elizabeth and Susan* on her return journey to Port Natal.

Isaacs seems not to have known about the dispute over the ship's registry. He protested that Francis had allowed the *Elizabeth and Susan* to sit at anchor in the port and to return to Natal only a few months earlier. Indeed he had, but only after the governor had ordered that Francis not interfere with the vessel's departure. Bourke wanted to be rid of James King and the Zulu emissaries. Now that King was dead, Francis pounced. He assured Isaacs that when the confiscated ship was sold, he would generously forgo the proceeds to which his position as port officer entitled him. King's surviving family likely never saw a penny; neither did Isaacs, who concluded that Francis was nothing but a "wholesale dealer in unredeemed pledges." Shortly afterward, Francis narrowly avoided suspension for having removed ivory from the *Elizabeth and Susan*, possibly in a kickback scheme. Young Charles Maclean, with his captain dead and his legacy appropriated, had experienced enough colonial chicanery and sought a passage back to London. Maclean lived his life at sea and eventually became a ship's master and a fervent antislavery advocate.[14]

Francis's confiscation of the *Elizabeth and Susan* cemented Isaacs' contempt for "the power which official duties give."[15] He had been frustrated once too often by the insolence of office. Isaacs resented the way colonial functionaries obscured their misdeeds behind a bureaucracy buttressed by titles, patronage, and red tape. But Isaacs also absorbed the government's lessons in how to wield the power of pen and paper, much as he had learned how to employ force and brinksmanship from the Zulu. He firmly grasped that there was profit in power, and power in profit.

Isaacs and Farewell had reconciled to some degree, united in their belief in Port Natal's future. Their enthusiastic reports about the region's beauty and fertility circulated throughout the Cape Colony and stirred considerable interest in annexing the territory for Great Britain. Saxe Bannister, the activist lawyer who a few months earlier had charged

Governor Bourke with a cover-up of atrocities committed by colonial forces against the amaNgwane, began formulating a "humane policy" of colonization. Urged on by Farewell, Bannister rallied a "mercantile body at Cape Town" to support settlement, increase the ivory trade, put a stop to the lawless Boers' "modified slave-trade," and remove "obstacles in the way of . . . civilizing" the Indigenous peoples. In the detailed outline he submitted of his plans to colonize Natal, Bannister did not mince words, The document's first line charged: "The conduct of the British government in the colonies is not sufficiently just to the natives." He insisted that African peoples possessed "a sufficiently clear acquaintance with principles of justice to see the inconsistency and injustice of our course towards them." Bannister joined ranks with those Native peoples subjugated by British rule to ask the secretary of colonies: "What sort of government is that which will not avenge the wrongs committed by its own subjects?" Unlike Farewell and the other whites, including Isaacs, Bannister appears to have been wholly motivated by humanitarian aims. The first of these aims, "civilizing" African peoples, was inextricably bound to the second, abolishing slavery. And through the pursuit of these goals, British territorial expansion increased apace.[16]

The governor stiffly responded to Bannister that he "could not perceive the advantages of the proposal" to annex Port Natal.[17] Farewell likewise found his way blocked. Representatives of the colonial government distrusted the motley collection of ungentlemanly naval officers (Farewell and King), a reputed thief and known ruffian (Fynn), a Jewish fortune-hunter (Isaacs), and a rash malcontent (Bannister) who submitted their briefs on the future of Port Natal. And their wavering loyalties—to King George IV, to King Shaka, to the rights of "savages," and to their own commercial speculations—seemed to prove their untrustworthiness.

One officer who met with Farewell and others at Natal claimed that "truth seems to be utterly unknown to the whole party." He singled out Farewell in particular, noting that he "does not speak out like an upright honest man and, moreover, he looks very much like a drunkard." The newly installed colonial governor, Galbraith Lowry Cole, another wounded veteran of Wellington's campaigns against Napoleon, informed his superiors in London that Farewell's statements about the desirability of settling Port Natal could not be credited. "As far as H.M.'s Government is concerned," Murray instructed Cole with finality, "there is no intention of forming any settlement" in the vicinity of Natal. Cole was to report Murray's decision to the would-be colonizers in no uncertain terms. An entry in Cole's correspondence records his succinct assessment of their schemes for Port Natal's development: "Harbour insecure & Country unfit for settlement." This reluctance to embrace new territory under British sovereignty is less surprising than it might appear to many. Throughout the first half of the nineteenth century, British statesmen often preferred "informal empire" to formal claims, especially in Africa.[18] Commercial beachheads were far less costly to extend, maintain, defend, or disown than recognized political possessions.

Farewell was disillusioned by Whitehall's opposition but did not relent. He traveled to Cape Town to haunt Cole's offices and plead his case. The new governor had by now become convinced that Farewell and Fynn had contributed to the lawlessness beyond the colony's borders, and he resisted any scheme they or their advocates proposed. He did, however, grant Farewell one hundred pounds of gunpowder in March 1829 in order to be rid of him.[19] The obstinate Farewell doubled down, deciding to return to Fort Farewell and annex the area in his own name, despite the explicit refusal of governmental support. Unable to sail back to the Zulu coast with the *Elizabeth and Susan* now impounded, Farewell was forced to travel overland to Dingane's kingdom, as Cane and Mbozamboza had done. Organizing such an expedition took time, and Farewell remained in the Cape Colony until September.

Isaacs left Farewell behind and arranged to sail for St. Helena. Disgusted by his treatment at the hands of British officials, reeling from the upheaval in the Zulu Kingdom, still grief-stricken for James King, and ailing from his years as a castaway, Isaacs resolved to recover his strength at the home of his uncle Saul Solomon. He may have booked a passage on one of the fast-sailing, teak-timbered, copper-bottomed schooners that plied the route between Cape Town and Jamestown. By 1829, St. Helena had shrugged off its infamy as Napoleon's penitentiary. The East India Company resumed control of the island but swiftly saw its economic vitality throttled. When the military garrison departed after Bonaparte's death, the rapid drop in population led to a decline of commerce, except for the trade in flesh. Impoverished women turned to prostitution. The transgender Dr. James Barry, Governor Somerset's reputed lover in Cape Town, was posted to Jamestown and later described with dismay the venereal disease rampant among residents.[20]

With fewer visitors to fleece, Saul Solomon's business empire fell on hard times. But Solomon welcomed his long-absent nephew and again provided him with a home and employment. Isaacs convalesced in the "peculiarly fine air" of St. Helena, where only eight deaths due to consumption—the "white plague" that caused about 25 percent of all fatal illnesses in London—were recorded for 1829 on St. Helena.[21] Life was good in Jamestown, so long as you were not one of those men or women yoked in servitude under the East India Company's legal exemption from the Slave Trade Act.

As Isaacs regained his strength (and presumably discarded his civet cap for more genteel dress), he found his uncle eager to discuss economic opportunities in the Cape and even to consider relocating his business empire there. One day, while relaxing with Solomon and a guest, the captain of an American vessel, Isaacs related his adventures in Natal. The American captain pressed Isaacs for details of the coast and the possibilities of trade with the Zulu. Isaacs was pleased to furnish him with "information respecting that unknown and unfrequented port." He heartily approved of the American captain's "enterprising spirit" and willingness to seek new outlets for trade.[22]

In these accolades we can hear echoes of Isaacs' distaste for British nobility, not to mention his interest in exploiting emerging markets. Isaacs knew that the American captain would "lose no time in making [the American] government acquainted" with the news, and that a nation barely fifty years old "would greedily and thankfully receive every possible information that might enable it to extend its commercial intercourse." Isaacs at first made out that the valuable intelligence had innocently "escaped [his] lips over a cup of tea." Later he confessed that he divulged the information because he was still embittered at the colonial government's high-handed dismissal of Shaka's ambassadors, the unjust seizure of the *Elizabeth and Susan*, and the refusal of his repeated requests to submit his eyewitness reports on the Zulu to the governor.[23]

While Isaacs wined and dined in Jamestown and regaled his extended family and their guests with stories of his adventures, Farewell made his final arrangements for his overland trek to Port Natal. His journey took him through the western edge of Zulu lands, where Nqeto, a chief in open rebellion against King Dingane, held sway. Nqeto (called Catoe or Kato by Isaacs) had been an ally of Shaka. Farewell offered to help Nqeto form closer relations with the colonial authorities. "I apprehend little danger from [his] nation," he confided. Farewell was accompanied by several Europeans, including John Cane, who had already made the overland journey at Shaka's insistence, a number of African porters, a dozen or so horses, and a few ox-drawn wagons containing firearms, gunpowder, clothing, and more than two tons of beads for trade. Farewell knew that Dingane would be pleased with these treasures. The new king had also requested such items as a scarlet cloak, brass forearm guards, "tea trays figured with lyons or tigers," ivory snuffboxes, white and blue crockery, toothbrushes, razors, a shaving brush, and a looking glass.[24]

Farewell was apprised of the hostilities between Nqeto and Dingane, and Nqeto knew that the lieutenant was travelling with wagonloads of gifts for his rival. But Farewell and a small party nonetheless turned off the main trail toward the rebel chieftain's *umuzi* at the end of September 1829. Cane remained behind at a crossroads with a cadre of Hottentots to guard the wagons. Nqeto and Farewell exchanged information and presents into the evening, and the renegade chief offered Farewell every hospitality. Rather than return to his wagons, Farewell and his men pitched a large tent on level ground near Nqeto's *umuzi* and settled down to sleep.

Near midnight, a detachment of Nqeto's men surrounded Farewell's tent, slashed the supporting ropes, and when the canvas collapsed over the occupants, attacked them with assegai. The career of Farewell, one of the most inaptly named men in colonial history, was at an end; he and several others were speared through, bludgeoned, or hacked to death. One survivor rushed back to Cane to warn him and the others of Nqeto's treachery. The Hottentot guards abandoned Cane, who likewise fled into the bush. Nqeto's warriors looted the wagons and brought the contents back to Nqeto, who had feared—with ample reason—that Farewell would offer valuable intelligence on his rebel forces to King Dingane.[25]

Farewell left behind a widow and son who learned of his death only months later. The news of the massacre did not reach Isaacs for some time either. When his health improved, Isaacs began seeking a passage back to Port Natal to protect his commercial interests there. Just as Isaacs had anticipated, the captain of the American vessel had returned to the United States and passed on word of trading opportunities along the Zulu coast. Soon another American-flagged ship, under the command of a Captain Page, arrived in Jamestown. The fast-sailing brig *St. Michael* had been outfitted by a significant New York shipping concern, P. J. Farnham and Co., based on the detailed information Isaacs had supplied a few months earlier. Captain Page disembarked, sought out Isaacs at his uncle's, and offered him "handsome emoluments" to proceed to Port Natal as his guide and supercargo.[26] They weighed anchor on 18 February 1830.

St. Michael reached Natal Bay at the end of March. Captain Page and a handful of sailors boarded a longboat to sound the treacherous bar at the bay's entrance. Surging waves capsized the longboat, plunging Page and the others into the churning surf. They managed to cling to the keel of the overturned vessel as it bobbed in the current. One crewman was swept away and struggled to keep himself afloat, swimming vigorously for the outstretched arms of his friends until a shark seized him and dragged him into the depths. Isaacs watched helplessly on deck as the horror unfolded. After two hours of struggle, he saw the remaining survivors waving from a sandy outcrop. There the drenched sailors spent the night while the *St. Michael* stayed safely out to sea, away from the riptides and fierce winds that pounded the narrow mouth of Port Natal.[27]

In the morning, Page and his crewmen rowed back to the *St. Michael* in the recovered longboat. They brought with them a shipwrecked Portuguese sailor who had survived the sinking of the *African Adventurer*, a slaver that had gone down near the Umgeni River nearly three months before. Portugal did not outlaw the slave trade in its colonies until the late nineteenth century. The sloop had departed the Mozambique coast laden with 160 slaves and enough supplies for a two-day journey but had gone off course. For three weeks the *African Adventurer* drifted until the crew spotted the Natal coastline. By that time, those on board had endured days without water. Slaves cramped below or chained on deck beneath the brutal summer sun perished in agony. The dead were thrown overboard—as were many of the living, in order to conserve the meager rations. Only thirty of these wretched captives remained alive when the ship ran aground. The wreck of the sloop was later surrounded by bleached heaps of the bones of those who had survived the hellish voyage, only to succumb on shore.[28]

The Zulu did not participate in slavery and treated the commerce in human lives with repugnance. Those who did deal in slaves, like the Portuguese inhabiting a distant outpost to the east of the Zulu Kingdom, were disdained. Dingane was said to have asked pointed questions about the Portuguese treatment of slaves and had been reluctant to help the surviving mariners of the *African Adventurer*. Isaacs recalled their victims' ordeal in dismay, recording that "such a recital of human suffering, from the detestable custom of slavery, is enough to melt the mind to pity." He mourned "that such an execrable traffic should be permitted to be carried on by a nation [Portugal] which might be immediately annihilated by the British power."[29] Power exists to be applied, Isaacs suggested, and power ought to protect freedom, even that of Africans.

Nasopongo, formerly James King's assistant and now Isaacs' "faithful boy," greeted him affectionately when he disembarked from the *St. Michael* in Port Natal. Nasopongo was no child: whites in South Africa often infantilized their adult Black servants by referring to them as "boys" or "girls." Some still do. From John Cane, Isaacs learned the details of Farewell's murder. Fynn soon arrived and escorted Isaacs to his kraal, where the friends celebrated their reunion and discussed how best to resume their "commercial operations" under King Dingane, who they hoped would honor the territorial grants and trade rights extended to them by Shaka.[30] They selected merchandise to present to Dingane from the cargo packed aboard the *St. Michael* and received an invitation for an audience with the king.

Dingane had moved his *ikhanda* from kwaDukuza to uMgungundlovu ("the place of the elephant"), about ninety-five miles to the northeast. Fynn, Isaacs, and fifty porters carrying gifts set out for uMgungundlovu mid-April, two weeks after Isaacs' return. Isaacs borrowed an ox to ride, as his feet had grown tender during his convalescence at his uncle's home. The ox was soon attacked by wild dogs, and Isaacs had no choice but to walk and endure the pain of his long march one footfall at a time. Squalls bore down on the column, which was weighed down by the gifts they carried.[31]

They arrived in uMgungundlovu on April 29, 1830. Dingane's *ikhanda* stretched in an ellipse across a valley along the White uMfolozi River. Like kwaDukuza, uMgungundlovu contained within its roughly concentric palisades and windbreaks enough beehive huts for thousands of residents, barracks for warriors, royal residences and "seraglios" housing as many as 1,500 women, and numerous cattle enclosures, abattoirs, and milking compounds.[32] Woodsmoke filled the air. Isaacs wandered through a crowd of men on their haunches milking cows, women beating and smoothing the earthen floors of their huts, warriors sharpening assegai, children racing in between the legs of adults grinding corn, and cattle lowing from every direction. In his newly constructed residence, Dingane graciously received Isaacs.

Isaacs described the king as exhibiting "great muscular strength" and possessing "keen, quick" eyes that noticed every movement and gesture of his guests. Dingane presented a "commanding appearance" and was "at least six feet in height," Isaacs wrote, and was "of a dark brown complexion, approaching to a bronze colour." He was impressed that the king was careful to "weigh every word before he utters it," but he also judged Dingane to be vain and self-indulgent.[33] A later visitor

FIGURE 5. "King Dingane Reclining in His Residence," ca. 1835, after an illustration by Allen Gardiner, from J. G. Wood, *The Uncivilized Races of Men in all Countries of the World* (J. G. Wood, 1878).

sketched Dingane as pot-bellied and short-limbed, "tall, corpulent, and fleshy." The Europeans entered the royal courtyard and squatted among Dingane's people. Fynn and Isaacs brought with them one of the survivors of the *African Adventurer*, a Chinese man, perhaps a crew member, whose "long black tail"—a queue—was much admired by Dingane and his people.[34]

The king addressed Isaacs fondly: "uDambuza mtabate" (I see you). Isaacs replied with the customary "Yebo Baba" (yes, Father). Dambuza-Isaacs and Mbuyazwe-Fynn then displayed their gifts, which moved the crowd to exclamation. Dingane doled out clothing and bolts of cloth to his favorites but ordered the beads and other trinkets removed to his inner residence. The king had not been expecting Isaacs, he explained, and was caught off guard by the bounty he and Fynn had brought. He then promised them whatever ivory he and his people might collect. "See the mountains and forests," the king proclaimed, "they are all mine, they contain innumerable elephants, and my rivers the hippopotami. I have given up going to war; I mean to cultivate peace . . .

[and] I shall then hunt the elephant and the hippopotamus, which will be an amusement for my subjects and enable me to remunerate my friends." Rather than make war against regional foes—the Zulu idiom was to "eat up" their enemies—Dingane aimed to sate himself with luxury. Isaacs was pleased to see that the new king sought his people's repose.[35]

Before Dingane sent Fynn and Isaacs back to Port Natal, he gave them sixteen tusks, sixteen head of cattle, and a score of hides—all that he could lay his hands on at that moment. He moved to appoint Fynn "King of Natal," bestowing on him the status of tributary chief. Fynn demurred at assuming this honor, perhaps because he did not want to be caught between the Zulu King and His Majesty. Dingane then turned to Isaacs and proclaimed: "To whom does all that country belong?—have I not given it to you?—and are not the people who inhabit it yours?" Dambuza-Isaacs had no hesitation in being made a minor king. He took Dingane's rhetorical flourish as a sign that "the original grant of the tract of country before ceded to me by his predecessor Chaka was . . . recognized by Dingane and confirmed to me." Whether this was Dingane's intended meaning is unclear. Back at Port Natal, Isaacs and Fynn formalized their commercial partnership and agreed to amalgamate their territorial grants.[36] Isaacs began staking out a site for a homestead in Natal. Fynn took his band of locusts—the *iziNkumbi*—into the interior to seek out more ivory.

In early June, Dingane's people arrived in Port Natal with an additional fifty tusks and scores of cattle hides, which were loaded onto the waiting *St. Michael* along with fifty tons of hand-cut local timber that would be crafted into ships' knees—curved bracing joints—at Jamestown's shipyard. Captain Page departed Port Natal with this cargo shortly thereafter. Among his passengers was one unnamed Zulu man who Page hoped would catalyze interest in future commercial traffic between America and Natal.[37] What became of this man is unknown. Perhaps he arrived in the United States, the first Zulu visitor to the young republic. If he made it to New York or Philadelphia, he would have learned of slavery in the Southern states and witnessed the stirrings of the organized abolitionist movement in those cities. Or perhaps he disembarked at another port—St. Helena or Havana—or died at sea. If he ever made it back to the Zulu Kingdom, his arrival, like his identity, went unrecorded.

Captain Page planned to return in about a year, bringing more of the gifts that had charmed Dingane, as well as woolen and cotton clothing to barter for ivory and hides. Those Zulu who had obtained European clothes "left off wearing hides to cover their nakedness," Isaacs observed. He boasted that he had "opened a trade in hides, and when other articles were imported," the Zulu "found the want of [those] goods, so they found the means to obtain them" by hunting elephant for ivory. The export of ivory in turn generated more imports of clothing, beads, and trinkets. As Isaacs explained, he desired nothing less than "to instil into them a knowledge of artificial wants, and a desire to satisfy these wants." This succinct definition of capitalism—creating appetites for consumption—encapsulates the destruction globalized trade can wreak on Indigenous cultures. As one early traveler in Natal remarked, the Zulu could not "foresee that the admission of a few mercantile adventurers may perhaps ultimately lead to the subjugation of [their] kingdom and posterity."[38] Influential colonial reformists, however, believed that free trade liberated peoples rather than subjugating them.

One of those reformists, Edward Gibbon Wakefield, edited a popular 1835 edition of Adam Smith's *The Wealth of Nations*. He argued that "the wants of savages" are few, but as wants multiplied, so too did the production and exchange of goods in order to fulfill those desires.[39] This dynamic of consumption created civilization. Wakefield believed that the gospel of trade would set all races free and augment the efforts of missionaries and British philanthropists in the colonies. Isaacs and Fynn may not have entertained such humanitarian justifications for their trade, but in any case, business between the Zulu nation and Messrs. Mbuyazwe-Fynn & Dambuza-Isaacs was booming.

Isaacs and his African retainers remained busy hunting elephant and setting up a homestead that featured orderly agricultural fields, storehouses, and dwellings. He made several visits to Dingane, and the king received him with an open hand. The two spoke at least once about Zulu theology. The *zulu*, or heavens, Dingane explained, were formed by the accumulation of smoke and dust. Isaacs propounded the Judeo-Christian notion that the "vast expanse of the skies . . . were known to be the work of a Creator." Dingane shrugged off the disagreement with perhaps the most reasonable response possible to any doctrinal dispute: "You *umlungus* have your ideas, and we have ours." Missionaries penned many of the accounts of the period of first contact, and naturally their works are replete with references to Christian theology. Even the impious Fynn devoted space to the expressly Christian content of these conversations.[40] Isaacs, however, remained reticent about religion.

The scattered references to God in *Travels and Adventures* are half-heartedly deist and suggest a sort of impersonal, even embarrassed approach to faith.

Conversion was rare among Georgian Jews. For the most part, the emergent Jewish middle class achieved their desired integration through acculturation to British mores rather than through spiritual means.[41] In Natal, Isaacs' social acceptance was predicated on adopting the attitudes and behaviors of the other white merchant-adventurers. So long as Isaacs could scheme, sail, ride, march, fight, hunt, herd, kill, and labor, he remained the equal of the Christian Farewell, King, Fynn, and Cane. Life beyond the bounds of "civilization" rendered irrelevant the second-class status Isaacs would have held in London as a Jewish urban savage. Of course, Isaacs' becoming a man like any other Englishman owed much to his becoming "'white," a status he achieved in relation to the Hottentots and Xhosa, if not the Zulu. This transformation could perhaps have been accomplished only on the fringes of empire, where social and class hierarchies were mutable but racial hierarchies more entrenched. Ironically, Isaacs improved his economic and social status precisely by adopting those Zulu values and behaviors that were thought to define masculine British values and behaviors.

During one visit to uMgungundlovu, the King requested Isaacs' presence in his royal courtyard and asked him to bring along his musket. Isaacs obediently loaded his weapon and arrived to find Dingane seated opposite "two fine-looking women." The women's downcast expressions caught Isaacs' eye as he saluted the king and took his place among the assembled chieftains. He was told that the women were widows of the defeated rebel Nqeto, the chief who had ordered Farewell's murder. Dingane sentenced the women to death and motioned several of his men to take them "away towards their home," as the execution site was derisively known. As soon as the wives had been escorted away, Dingane ordered Dambuza-Isaacs to "go and shoot them." Horrified, Isaacs refused and pleaded for their lives. Dingane was unmoved. "They killed one of your countrymen," the king thundered, "and I insist on their lives being taken by the musket." Again Isaacs resisted. Dingane's chieftains clamored for blood and demanded that Nasopongo take the musket as Dambuza's proxy.[42]

Isaacs handed over his musket and pistol to Nasopongo, who walked to the site where the doomed women waited under guard. Dingane fixed his eyes on their terror. Nasopongo approached and fired Isaacs' musket at one of the women from about thirty feet away. The shot passed clean through her breast, and she dropped dead. The second woman raised a woven mat to shield her torso and then wheeled away. Nasopongo raised the pistol and fired, wounding the woman in the back. The faint blue smoke of the shot hung in the air. The injured woman backed away while he laboriously reloaded the musket. Again he leveled his weapon, aimed, and fired. She fell to the ground dead amid the cheers of the chieftains. Dingane "indulged in the sight of this barbarous act" and presented Isaacs with twenty liters of beer to celebrate the executions. Isaacs ruminated that Dingane had exulted in bloodshed as much as Shaka had done. The king later entreated Isaacs to provide him with a musket, but he refused, or at least claimed to have done so.[43]

Two days later, the sky above uMgungundlovu darkened as clouds of locusts descended on the king's *ikhanda*. They fouled the Zulu huts, covered the pastures, and "consum[ed] all before them," Isaacs observed, "as in olden times, when they scourged the Egyptians, as a retribution for their monarch's disregard of the commands of his Creator."[44] Isaacs' meditation on the biblical plague casts Dingane in the role of a hard-hearted Pharaoh for his brutal execution of Nqeto's wives. With uncharacteristic humility, Isaacs refrains from according himself the role of Moses. But Isaacs failed to recognize the ways in which his own adventure capitalism, with its appetite for ivory and hides, and its spewing up of European clothes, quite literally denuded the Zulu Kingdom.

Fynn had meanwhile gone to the interior with a "body of armed natives," his own locusts, to hunt for ivory. He left Isaacs to take up residence near Mount Pleasant, where James King had been interred. There Isaacs drilled a small hunting party of his own retainers, putting them through target practice until most could hit something as large as an elephant from sixty paces 90 percent of the time.[45] While this may sound easy, we should recall that muskets of the day were not especially accurate.

Isaacs had other followers build him a beehive hut while he supervised the sowing of cornfields and gardens around his homestead. He recorded that the more orderly the settlement at Natal appeared, the greater the number of families of conquered people, Zulu expatriates, and outcasts who sought to join it. One of those who settled nearby was Hlambamanzi, the former favorite of Shaka. Hlambamanzi brought along his ten wives and numerous followers, perhaps to keep an eye on the Europeans for Dingane, or perhaps because his position under

Shaka put his life at risk among the ever-shifting alliances of Dingane's court.[46]

Isaacs described Mount Pleasant and the surrounding kraals as "hamlets with numerous inhabitants pursuing their avocations of guarding their herds and cultivating their patches of land" under white protection. In a cringe-inducing passage, Isaacs fancied that the Africans "looked up to [their] European commanders as being beyond comprehension; [they] feared their power and loved them because they protected [them] against tyrannical chiefs and kings." Isaacs, as an *induna* protected by Dingane, duplicated the social structure of the Zulu *umuzi* at Mount Pleasant, which rendered his settlement culturally legible and attractive to others in the region. He established a tribunal to function as a kind of senate, set up a rudimentary taxation system, created a pool of paid laborers, and concluded that these innovations were helping the mixed group of Hottentots, Zulu, and members of other satellite and subjugated groups achieve a "rational order of living."[47] Despite the paternalism of Isaacs' words, he nonetheless accorded the Africans of his settlement their human dignity, which cannot be said for many of the Europeans who sojourned in Africa either before or after him.

Of course, Isaacs in turn depended on his large retinue to house and feed him, do his dirty work, hunt for elephant and hippopotamus, and serve as a market for imported goods. He also enticed, or coerced, women from his flock into his bed. At least one of these women gave birth to the children he later abandoned. When he described an African bride as "lovely . . . meek, gentle, tender [and] elegant," and resembling a "black fairy, or something superhuman," perhaps he was recalling a woman he had cared for or even loved. Was it Macowzala?[48]

On the morning of 11 August 1830, Isaacs was startled awake by the retort of a musket shot. He leaped from his bed and raced toward the sounds of screaming. There he found "a poor boy . . . lying prostrate on the ground, his arm nearly shattered off, with a deep wound in his belly that exhibited his entrails, and the upper part of his thigh lacerated." One of Isaacs' ivory hunters had been cleaning his weapon when it discharged. Isaacs called for the local healer, who was certain the victim would not survive. Nonetheless, Isaacs resolved to amputate the child's arm. He and others held him down, dosed him with a potentially fatal forty drops of laudanum, and sawed through the remains of the boy's splintered arm from the elbow. Isaacs sewed up the ragged stump, cleaned and dressed the boy's other wounds, and posted three people to nurse him around the clock.[49]

Over the coming weeks, Isaacs was gratified to see the boy's improvement. By the end of September he had regained some of his strength. Isaacs described how the people of Mount Pleasant assembled to pay a ceremonial visit to the child. Isaacs had an indifferent record as a physician, but this time his patient would live. Isaacs was much relieved: the patient was his son Porter.[50] Isaacs does not record Porter's age, but he could not have been much more than four years old, given that Isaacs had arrived in Natal in October 1825. Within a few years, Isaacs would abandon Porter and the boy's mother. That Isaacs had himself endured a fatherless youth and been forced from his mother's side did not, apparently, make him more sensitive to the sufferings of the women and children he left behind. Instead, as with many of those who survive poverty and hardship, his own gritty resilience seems to have made him insensitive to the immiseration of others.

Throughout the rest of 1830 and into 1831, Isaacs superintended Mount Pleasant, cultivated his lands, and amassed ivory. He traveled throughout the Zulu lands opening trade with various chiefs. Though he occasionally suffered from dysentery, Isaacs possessed an iron constitution. He mounted pack oxen and later acquired a horse to ride, and he survived falls from both. He marched for hours, sometimes without food and typically barefoot, though at trail's end Nasopongo would be there to massage his swollen feet with animal fat. Isaacs endured rainstorms in open country, chilling temperatures, and extreme heat. He ate without utensils, "exhibit[ing] alacrity with thumb and finger," partaking of parboiled beef, milk curds, home-brewed beer, and green corn with the local peoples.[51] Considering the perils Isaacs faced from disease, lack of hygiene, contaminated water and food, injury, infection of even the smallest of wounds, and the harsh elements, his robust health was little short of miraculous. The dangers posed by the wild game he stalked—elephants, hippos, the occasional water buffalo—and from the beasts that stalked him—crocodiles, wild dogs, leopards—also testify to his fortitude.

Isaacs paid visits to Dingane and came to know the Zulu king as he had known Shaka before him. Unlike Shaka, Dingane was given to indolence and preferred the pleasures of the royal court to the glories of battle. King Dingane entertained his *abelungu* visitors with dances and songs performed by graceful, beautifully arrayed women and girls. A proud Dingane told Isaacs that he had composed the songs they sang and designed their beaded adornments. After one of these command performances, Isaacs recorded that "the Zoola females far exceed those of all the African tribes I had seen, in either dancing, singing, or agility; as well as in their personal charms." Isaacs' fond eye for women is evident from *Travels and Adventures*. He shared this regard for the female form, and the particular attractions of Zulu women, with Dingane, who was said to have luxuriated in the attentions of four hundred women.[52]

The king could be excessive in cruelty as well. During one visit, he demanded that Isaacs hand over a magnifying glass so that he could imitate the trick of concentrating the sun's rays to spark a flame. He held the lens over a young man's hand and singed his flesh. The victim writhed in agony but could not scream, on penalty of death. Like Shaka, Dingane was fascinated by European technologies and eager to receive information from beyond his own lands. And like Shaka, he saw the value of the whites' firepower and sought their assistance in dislodging a stubborn enemy from a rocky stronghold. This time Isaacs risked enraging Dingane by refusing to supply him with his squad of trained elephant hunters.[53]

If Dingane thought Isaacs unhelpful, Cape Colony officials suspected him of being downright seditious. Gossip and press reports made him out to be a sort of paramilitary adviser to the Zulu in the service of the United States. The American desire to obtain a South African port had been noted with concern well before Isaacs ever arrived, and so his partnership with a US shipping interest aroused concern. The British may not have wanted to annex Natal themselves, but neither did they want a competing power to gain a foothold there. Newspapers claimed that the *St. Michael* had unloaded a cargo of weapons at Natal and then "sailed for the United States for the purpose of bringing out men and supplies for the forming of a settlement there. Nathaniel Isaacs . . . remained in charge of arms, and is said to be instructing the natives in their use." Governor Cole was warned that the *St. Michael* "landed a quantity of muskets, cutlasses, gunpowder and salt, which have been left under charge of one of the crew named Nathaniel Isaaks. . . . This person is stated to be instructing the natives in the use of fire-arms." Isaacs publicly protested these "downright falsehoods," insisting that "the few firearms . . . that were on board . . . were afterwards purchased by me for my own use and protection; and although Dingaan and others frequently tried to get them . . . from Mr. Fynn and myself, never did they succeed."[54]

The charge that Isaacs trained Africans in the use of weapons was correct, but the idea that he aspired to an American coup in Port Natal was almost certainly nonsense. So too was the rumor that Governor

Cole would soon send a force to rout Isaacs and Fynn and put a stop to their alleged transatlantic conspiracies.[55] Inflated reports of the partners' success in the ivory trade may have been spread by jealous competitors at the Cape in order to encourage future military action against Dingane and his white vassals.

Cape Town's Europeans were not the only ones capable of political machinations. Hlambamanzi, the one-time slave to Boer outlanders and former prisoner of the Crown, was an adroit agent of his own self-interest. He had particular reason to resent the whites. They had enslaved him, humiliated him, flogged him, clapped him in irons, press-ganged him, and swapped him among themselves like so much merchandise. Hlambamanzi had found refuge among the Zulu and achieved a position of honor under Shaka. But when he returned to Port Elizabeth as interpreter for the king's first diplomatic mission, the colonial officials ignored him, regarded him as Shaka's spy, or dismissed him as a confidant of the shady James King, the one white man who had ever treated him with kindness. Following Shaka's assassination, Hlambamanzi lost his standing. His experience among the Boers and English left him contemptuous of their assumed superiority. Whites were just "little wild beasts" with muskets. And Hlambamanzi possessed his own musket, perhaps an earlier gift from Farewell.[56] He would not be overawed by their magic. The *abelungu* were men to be manipulated like all others.

While Isaacs scoured the countryside for ivory, Hlambamanzi set own plot in motion. He dispatched a messenger to Dingane bearing the news that John Cane intended to lead a British attack on the Zulu Kingdom. Isaacs learned of the rumor from a Zulu chief he had befriended, but knowing there was bad blood between Hlambamanzi and the surly Cane following a dispute over cattle, he was certain that colonial forces had no intention of invading and that Hlambamanzi was sowing discord for his own ends. So Isaacs raced on horseback to Dingane to counter Hlambamanzi's reckless claims and offer himself as a hostage to the truth.[57]

Isaacs arrived at uMgungundlovu at sunset on 17 April 1831, nearly a year after his first visit to Dingane. The king received him at the palace threshold and admired the speed of Isaacs' horse as he rode in. Dingane wore strands of pink and white beads that stretched from his neck to his knees, and brass bangles on his arms and legs that glinted in the light of the setting sun. Isaacs dismounted and prostrated himself before the king. He presented Dingane with gifts, including construction tools and a single white mouse, which the king praised. They discussed Hlam-bamanzi's allegation of an imminent British invasion; Isaacs assured the king that the reports were fabrications meant to discredit Cane.[58] The king's anxieties were eased, but Cane's delay in traveling to uMgun-gundlovu to explain himself earned him Dingane's suspicion and enmity.

Isaacs tried to intercede on Cane's behalf, explaining that Hlam-bamanzi sought to ingratiate himself with the king and appropriate Cane's livestock. Unmoved, Dingane told Isaacs that he had dispatched warriors to drive Cane from his kraal, seize his cattle, and expel him from Zulu lands for his disobedience. Again Isaacs protested. "If you see [Cane] in the bush," Dingane finally conceded, "tell him to come here with Jacob [Hlambamanzi], and . . . time will prove if false infor-mation has been sent to me; if it be false, Jacob shall not go unpun-ished." Dingane then gave Isaacs beef and beer, which he consumed beneath a shade tree while admiring the king's women.[59]

Afterwards, on the road back to Port Natal, news reached Isaacs that Dingane's warriors had laid waste to Cane's kraal. Dingane's instruc-tions had been to "cut them [Cane's people] into yard lengths like our beef." But Cane had gotten wind of the impending attack and managed to escape to Fynn for protection. The other inhabitants fled into the bush, taking whatever they could. From a hideout twenty miles from Port Natal, Fynn composed a hasty letter to Isaacs and sent it off with a trusted messenger, not knowing precisely where Isaacs was, or even if he had escaped uMgungundlovu unscathed. "Having some faint hopes that you are still alive is what induces me to write and inform you of our situation," Fynn's letter began, "I gave you up for gone. . . . I can assure you, we concluded we could never hope to see you again."[60] Somehow, the messenger located Isaacs.

Though wary of Dingane's shifting moods, Isaacs was perfectly fine. He had visited Cane's kraal and witnessed the devastation. "The first thing that attracted my notice was a few sheets of an Encyclopedia scat-tered along the path," Isaacs wrote. "The kraal had been burnt for fuel; the cat had been speared and skinned; the ducks were scattered lifeless about the place." The gory scene and Fynn's startling letter moved Isaacs to panic. Fortunately, the *St. Michael* had returned earlier than anticipated and now lay at anchor outside Port Natal. Isaacs spotted her, raced to the waterline, waded waist-deep into the sea, and waved a blanket and fired his pistols to attract the vessel's attention.[61] The brig entered the port without incident, and Isaacs was welcomed aboard by Captain Page.

Emerging from their hideout, Cane and Fynn arrived that evening and were taken aboard. Fynn described the chaos preceding the attack on Cane's homestead. When news reached him of the warriors' approach, he thought all the Europeans were at risk. Fynn divided his and Isaacs' property among their retainers, who concealed whatever they could. Fynn had pried open Isaacs' writing desk and rescued his papers, thinking that if Isaacs were killed, at least his journal would comfort his mother back home. Two tons of stockpiled ivory remained scattered in various secret caches. And though Fynn's kraal and Isaacs' Mount Pleasant holdings were untouched, they could no longer be sure of Dingane's protection. Instead they put their faith in the *St. Michael*'s artillery, reasoning that the sound of the cannons would strike terror into the Zulu and allow them time to collect their ivory stores.[62] Isaacs spent the next weeks shuttling between the *St. Michael* and Mount Pleasant. He sent his people to gather the hidden ivory and soothed their anxieties.

Messengers tramped from Port Natal to uMgungundlovu protesting Dingane's attack on Cane. The king expressed his sorrow at the hostilities, claiming that Cane had aroused his anger by defying his summons. Dingane maintained that Isaacs, Fynn, and their property had never been his targets and that they and their followers remained under royal protection. He invited Fynn to parley with him on behalf of the alarmed *abelungu*. Isaacs and Fynn were wary of Dingane's assurances, accompanied as they were by renewed requests to supply him with a cadre of men trained in musketry to dislodge an enemy chieftain from his redoubt. After conferring with Fynn, Isaacs agreed to loan the king a squad of armed men.[63] The lure of ivory was strong. Both partners were ready to compromise their safety, the lives of their men, and their loyalty to the Crown to corner the market in white gold.

A party of about eighty porters, with Fynn at their head, set off for uMgungundlovu, carrying a lavish array of gifts that had arrived on the returned *St. Michael*. Fynn hoped he could placate Dingane with the beads, brass ornaments, snuffboxes, iron pots, rugs, blankets, cotton cloth, and woolen items. Isaacs remained behind at Port Natal to supervise the collection and loading of ivory onto the *St. Michael*, and to prepare Mount Pleasant and the neighboring kraals against a possible sneak attack of the kind visited on Cane. Isaacs patrolled his lands on horseback armed with two pistols and a cutlass and was escorted in the field by his trained musketeers. But Dingane sent no *amabutho* against the Europeans. Instead he presented Fynn with about one ton of ivory and twenty head of cattle. Dingane looked with favor on the presents Fynn had brought, but something was missing. Where were the troops he had requested? Only eleven "musket boys" had arrived.[64] Dingane had expected more armed men, and it is likely that he had also anticipated receiving weapons.

Fynn informed the king that he would have to supply more ivory if he wanted more men or arms. Dingane reproached Fynn: "You are never satisfied. Last year the vessel [St. Michael] went away dissatisfied, and it is the same this year." The king then recited the litany of evils Hlambamanzi had witnessed and reported to him: "At first the white people came and took a part of [the natives'] land; they then increased, and drove them further back, and have repeatedly taken more land. . . . [A] few white people intended to come first and get a grant of land, as Mr. Fynn and Dambuza had done; they would then build a fort, when more would come, and demand land, who would also build houses and subdue the Zoolas, and keep driving them further back." Although Hlambamanzi's news of an imminent British attack was invented, he was accurate in identifying the Europeans' method of encroachment on Indigenous people and their land. Dingane demanded an explanation: was it true that a British force guided by Cane "was about to advance" against the Zulu?[65]

Fynn feared that Dingane and his chieftains had been swayed by Hlambamanzi's accounts of white perfidy. Soon Fynn's comrade Lukilimba, who had been reluctantly drawn into Hlambamanzi's intrigues, confessed that "he thought Dingane intended to destroy" the whites as soon as the St. Michael set sail again. Fynn was enraged at Hlambamanzi's schemes, at Dingane's alleged duplicity, and at Lukilimba's quiescence. And he was frustrated that all his commercial opportunities were vanishing before his eyes. Fynn reached for his pistol to take revenge on whoever was nearest at hand; he shot and killed the inconstant Lukilimba.[66]

Fynn found a stub of pencil and scribbled a message to Isaacs: "Please come and meet me as soon as possible. . . . [C]ome immediately. . . . I can't say any more just now." The note reached Isaacs at Port Natal at midnight. He saddled his horse before dawn and rode out to meet Fynn, whom he found on the trail coming from Dingane. "We have no further hopes in remaining here," Fynn informed Isaacs. Recounting his meeting with Dingane and Hlambamanzi's deceptions, Fynn concluded they were no longer safe among the Zulu. Only the cannons of the St. Michael protected them. The partners decided to split up. Fynn determined to take his retainers toward the western frontier, at the limit of

British control. Isaacs set sail aboard the *St. Michael*, but he pledged to return to Port Natal when tensions eased.[67]

Isaacs later learned that had he and Fynn waited a few weeks, they could have resumed their lives at Port Natal unmolested. Dingane had belatedly unraveled Hlambamanzi's intrigues and decreed that the interpreter should be put to death. As Cane told it, Dingane "ordered me to attack Jacob with my people and massacre him." Cane happily obliged and captured Hlambamanzi's cattle. Fynn returned to Natal in September 1831 and reconciled with Dingane, who had in the meantime rededicated himself to "cultivating the sweets of peace" after purging his court of chiefs who had opposed the Europeans' presence.[68]

But Isaacs had already set sail aboard the *St. Michael* and was to spend the next eight months accompanying Captain Page to a variety of ports. He dined with Portuguese officials in Mozambique and grieved over their abuse of slaves. At another port, a Portuguese harbor pilot intentionally misdirected the *St. Michael* toward a reef in the hope that it would founder and he and his accomplices could loot the ship. Isaacs called the Portuguese "savages" for plotting such treachery. Only a combination of skill and luck allowed Captain Page to escape that disaster. From there they sailed to Mahajanga on the west coast of Madagascar, where Isaacs complained of the sun, the humidity, and the clouds of mosquitoes. He was, however, impressed by the industrious locals who prepared "jerked beef" for export to Havana, where the inexpensive protein was said to be consumed by slaves. Isaacs feasted on a delectable shipboard supper of porpoise-brain meatballs, "well seasoned with pepper and prepared with onion."[69]

Captain Page pointed his ship northward, touching at Ndzuwani ("Johanna") in the Comoros Islands. There Isaacs met Prince Ramanataka, a military leader and exiled brother-in-law of the recently deceased Malagasy King Radama. Ramanataka had fallen afoul of the new regent, Queen Ranavalona I, who was infamous among Europeans for her bloody persecution of Christians. Certain that the queen intended to kill him in a purge, Ramanataka fled Madagascar with about two hundred armed troops and a fortune pilfered from the royal treasury. Isaacs described Ramanataka as "brave [and] very intelligent for a coloured man," and he commented more admiringly that the general had "adopted the English costume" and disciplined his loyal militia "after the custom of the British infantry."[70] The fugitive Ramanataka arrived on Ndzuwani and offered to put his troops at the service of the local ruler in exchange for sanctuary.

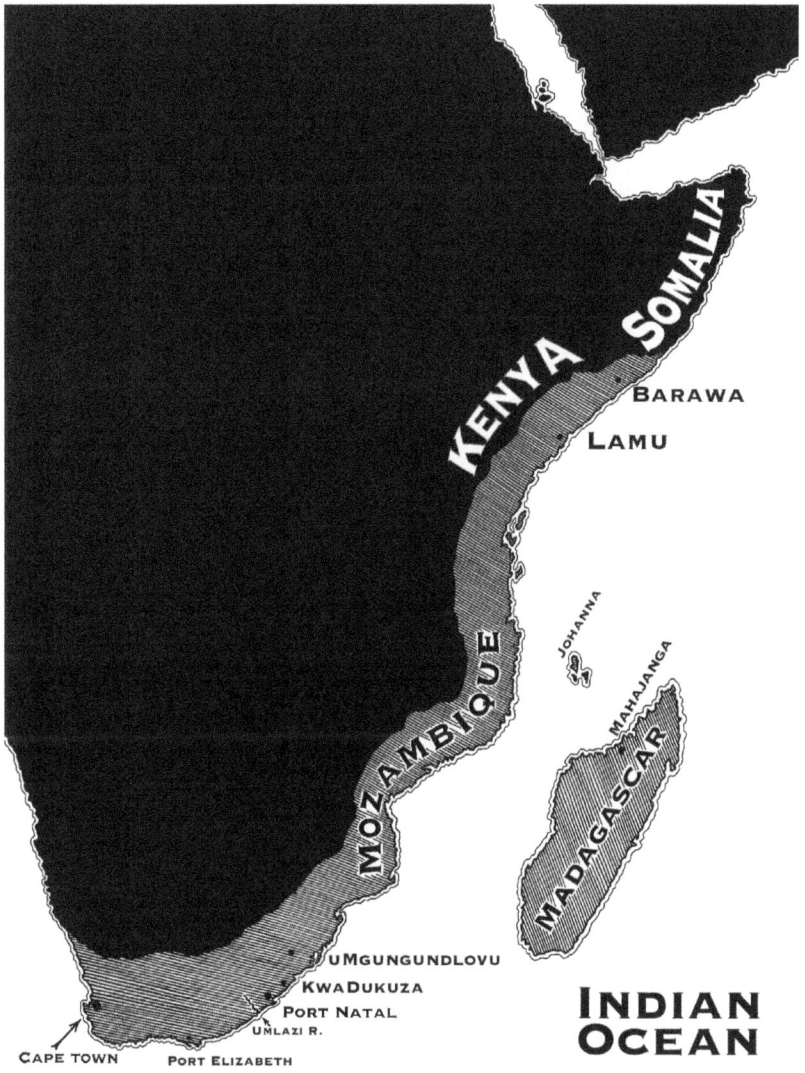

MAP 6. South Africa and the East African coast.

Though safe from pursuit, Ramanataka was frustrated that the small island offered little scope for his martial ambitions. He took an immediate liking to Isaacs, and after hearing him expound on Port Natal's numerous charms, the general proposed that he invade and occupy the region as a British ally. Ramanataka pressed Isaacs to sound out officials at the Cape to see if they would approve of his designs. Isaacs duly

informed Governor Cole that Ramanataka "wanted to charter the brig
I was in to convey his people to Natal ... and that British subjects
might (as they have done at Madagascar) reside at Natal under his protection." Isaacs further suggested, with more than a hint of reproof,
that should Governor Cole's own power prove "too limited to colonise
Natal," only about fifty British soldiers—presumably reinforced by
Ramanataka's men—would be necessary to defend Natal and demonstrate to "the Zoolas that the white people have a government which at
present they doubt." Isaacs wanted to control the chessboard in Kwa-
Zulu-Natal and deploy Prince Ramanataka's troops as his pawns. But
Cole rejected the overture, insisting that "he could not think of encouraging one barbarian against another."[71] (Never mind that such tactics
often defined British colonial policy of the era.) Isaacs was again disappointed, though probably not surprised, by the Cape government's lack
of vision and the dismissal of yet another of his schemes for appropriat-
ing Natal's resources.

The *St. Michael* sailed further north, stopping at Barawa on the Somali
coast. Isaacs despised the faded medieval port and its seedy inhabitants.
"To say anything complimentary of these people," he noted acidly,
"would be a libel on their characters." He was, however, impressed with
the American whalers and traders he met along the East African seaboard. Isaacs praised the "enterprising American[s], whose star-spangled
banner may be seen streaming in the wind, where other nations ... would
not deign to traffic." He was also won over by the shipyard skills demonstrated by the people of Lamu in today's Kenya. The copper sheathing of
the *St. Michael* had stripped off in several places during her voyage, but
the locals repaired the vessel with a mixture of lime and grease that Isaacs
found to be an "excellent substitute for copper." In Lamu he also listed
the items that could be had "at reasonable prices": medicinal roots, animal hides, tortoiseshell, ivory, dates, and coffee. When the *St. Michael*
finally docked in New York in April 1832, most of these items appeared
in newspaper notices listing the cargo offered for sale.[72] But Isaacs was
not aboard ship when stevedores unloaded her exotic cargo in New York.
In March he had disembarked in Jamestown, where Saul Solomon and
his family welcomed him once again.

Isaacs set about copying extracts of his journal and sent them off to
Cape Town for publication in his ally John Fairbairn's *Commercial
Advertiser*. These excerpts were designed to pressure the colonial gov-
ernment into extending British sovereignty to Port Natal. If successful,
Isaacs was certain he would prosper, by virtue of the land grant made to

him by Shaka and upheld (he believed) by Dingane. The first of these accounts was dispatched in April 1832, with a notice from Isaacs that he would shortly be returning to the colony.[73]

By autumn Isaacs was indeed back in Cape Town, now lodging at the South Africa Society House, less than a mile from Government House. There he seems to have met with Cole, although the governor did his best to ignore the persistent young man and his petitions. In a letter, Isaacs warned Cole that "the Americans have [Natal] in view, that is, to open a trade there," and he stoked colonial fears of competition by adding, "Had I gone there [to the United States], instead of coming here, Natal at this period would have had the American ensign flying on its shores, and even at this moment I should not be surprised if the vessel that took me away has fitted out to return there." Then, buttressing prospects of imperial and commercial advantage with a legal and ethical argument, Isaacs observed that the annexation of Natal "would be the means in a great measure of checking the illicit traffic in slaves, that is at this day carried on in an extensive way . . . all along the eastern coast." Isaacs' letter preserves in a neat and fluid hand the sincere hope that "His Excellency" would invite him for another meeting to discuss matters.[74] His hope was in vain.

The aristocratic Cole did not have time for the excitable young merchant-adventurer. Cole did not need Isaacs to present him with a brief for settling Natal. Unbeknownst to Isaacs, the governor had forwarded similar views to the colonial undersecretary more than a year earlier. Cole shared the belief that an American settlement in Natal would prove to be "embarrassing" for the colony, though he did not admit as much to Isaacs. This sort of political double game was the rule rather than the exception when it came to questions of territorial control in South Africa.[75] Yet no sign was forthcoming from London in favor of annexation, and Cole continued to officially ignore Isaacs' entreaties.

A desperate Isaacs again wrote to Cole, asking "that in the event of Natal being colonised . . . it may please your Excellency to allow me the preference of a certain piece of land. . . . having built on the land, and cultivated, and cleared the soil I now ask for in the event of that place becoming a British Settlement, and that this Government will be pleased to allow me the beforementioned land." This time Isaacs' plea did receive a response from on high: "His Excellency the Governor does not see how he can comply with the Memorialist's request as the Land is not within this Colony."[76] One can almost hear Isaacs smacking his forehead in disgust at obtuse officialdom.

Isaacs had no better luck in getting a response from his friend Fynn. In December 1832 he complained that he had written his former part- ner five letters but had received no word from him in return. Isaacs added that he had decided to sail to England, as he had failed to locate a suitable trading vessel at the Cape but had "received good accounts from home and there's no doubt but that my friends will furnish me with the needful on my arrival there." He promised Fynn that he would use his connections in England to "bring the subject of Natal before the House [of Commons] for discussion . . . and should I succeed in getting Natal settled it will be a fortune for you as well as for myself." Isaacs then assured Fynn that he would return after six or seven months to reunite with him and the ragtag clan settled around Mount Pleasant.[77]

PART TWO

From England to West Africa and Back

This was always the way with white men, they first sent quiet people to do good; then merchants; then as their numbers increased, they built forts and brought guns; and, at last took away your country.

—Namina Lahay, ca. 1841

CHAPTER 5

Feverish Trade, 1832–1837

Isaacs arrived back in Canterbury "a big bronzed and bearded man" of twenty-four.[1] His experiences at sea and in the Zulu Kingdom had surely surpassed any of his youthful fantasies, and he had plenty of rousing stories to tell his family. But in England virtually no one had heard of King Shaka or of the Zulu people, or much cared about the fate of an obscure corner of Africa that was not even on the map, quite literally: no map of the interior of eastern South Africa yet existed. Isaacs would soon change that.

Great Britain had undergone seismic transformations during Isaacs' decade-long absence. Arthur Wellesley, the Duke of Wellington, had resigned his position as prime minister in 1830, in part due to his staunch opposition to constitutional and economic reform. Once hailed as the savior of Europe for defeating Napoleon, Wellington now had to suffer the indignity of mobs of the dispossessed and discontented smashing the windows of his stately home on the edge of London's Hyde Park. The opposition leader, Charles Grey, solidified his role as prime minister at the end of 1832 after steering the passage of the Reform Act, which extended voting rights to increase the electorate by about 45 percent (though it excluded women).[2] The poor who received relief were now explicitly disenfranchised, but this legislative compromise may nonetheless have forestalled violent revolution in the United Kingdom.

Jewish civil rights had expanded as well. The debate as to whether native-born Jews could legally hold title to land in England was resolved in their favor in 1830. That same year, Jews were permitted to operate retail establishments in London. Yet Jews remained the only religious group denied full political rights in Great Britain. Jews could not be admitted to the bar. Unbaptized Jews could not be elected to Parliament. [3] And slave holding—though not slave trading—was still legal in portions of the empire, including the Cape Colony.

Remembered today as the namesake of a bergamot-flavored tea, Charles, second Earl Grey, should more rightly be recalled for his role in helping pass the Slavery Abolition Act in August 1833. Yet Grey was a colonialist, and his nepotistic administration favored a systematic approach to empire building. Promoting the settlement of Great Britain's colonies offered a solution to economic and social unrest at home. Given persistent domestic turmoil, the Colonial Office emerged as a significant arm of government. A coterie of colonial reformers, influenced by Jeremy Bentham's utilitarian philosophy, petitioned Grey's inner circle with plans that called for "waste land" to be settled by British emigrants. [4] Their passage out to these overseas possessions was to be funded by the capitalization of a joint stock company, according to a proposal circulated by Edward Gibbon Wakefield.

Wakefield was a well-educated, handsome rake with a Georgian pompadour. He was also a kidnapper who tricked a fifteen-year-old heiress into a forced marriage (the second time he had married an heiress in order to embezzle her money), an inmate of Newgate prison (owing to the kidnapping), and a bold prison reformer (owing to his incarceration). Later, he became an ardent abolitionist, an opponent of convict transportation, the editor of an edition of Adam Smith's *Wealth of Nations*, an adviser to government bodies on colonization, and a member of Parliament in New Zealand. In the early 1830s, Wakefield used his flair for the dramatic to propagandize for his reformist colonial policies. His ideas captivated aristocrats, an educated elite, and other highly placed friends who could sway Prime Minister Grey's administration. The young John Stuart Mill supported Wakefield's views. Wakefield's collaboration with Jeremy Bentham during the last year of the philosopher's life converted the prime minister's undersecretary for the colonies to the cause of systematic colonization. Wakefield's theories influenced the imperialist policies of Great Britain in varying degrees for decades. [5]

Wakefield's proposal called for those with capital in England to purchase colonial land for investment and lease it to settlers who would be drawn from the "excess" pool of restive British labor. The settlers' agricultural and small-scale industrial labor would be concentrated for development around centrally planned towns. Land prices would increase as the colony developed. Capital and labor would collaborate for mutual benefit, but wealthy landowners would remain a privileged class. Though development would be regulated to prevent abuses against Native peoples, they would nonetheless be enmeshed in a racialized hierarchy analogous to plantation economies. Indigenous peoples, who would be converted to Christianity, would provide cheap labor as well as a market for agricultural surplus and imported goods. The reformers who approved of Wakefield's schemes believed they could establish flourishing little Englands across the empire and contribute to the onward march of civilization.[6]

Grey's support for colonization based on Wakefield's theories of political economy would seem to have little to do with the fortune-hunting Isaacs. Yet Isaacs tied his dreams to those reformist advisers who could influence Grey and other powerful government figures. Among them was the prime minister's son, Viscount Howick (Henry George Grey), who served as undersecretary of state for war and the colonies from 1830 to 1834. He was succeeded, rather confusingly, by Charles Grey's nephew, Sir George Grey. The cadre of utilitarian-influenced radicals and family members who orbited Grey during his ascent and tenure as prime minister offered Isaacs a rare opportunity to promote his free-trade vision for Port Natal within the mahogany confines of Whitehall.

In early September 1833, Isaacs sought a personal interview with the secretary of state for the colonies. Isaacs warned that should the British not express sufficient interest in colonizing Natal, then "the subject will shortly be taken up as a matter of combined enterprise by the Americans." Isaacs' petitions went unanswered. He compounded his temerity by composing and circulating a prospectus for a proposed South African Company for Commerce, Colonization & Agriculture at the Bay of Natal, with ten thousand shares to be offered at £50 each. "The objects of this company are to open a trade with Ports to the Eastward of Natal and particularly with the Zoolo Country," the document explained, noting that "Capt King and Mr Isaacs . . . have obtained the concession of a large tract of land amounting to many millions of acres . . . accompanied by a grant of exclusive rights of commerce." A similar, though more substantive *Plan of a Company to be Established for the Purpose of Founding a Colony in Southern Australia*, authored principally by Wakefield, had previously made the rounds of government and appeared in print in 1832. Wakefield's plan aimed at the creation of a joint stock company to promote settlement in Australia, likewise to be capitalized with ten thousand shares offered at £50 each.[7] The economic foundation of Isaacs' prospectus almost certainly relies on figures supplied to the government by Wakefield a year or so earlier.

Like Wakefield's plan for Australia, Isaacs' prospectus describes the products available in Natal, its climate, and the promise of trade and future development. The prospectus maintains that "the proposed enterprise is neither a visionary scheme nor the offspring of a sudden and immatured conception." To prove his seriousness, Isaacs noted that he had wangled an interview with Lowry Cole, the former governor of the Cape, who had returned to England. Cole was said to have reviewed the prospectus and "doubts not of the Government rendering every assistance and patronage to the projected establishments."[8] Cole had earlier sent private communiqués to London in support of British colonization of Natal, though he had done his best to thwart haphazard and informal annexation of the sort implemented by King, Farewell, and Isaacs. Now, back in England and free from the burdens of governance, and with King and Farewell both dead, Cole seems to have thrown his support behind Isaacs' more mature scheme.

Isaacs' prospectus claimed that several businessmen at the Cape and in St. Helena "have signified their intention of becoming partners in the adventure." Saul Solomon was likely among them. According to Fynn, Isaacs garnered the support of "a long list of influential merchants," including the East India trading firm of Maynard & Company, which had begun exploring commerce in Port Natal as early as 1831. But when word reached London that a disorganized handful of merchants had already elbowed their way into Port Natal, the directors of Maynard withdrew their support, as did the heads of other leading firms. Isaacs' South African Company had promised to "offer a new era for British capital and enterprise," but it had collapsed before it had even begun.[9]

Perhaps to accompany his prospectus, Isaacs commissioned a map of Natal and the surrounding region. That map, titled "Africa: South-Eastern Coast," indicates Isaacs as the creator, though the work reveals the hand of a more expert draftsman. Whoever revised Isaacs' sketches did not do an especially thorough job: the map exhibits two spelling mistakes ("heigth" and "abundally"). Nonetheless, Isaacs' map—probably based on sketches he had made during his travels through Zulu lands—presents the earliest known cartographic projection of Zulu territory.[10]

An area tinted reddish-pink—the traditional hue representing the British Empire in the nineteenth century—appears at the center of the

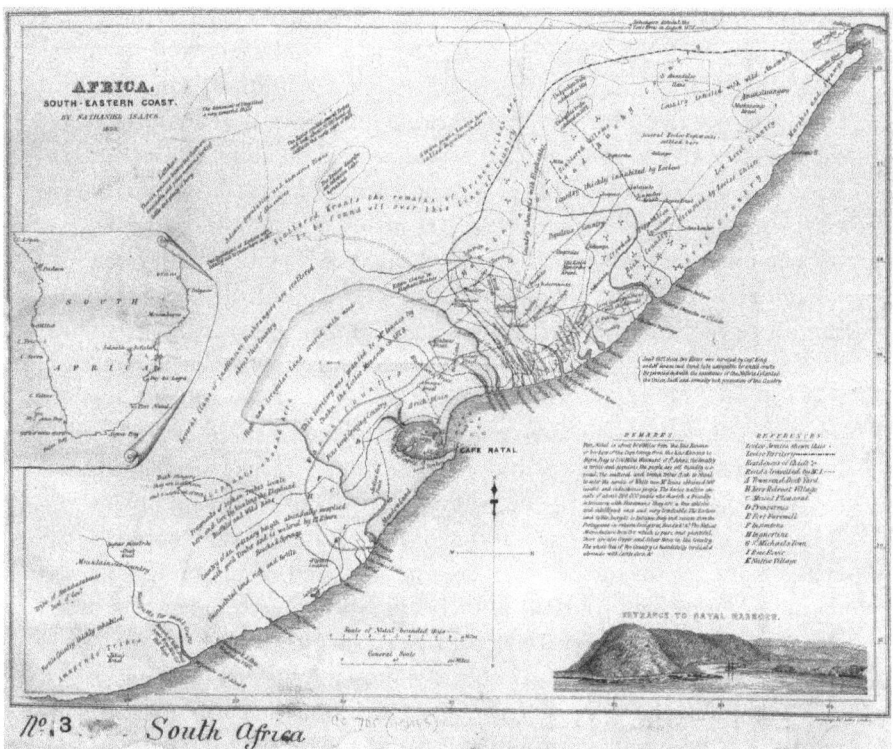

FIGURE 6. Map by Nathaniel Isaacs, "Africa: South-Eastern Coast," National Archives (UK), ref. CO700/South Africa 3.

map bearing the legend "This territory was granted to Mr. Isaacs by Chaka the Zooloo monarch in 1828." On his map, "Fort Farewell" and "Townsend Dock Yard" are both located within Isaacs' imperial zone, but his kraal at Mount Pleasant rests just beyond its rosy borders. Why Isaacs (or errant cartographers) excluded his private holdings from the region is a mystery. For Isaacs and his contemporaries, land ownership signified power; most men who were not landowners in England— including urban workers and merchants—were disenfranchised until passage of the Second Reform Act of 1867. Land remained the single greatest source of British family wealth by far until the 1880s.[11] Isaacs' preoccupation with documenting his land grants therefore makes perfect sense, even if the precise borders he or his cartographer recorded do not.

Beyond the boundaries of Isaacs' land grant lies an area bordered in blue and denoted as "Waste Country." This designation—a land without inhabitants, *terra nullius*—accords with Wakefield's usage of the phrase. Yet the rhetoric of Isaacs' map is self-contradictory: he locates a "native village" and a smaller *umuzi* within the same region. Isaacs cannot therefore simply be accused of denying or wishing away the presence of the Zulu and other Indigenous inhabitants. In fact, his map records dozens of Zulu *imizi* and encampments and includes legends such as "populous country," "crowded population," "rich country occupied by Zooloo," "scattered villages," and "country thickly inhabited by Zooloos." Nor does Isaacs ignore non-Zulu peoples. He is careful to name and locate a number of "tribes" across the map and indicate their standing in relation to the Zulu. A prominent block of text reports that "the Zooloo nation consists of about 200,000 people who cherish a friendly intercourse with Europeans[;] they are a fine intelligent and athletic race and very tractable." If his map was meant to encourage settlement of Natal along the lines of Wakefield's plans for Australia, Isaacs' documentation of large Indigenous populations negated any claims of the land being uninhabited. This may have been the undoing of his proposal. Both his prospectus and his map proved that "the Zoolo Country" was only *terra incognita* and not the *terra nullius* that was far more desirable to colonial reformist thinking.[12]

Isaacs must have been frustrated with the lack of government response. Where he had succeeded in finding favor with the kings of the Zulu, he had failed with all the king's men in Britain. His entreaties to officials in Cape Town and in London had been brushed off time and again. He had tried to clothe his naked ambition in the trappings of patriotic imperialism: the annexation of Natal, his correspondence, his map, and his prospectus all argued that he could enrich Great Britain through its favored doctrine of free trade. But he simply did not possess the patronage or the right pedigree to sway the aristocrats and veteran officers of Napoleonic conflicts who dominated the British colonial administration. Their apathetic response to reports of the promise of trade in Natal allowed other merchants time to infiltrate the area that Isaacs believed to be his deeded property. In consequence, wealthy individuals and established companies, such as Maynard, who at first supported Isaacs' venture, now turned their backs on it. Isaacs' map had been a means to publicize a land ripe for annexation, where "elephants abound" and "the whole face of the country is beautifully verdant."[13] Instead, Isaacs found himself humbled in dreary England, fearing claim jumpers in Port Natal, and bereft of any significant public or political support.

Isaacs remained convinced that his prosperity depended on the lands granted to him by King Shaka. He would need to leverage his knowledge of the Zulu coast to persuade the government of the wisdom of formally extending the Cape Colony's frontiers. Like Benjamin Disraeli's Jewish messiah-king in the novel *The Wondrous Tale of Alroy* (1833), Isaacs possessed a "soul that pant[ed] for Empire." But how to convince those who held power in London to plant the British flag in Natal? Isaacs' one advantage was his intimate familiarity with the Zulu Kingdom. But even this competitive edge was dulling quickly. James King's former antagonist, Captain William Fitzwilliam Owen, had just published his two-volume *Narrative of Voyages to Explore the Shores of Africa, Arabia, and Madagascar*, which made passing mention of the "warlike, but restless Zoolos."[14] Isaacs realized that he could outdo Owen's lackluster narrative with his own eyewitness accounts of Zulu life, culture, and warfare. He hastened to revise the journal he had kept of his travels among the Zulu and turn his notes and those of King into a popular book for a readership hungry for knowledge of exotic locales.

Isaacs had expected that Henry Francis Fynn would rise to the task of publishing such a book. In late 1832, before he left Cape Town, Isaacs had encouraged Fynn to revise portions of his diary for publication. "I am most anxious," Isaacs wrote, "to see your work out—when do you intend to publish?—the sooner the better, and endeavour to exhibit the Zooloo policy in governing their tribe, I mean show their Chiefs, both Chaka and Dingane's treachery."[15] Isaacs had done just that in the excerpts from his own journal that were published in Fairbairn's *South African Commercial Advertiser* in 1832. Isaacs hoped that with Fynn's work in print too, the push to annex Port Natal would gather momentum.

Isaacs even offered to help Fynn promote the "book you have in hand on the History of the Zooloos." Fynn does not seem like the kind of man to have written any sort of book, yet Isaacs had watched Fynn tote a manuscript through the bush for years. Fynn composed his observations under difficult conditions, in part using a natural ink he obtained from crushed flowers, and he protected his field notes from the elements by wrapping them in an ear he had cut from the corpse of one of the elephants his men had brought down. This original manuscript was later mistakenly buried in its leathery cover alongside the body of Fynn's brother. According to custom, a deceased man was laid to rest with his personal effects, and the Africans present at the death of the younger Fynn thought the precious diary belonged to him. The elder Fynn, who was away at the time, could not disinter his brother's body without facing charges of sorcery, which might end in his own death. There was nothing for Fynn to do but re-create his manuscript from memory, which he did over the course of many years.[16]

But in 1832, Isaacs was still impatient to see Fynn's proposed history in print. He even instructed Fynn on how to improve his manuscript with lurid portraits of Shaka and Dingane: "Make them out as bloodthirsty as you can, and endeavour to give an estimation of the number of people that they have murdered during their reigns, and describe the frivolous crimes people lose their lives for. Introduce as many anecdotes relative to Chaka as you can, it all tends to swell up the work and makes it interesting."[17] This advice suggests Isaacs' own inclination to sensationalize the violence he witnessed during the Zulu kings' reigns. Nonetheless, his words do not fundamentally discredit the observations contained in *Travels and Adventures*. Zulu informants also highlighted Shaka's violence and perverse cruelties—though their testimonies may have been influenced by Isaacs' written record, which appeared before Zulu oral traditions were collected. While we cannot reconstruct the precise numbers of deaths due to war and execution under Shaka and Dingane, on balance I believe that Isaacs exaggerates them. For this reason I have avoided dwelling on the most lurid elements of his narrative. That said, Isaacs' descriptions of the Zulu accord with the historical record overall and feature numerous passages that record his sincere appreciation for the Zulu people and their culture, along with notes of paternalism and ambivalence.

Back in London in 1834, Isaacs realized he could not count on the wayward Fynn—who seems never to have answered his letters—to publish his projected history. Isaacs came to believe that he would have to be the one to chronicle the Zulu and their kings. The work he envisioned would attract readers by its graphic depictions of exoticism, bloodshed, and derring-do. He further aimed to provide the first amateur ethnography of the Zulu people and an analysis of their "Zoolacratical" government. Such a book would allow Isaacs to publicly stake his claim to the land granted him by Shaka and Dingane while also placing pressure on the authorities, in the hopes that "the government of Great Britain may view the advantages which the port of Natal offers for the extension of commercial enterprize." With these aspirations in mind, Isaacs made his way to the office of the publisher Richard Bentley, likely because Bentley had published both Owen's *Narrative of Voyages* and a high-profile economic work by Wakefield (*England and America*, 1833). Bentley was not yet the household name it would become with the launch of the periodical *Bentley's Miscellany* in 1836 under the editorial direction of Charles Dickens. Even so, Bentley's firm was well regarded. In January 1834, Isaacs signed a contract with Bentley to publish *Travels and Adventures*.[18]

Isaacs would have followed news of the passage of the South Australia Act that summer. The act provided for the establishment of a colony along the lines of Wakefield's proposals. Then in June 1834, nearly two hundred leading citizens of Cape Town pressed for Port Natal's annexation, citing "the various documents . . . transmitted to England by the Colonial Government, particularly to that which has been received from Mr. Isaacs."[19] Isaacs was therefore not forgotten at the Cape, and his efforts to expand trade were lauded. He must have dreamed that when his book appeared, readers would pressure Parliament to legally sanction his homestead on Mount Pleasant as the capital of a new colonial outpost in South Africa. But until these schemes could be implemented, Isaacs would need to earn a living.

From at least 1834, he served as an agent and junior partner in George Clavering Redman's shipping firm.[20] Redman's ships called at West African and Indian ports, spending months at sea, and returned to disgorge their cargo in London at the East India Docks in London's East End, abutting the Thames River. These Georgian-era dockyards occupied hundreds of acres of waterfront with the massive warehouses and pulley-cranes needed to load and unload thousands of vessels a year. The half-terrestrial, half-floating world of maritime London teemed with clerks and merchants, laborers and beggars, sailors and soldiers, bankrupts, thieves, publicans and prostitutes, customs officers, crews of chattering Europeans, drawling Yankees, and West Indians, Africans, Chinese, and Indians. Brothels and lodging houses, butchers and bakers, sailmakers and slopsellers all crowded the cobblestoned streets. The sour smell of beer, the ammonia of urine, the stench of horse manure, and the musk of tobacco blended with the perfume of rum and the dank smell of the riverbank's mud. Fragrant coffee, tea, and spice aromas wafted from one warehouse; from another came the sickening smell of curing hides. By the docks, the stink of sulfur and piney fumes of turpentine vied with the mushroom funk of dry rot.[21]

Isaacs thrived in this atmosphere of bustle and toil. By December 1834 he was once again at sea, heading to take charge of Redman's affairs at the busy terminus of Bathurst, a British outpost on St. Mary's Island at the mouth of the River Gambia. Young Nathaniel had set off for St. Helena to learn his uncle's business at the age of fourteen; now it was time for him to take on an apprentice of his own. This was twelve-year-old Benjamin Moss, a Jewish boy from Deptford, west of Greenwich. Ben Moss was also related to Saul Solomon, though more distantly: he was the nephew of Saul's brother's wife, Hannah Moss, who had once resided with her husband in Jamestown.[22]

The opportunities available to a Jewish boy on London's outskirts could not compare to the rewards promised by trade with West Africa. And given the temptations of the nearby metropolis, with its freely flowing drink and its myriad corruptions, Samuel Moss might have considered young Ben safer among the loutish tars of a merchant vessel than along the Thames, where "cloaked marauders" materialized from the shadows. He may also have feared that if his son did not rise above his current station, he might be forever discriminated against and marked as an "urban savage" because of his working-class Jewish origins.[23]

In the 1830s, Londoners were fearful of the city's underworld denizens, which included not only a catalogue of miscreants—sharpers, swindlers, coiners, duffers, puffers, gamblers, housebreakers, pickpockets, river pirates—but also a special breed of rogue defined not by his actions but by his "racial" identity: the Jew. Self-proclaimed experts alleged that "under the pretence of purchasing old clothes," as many as 1,500 Jews made their rounds in the city, "prowl[ing] about the houses of men of rank and fortune, holding out temptations to their servants to pilfer and steal small articles." By the time the journalist Henry Mayhew interviewed London's itinerant poor in the 1840s, Jews were stereotyped as canny ragmen and swarming costermongers—fruit and vegetable vendors, hard-bargaining sea-sponge sellers, operators of rigged games, shrewd peddlers with "negro" hair, and street urchins. Mayhew was at pains to note that London's Jews were members of the "wandering tribes" and were therefore lower on the hierarchy of civilization than proper Anglo-Saxons with fixed residences. He explicitly compared vagabond Jewish "street folk" to nomadic tribes in South Africa.[24] Thus a growing awareness of South Africa's peoples filtered into British consciousness through published reports like Isaacs', and these descriptions helped influential journalists like Mayhew define Jews and other subalterns, like the Irish, as only superficially white. English Jews such as the Isaacs, Solomon, and Moss families might have longed for acceptance, yet even in the years of reform they were scorned as parvenus, or worse.

The famed author Benjamin Disraeli entered Parliament in 1837. Despite his celebrity and his baptism, even he could not avoid slander against his Jewish origins. Fellow statesmen and writers maligned Disraeli throughout his long career, including his tenures as prime minister, as a "Jew adventurer."[25] The two words were used to reinforce one another. A Jew was a member of a suspect race, often thought to be dishonest, motivated by self-interest, and having a preternatural aptitude for commerce. The term *adventurer* possessed a similar valence, suggesting an appetite for financial speculation, a gambler's temperament, and a mercenary approach to self-advancement. Those who applied the term to Disraeli were implying that no matter how great his achievements, he remained fundamentally inferior to a Christian-born gentleman. In such a poisonous atmosphere, working-class Jews could hope for little security.

A promising apprenticeship—which could often resemble indentured servitude—offered industrious boys and their families a chance to attain some measure of economic stability. As Samuel told his son before he left with Isaacs, "There would have been no prospect of you ever doing any good for yourself in Deptford, therefore I hope you do not regret leaving us."[26] Still, the geographic separation from family and the length of sea voyages made such partings difficult. Although letters delivered by ship often missed their recipients or crossed in transit, Samuel Moss and his son managed to exchange at least thirty-two letters during Ben's thirty or so months in Isaacs' service. Occasionally, Isaacs would pen a postscript to one of these letters, giving a brief update or sending greetings to Ben's affectionate father. Their correspondence offers a remarkable window on mid-nineteenth-century maritime trade and its hazards. These letters, preserved in a family archive, also provide an intimate and more objective perspective on Isaacs' character than does the bluster of portions of *Travels and Adventures*.

"Mr. Isaacs has been exceedingly kind to me when laying on my sick bed from the effects of Sea Sickness," Ben wrote in January 1835 after arriving in Bathurst. Since he also complained of constipation on the voyage out, his father dutifully dispatched him a package of laxative powders. He instructed his son to "look to [Isaacs] as your father—attend to his directions & do nothing without his approbation." Having lacked a father for much of his life and having abandoned his own children in South Africa, Isaacs might have been expected to be an inept father figure, but it appears that he warmed to the role. Ben informed his father that Isaacs had "acted as my Physician and Nurse and more so as a father by attending to my wants." A postscript from Isaacs reassured Samuel that Ben "is a fine boy and one that bids fair to be a clever man."[27] Perhaps Nathaniel Isaacs saw something of his own younger self reflected in the devoted gaze of his apprentice, a Jewish boy escaping deprivation.

That Isaacs took the time to write to Ben's father at all reveals his affection for the Moss family. Isaacs was a hard-nosed businessman trying to keep track of scattered vessels and their cargoes, and he worked under difficult conditions in West Africa, all while enduring a climate that had proved deadly to many Europeans. White men wilted beneath the strong sun in the dry season and sickened during the downpours of the rainy season. Tropical fevers felled new arrivals and transient sailors.

The town of Bathurst lay on the southeastern edge of St. Mary's, an island about four miles long and not much more than a mile and a half wide.[28] Mangrove swamps edged the British settlement, while baobabs stood like portly sentinels on the higher ground. Seasonal storms flooded parts of Bathurst, which barely rose above sea level. Still, as Isaacs knew, one could grow rich while planted on this marshy coast.

Sephardi merchants fleeing the Inquisition had created scattered Jewish communities along the shores of West Africa, though by the eighteenth century this little-known diaspora had disappeared. Portuguese colonial aspirations in the region were eclipsed by those of Britain and France. Beginning in the 1820s, British traders flocked to Bathurst. Africans eager to profit from commerce brought "beef, mutton, poultry, fish, fruit, milk, butter, palm-wine and . . . African vegetables" to sell to the whites, and they bartered gold, ivory, animal hides, and mahogany for imported European goods. Bathurst's handsome colonial buildings defied the weather and winds that buffeted the shoreline. Merchants erected fine homes "built with stone or brick, and roofed with slates or shingles."[29] Though the port flourished, changeable political conditions and local skirmishes, along with the threats of theft, piracy, and seaborne disaster, remained constant worries. Shortly after Isaacs arrived in Bathurst in 1835, he learned that Redman's vessel *Eliza* had sustained serious financial losses at Portendick, a seasonal trading post hundreds of miles to the north, in today's Mauritania.

The *Eliza*'s captain had advanced goods to seminomadic chieftains who controlled trade on that coast in exchange for valuable cargo forthcoming from the interior. But before the *Eliza* could be loaded, two French naval brigs arrived and pounded the shoreline with heavy fire, while smaller boats with swivel guns shot from closer range, scattering the local traders and endangering the *Eliza*'s crew. At least one man— Baboukas, the nephew of a prominent local chief—was killed on shore. A French officer and thirty sailors brandishing pistols and cutlasses then

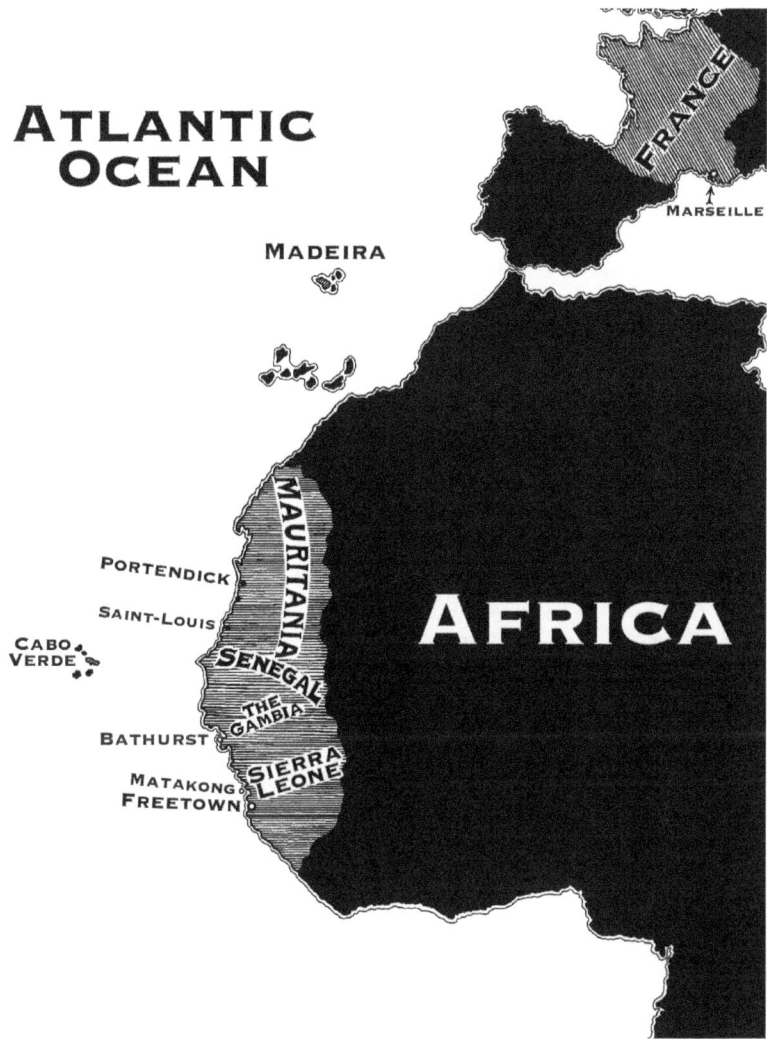

MAP 7. West African coast.

boarded the *Eliza*, took control of the ship, and forced it out to sea under armed escort.[30] Redman and Isaacs—not to mention the murdered Baboukas—had become victims of a struggle between French forces and the Muslim Traza (or Trarza) people.

France sought a monopoly over trade from the Saharan interior, but because the British were permitted by treaty to trade at Portendick, French forces used subterfuge and force to direct all commerce southward to their port of Saint-Louis (now Ndar) in Senegal. There a quasi-governmental syndicate dominated the ports and squeezed out British merchants. The Trazas—who were typically referred to as "Traza Moors"—rejected French restrictions on their movements and commercial dealings and skirmished repeatedly with colonial troops. French authorities in Saint-Louis suspected British merchants of arming the Trazas to foment instability.

As it turned out, there were no Trazas on shore at the time of the French onslaught, as attested to by Baboukas's uncle, who represented an entirely different clan. He swore revenge against the French, intending to "massacr[e] every Frenchman that might happen to be wrecked upon his Coast." The threat was not an empty one; wrecks were common in the region, whose coast is littered with sand bars, rocky shoals, shifting mud banks, and jagged reefs. One of the most horrific maritime disasters in modern history occurred nearby when the *Medusa*, a French vessel, ran aground thirty miles off the coast on its way to Saint-Louis in July 1816. Too few lifeboats meant that 150 passengers and crew—the majority of those aboard—were forced to take to an improvised raft that foundered in the breaking waves. Men battled one another to clamber to safety in the middle. Many were swept off and devoured by sharks. Others succumbed to the sun or starvation. Some reportedly resorted to cannibalism. Théodore Géricault's massive painting *The Raft of the Medusa* (1818–19) evokes the desperation of those left to perish. Meanwhile, those fortunate enough to have found space in one of the *Medusa*'s lifeboats made it to shore and trekked overland from Portendick south to Senegal. Along the way they were assisted by the Trazas, who later demanded compensation from France for providing them succor. The refusal to pay likely did not endear the French to the Trazas.[31]

Isaacs, however, engaged in friendly relations with the nomadic peoples who seasonally came to trade at Portendick. In 1834 Isaacs lived among them for two months at the height of summer. He gathered information about the goods they sought from England and learned much about the fractured politics of the region. Isaacs kept an account of the items the Trazas desired, and these goods—including some that had to be special-ordered, such as a handwritten Quran, a dozen red leather-bound books with silver clasps, and one hundred tent poles—were subsequently sent out on the *Eliza* and other ships belonging to Redman. Isaacs recorded the names of forty-two different clans who peopled the Sahara's sea of sand. During Isaacs' residence among them, fifty devout Traza men contracted with him for passage to Mecca aboard a ship he partly owned with Redman. Isaacs transliterated their names in a formal document signed by witnesses at Portendick.[32]

Soon after he learned of the disruption of the *Eliza*'s trade at Portendick, Isaacs entreated the Gambia's colonial governor to restore order and repel French aggression. He asked the governor to dispatch a British man-of-war to escort another of Redman's ships, which he intended to send north to Portendick to recoup the losses sustained through French belligerence. Redman, for his part, assessed his deficit to have exceeded £4,000. Yet, Isaacs noted, even full monetary compensation could not "atone for the loss of our dignity & the dishonor of our flag."[33] Here Isaacs was playing on the Crown's sensitivity to other European powers' territorial claims and interdiction of free trade.

Ben seems to have been unaware of these political contests. His brief letters focused on events along the commercial beachhead of Bathurst: "The inhabitants of this Port are a kind and familiar people and I have been introduced to a great number by Mr. Isaacs." Elsewhere, Ben wrote that he looked forward to seeing his "old schoolfellows" and telling them stories "about the Coast of Africa and all its black inhabitants." He felt sure that his father too would enjoy tales of "the 'black ladies' and when I relate to you their habits and customs & c. I am sure you won't refrain from laughter." These included "women [who] wear beads round their *posteriors*."[34] From these boyish comments, we learn that Ben closely observed local customs—even if he did treat them, or at least the women, with a fascinated condescension.

Ben's letters also express his gratitude to Isaacs for his introduction to the import-export business. "I am paying every attention to Mr. Isaacs possible," he apprised his father, "and I am now learning the way of trade." He wrote to a cousin: "I am ready to go on another voyage please God but I must not forget to mention Mr. Isaacs' kindness towards me for I find him a very good friend to me, and his kindness I never will forget." Ben's reflexively pious expression—"please God"— remains characteristic of the speech of some British Jews even today. It stems from an anglicized translation of a Yiddish idiom, *mirtzeshem*, which is itself derived from the Hebrew *im yirtze ha'shem* ("if it please God"). British Christians also used the phrase "please God," but as a benediction, for example, in the invocation "may it please God." The syntactic interpolation or conversational tic of "please God" remains a distinctive Anglo-Yiddish colloquialism. In the metropolis, Yiddish underwent a kind of baptism, its more piquant character washed away to leave behind a blander English much as Ben's original surname,

Moses, lost more than just a syllable when it was transformed into Moss.[35]

Back in "dull & miserable" Deptford, Samuel Moss related news from home. Ben's aunt Esther had moved to Bevis Marks Street, the site of Great Britain's oldest synagogue and the epicenter of Jewish communal life at the time. An uncle by marriage, George Bagshaw, had suffered some kind of paralysis. Ben's older sister later recalled Bagshaw as a "very well-informed and educated man."[36] He was likely not Jewish, demonstrating that whatever traditional affiliations the Moss family retained, intermarriage—at least among the upwardly mobile—was an accepted part of Anglo-Jewish life.

About six months into his apprenticeship, Isaacs left Ben in the care of another well-known merchant at St. Mary's while he sailed north to Portuguese-held Guinea-Bissau. Isaacs did not want Ben to accompany him during the unhealthy rainy season. So Ben busied himself collecting parrots—one gray and one green—in the hopes of bringing them to England as curiosities. Within two weeks, however, Ben grieved over the deaths of both his captive birds. Isaacs had returned by then and planned to head up the River Gambia to take on a load of mahogany. At times, Isaacs traveled as far as 250 miles upriver—a substantial distance in the unfamiliar interior.[37] Most West African traders preferred to remain in their coastal outposts and let Indigenous merchants come to them.

Isaacs trusted Ben to attend to business affairs during his absence. Ben wrote that he was keen for the experience: " I . . . am as happy as I could wish to be, Mr. Isaacs behaves uncommonly well to me and think it my duty to do for him everything that lays in my power as he has acted so very kindly towards me. . . . [And] this will be good for me in the course of time." He later wrote: "I have obtained a little insight into business and wishing to increase that knowledge I think it best to remain [in Bathurst]." He worked long hours at Isaacs' side once his master returned. Isaacs "act[s] as agent for the vessels that daily arrive here from England—belonging to Mr. Redman," Ben explained. At one point, Ben recalled that "Mr. Isaacs is so busy I am afraid to talk to him."[38]

Isaacs was indeed distracted. He had been working frantically to minimize the cascade of financial damages resulting from the illegal French blockade of Portendick, which made trade with the Trazas impossible. He sent the schooner *Marmion* to the Mediterranean, the brig *Meta* to Cabo Verde with a load of sorghum, the cutter *Prince Oscar* and the repatriated *Eliza* to Sierra Leone, and the brigantine *Matchless*, laden with rice, to Antigua in the West Indies—all desperate efforts to sell perishable cargo, even at a loss. Prices plummeted in all ports of call because of the number of other ships also trying to offload merchandise intended for Portendick. Tornadoes and heavy rains battered the ships at anchor. Rampaging Portuguese soldiers forced Isaacs' crews to abandon one port. Then the entire European crew of the *Meta* became ill, and the captain died.[39]

Despite hectic business duties and logistical headaches, as a spirited teen, Ben made time for diversions. He visited many islands during his months in West Africa, "all very fine . . . but none to equal old England." For a period, he was "rather indisposed" due to an overindulgence in guavas. He reported that he and Isaacs remained healthy, though, even when the *Meta*'s captain and crew sickened. In a letter home to an uncle, likely George Bagshaw, Ben sounded a note of triumph: "I am now learning the languages of the country and . . . have learned enough to understand and be understood." Ben had made a preliminary study of "Jolof" (Wolof). He was also pleased with the "new hat [and] a pair of Shoes" that Isaacs had given him and was appreciative that his boss had bought him bespoke "clothes suitable to this Country." Ben referred to himself as looking like a "dandy . . . on the Coast of Africa"; others gave him the moniker "Big Ben" for his smart tropical attire. This may have been a teasing nickname, as Ben could not have been very tall. In a letter to Samuel Moss, Isaacs remarked that no shoes small enough for Ben could be found in Bathurst.[40]

Isaacs' fondness for the boy was clear in all he did and wrote. In December 1835, after a year abroad, Ben had the opportunity to return to London, and Isaacs prepared for his apprentice's departure. But "as the time drew nigh," Isaacs apologized to Samuel, "my resolution failed, and I could not part with him, for to tell you the truth I never was more attached to anyone than I am to him." He continued: "I will promise that Ben shall be no expense to you, and I will do my best to bring him forward in order that he may assist in supporting those to whom I feel much indebted." The debt he refers to is surely his enduring sense of obligation to Saul Solomon and family for starting him on his own merchant career. Isaacs concluded his letter with the words "faithfully yours" and signed his name with customary vigor.[41]

In early 1836, Isaacs again left Ben in charge in Bathurst for several days, giving him clear instructions on preparing, weighing, and loading cargo. Isaacs also tutored Ben on dockside sales: "Sell all you can for *Cash* but give no credit—only to respectable Europeans. Suffer no one to have the Keys but yourself." Ben acquitted himself well, and Isaacs was quick to praise him to his father: "You will doubtless be glad to know that your Son Ben continues to improve as much in business as he does in appearance. I can assure you he merits all I can say of him. He is an attentive, willing, industrious good lad and bids fair to be an ornament to society." Ben, in turn, told his father that Isaacs would "*make a man of me.*" Though Ben missed his family, his sojourn in the Gambia with Isaacs had been worth it. "I have now a good insight into business and have learnt more during the last 6 months than I should of learnt at home for two years," he assured his father.[42] Soon he and Isaacs would sail north for Portendick and from that site of conflict continue toward home.

Isaacs had been absent from London for well over a year, but he was nonetheless gaining a reputation there. *Travels and Adventures* was published in mid-1836. The two-volume set cost a guinea (twenty-one shillings), putting it into the category of a luxury item. Isaacs' work was published not by Bentley's firm but by the less prestigious Edward Churton Company. The contract with Bentley had been canceled, though whether the change was initiated by Bentley or by Isaacs is not clear.[43] Bentley was known for insisting on holding the copyright to his authors' works, so it is possible that Isaacs, ever the businessman, balked at such a relationship. Like many publishers, Churton sold books by subscription, a commission-based arrangement that may have been more appealing to Isaacs than having his work owned outright by Bentley.

While sales figures are unknown, *Travels and Adventures* likely sold well because of the publicity it received from Churton's advertisements and because the Sixth Xhosa War (1834–36) thrust South Africa into the headlines. Isaacs' editor at the *South African Commercial Advertiser*, the humanitarian John Fairbairn, was one of those blamed by reactionaries for inciting the Xhosa to attack. Settler backlash against Fairbairn mounted as critics mined his newspaper columns for inaccuracies or editorials that appeared to give moral support to the enemy.[44] *Travels and Adventures* found its way into stores and libraries just as demand for accurate reports from the Cape Colony peaked. Reviews of Isaacs' work soon appeared in important magazines.

Some publications, like the *Athenaeum* and the *Metropolitan*, criticized elements of Isaacs' style and the rambling accounts of his travails among the Zulu. A twenty-nine-page essay in the *Quarterly Review*, known for its long-winded reviews, blamed "some booksellers' hack" for inflating the narrative of Isaacs' journal with a hodgepodge of literary quotations. Others, like the *Court* and the *Gentleman's Magazine*, publications aimed at a mass readership, acclaimed *Travels and Adventures* for its entertaining and at times "absolutely harrowing" descriptions of Zulu ferocity. The *Mirror of Literature, Amusement, and Instruction* devoted a front-page review to Isaacs' work and reprinted two of William Bagg's illustrations. The *Athenaeum* called attention, albeit obliquely, to Isaacs' unspoken Jewish heritage. An anonymous reviewer observed that Isaacs "discreetly avoids using the word Christianity," and acknowledged that the author's deistic sentiments amounted to "abstract notions of religion, which are equally suitable for Jew and Gentile."[45] Attentive readers may have taken the hint that Isaacs numbered among the former.

The reviews in these publications note how Isaacs' descriptions of the land and peoples amounted to an argument for British colonization of Natal. They split, however, on whether or not the government should form a settlement there. But the reviewers generally agreed that Isaacs had offered, in the enthusiastic words of the *Monthly Review*, "as valuable an account of the Zoolus, as any which has yet been given to the world."[46] In fact, Isaacs gave the English-speaking world its first sustained eyewitness account of Shaka and Dingane and the most detailed report available on the culture of the Zulu Kingdom. Isaacs' volumes provided the Zulu kings and their subjects with a kind of extended praise poem (*izibongo*), though one aimed at Europeans and necessarily written rather than oral. All subsequent scholarly works of Shakan historiography remain indebted to Isaacs' *Travels and Adventures*, even when they question his accuracy and decry his prejudices.

Another account of the Zulu, the missionary Allen Francis Gardiner's *Narrative of a Journey to the Zoolu Country*, was also published in 1836. Gardiner's recollection of his brief 1835 visit to King Dingane provides a chronology of his evangelism. With the Bible as his guide, Gardiner believed he had found evidence of Jewish rites among the Zulu, even describing one youngster as possessing "a Jewish expression of countenance." Here we see an oft-repeated pattern of colonial encounters with alien peoples described in terms of a domesticated example of British otherness: Jews. Gardiner periodically interrupts the narrative of his journey with original verses of thanksgiving. His execrable "Zoolu's Prayer" features a Zulu speaker who exults, "We love to hear the white man tell / How Jesus ransomed souls from hell."[47] Whatever Isaacs' stylistic faults, at least he never wrote doggerel from a penitent Zulu's point of view. Gardiner's work received significant attention and was at times reviewed alongside *Travels and Adventures*.

Both works were enlisted in a wave of public relations efforts, missionary schemes, and behind-the-scenes political machinations on behalf of annexing and settling Natal in the 1830s.

Those opposed to extending British influence in South Africa entered the fray as well. The *Athenaeum*, a month before its review of Isaacs' book, lashed out at Gardiner's hopes for colonizing and evangelizing the Zulu people. "A more absurd and mischievous scheme could not be proposed than that of usurping the dominion of the country round Port Natal," a reviewer concluded. Yet just as Gardiner's and Isaacs' volumes appeared in print, Isaacs' shipwrecked compatriot John Cane chaired a meeting of the settlers, traders, and missionaries then encamped around Natal for that precise purpose. Records indicate that a man by the name of Isaacs attended Cane's plenary council, though speculation by historians that this was Nathaniel Isaacs is incorrect. On the day Cane convened his meeting, Monday, June 20, 1836, Isaacs was about six thousand miles away treading the sandy shoreline of Portendick.[48]

Isaacs sailed from Bathurst to Portendick aboard a British gunship, but he dispatched Ben separately aboard Redman's *Matchless*. Ben kept a journal during this difficult voyage. First he bunked with Bah Fall, a Black man from Portendick who plugged his nose with snuff and intimidated Ben. He then suffered through a weeklong bout of seasickness during which he longed for Isaacs' comforting presence. Once recovered, he gaped at a frenzy of sharks that circled the *Matchless*, and he stared mournfully out at the Sahara's "vast plain of sand." He observed "a number of Moors" who pitched their tents and tied up their camels in the arid wastes. Water was rationed to a quart a day.[49]

Ben and Isaacs had come to this dangerous and "dismal part of Africa" to purchase a rare and valuable commodity: gum arabic. A sap exuded from certain species of acacia trees, gum arabic was one of West Africa's most valuable exports after slavery was outlawed. British manufacturers used vast quantities of the resin as a dye fixative in the expanding textile industry. Gum arabic could be purchased only at a few ports around the world. One of them was Portendick, where groves of thorny acacias flourished in sandy soil about one hundred miles distant, in an area under Traza dominance.[50] Workers scored the acacias' bark with knives to collect the amber-colored balls of gum. Many of these men were enslaved. Though slave trading had been forbidden by both the French and the British, societies in the Sahel continued to hold slaves. Typically, slaves who harvested gum toiled during the months of the scorching harmattan winds that scoured the Sahara. Sometimes they would eat the gum they managed to harvest to sustain themselves.

Europeans had traded for gum at Portendick for hundreds of years, but seagoing merchants had to contend with the notorious coastline, the lack of a safe harbor, and the shrewd—and sometimes violent—nomadic traders. By the 1830s, any infrastructure that had once existed at Portendick had been swallowed by sea and sand. According to Isaacs, all that marked the trading post were "two palm trees . . . without branches standing close together." The coast was "uniformly barren," and its meager supplies of freshwater proved a further hazard. The risk was great, but so was the reward. An agent like Isaacs could sell gum arabic back in England for many times what he paid for it. One British merchant at Portendick concluded that his net profits were approximately £1—about what a skilled workman earned in five days—for every thirty-five pounds of gum sold in London. In 1835 Isaacs had acquired more than 130,000 pounds of gum at Portendick. He could not afford to lose out on such a lucrative trade because of another contretemps with the French. Fortunately, his request for an armed escort had been granted by the Gambia's governor, who was pursuing a diplomatic agenda. Now sailing aboard the ten-gun *Pantaloon* as the governor's guest, Isaacs rendezvoused with the *Matchless* and was reunited with Ben.[51]

Isaacs introduced his apprentice to the Traza gum traders encamped on the coast, who assured Isaacs that supplies were "plentiful in the interior" and would soon be brought to shore from caches they had buried in the sand for safekeeping. After weeks of delay and constant promises to this effect, however, Ben no longer believed the locals and had formed a low opinion of them and their country: "There is nothing in the people or country to admire, the one is filthy, base and ungrateful, the other barren." Isaacs too described a disheartening scene: "Portendick has no regular establishment and the greater part of the year it is entirely deserted." Ben broke the monotony by washing in the ocean, riding a donkey, reading, writing, or fishing. As the days passed, Isaacs slowly accumulated supplies of gum, though Ben noted that he had been forced to pay high prices to obtain it. Isaacs bartered or advanced bolts of blue cotton cloth, knives, locks, looking glasses, snuffboxes, bullets, pistols, iron pots, blank books "with Morocco covers," sugar, molasses, sorghum, and other goods to tempt the locals further.[52] But gum was slow to arrive, and June dragged on into July.

On July 3, the Traza sultan, Mahomed Lahabebe, arrived in Portendick. Ben thought "he brought the whole of his nation with him—there were Princes, Soldiers, Camels." The *Pantaloon* and *Matchless* fired salutes to welcome him and his entourage as they appeared on shore in their flowing robes. The sultan was eager to sign a treaty with the governor of the Gambia and offered the British a tract of land on which to establish a port in exchange for protecting his people from French aggressions. Accordingly, the sultan and governor signed a convention outlining the deal. He also expressed a "great desire to have an English tent" as a gift. Isaacs and Lahabebe discussed business, and the sultan soon "dispatched his Princes in different directions to use their influence" to bring quantities of gum to the coast. Ben too had a chance to meet the sultan and his "domestics," possibly slaves. Trading remained desultory, but Ben reported that they had about twelve tons of gum stowed aboard the *Matchless*.[53] The sultan and his retinue then returned inland, and the *Pantaloon* sailed back to Bathurst carrying the governor, who would report to London on Lahabebe's offer of territory in exchange for defense.

Isaacs, now at the mercy of any patrolling French gunboat, waited nervously for caravans of gum to arrive. When they did not, he deployed the ancient bargaining technique of feigning disinterest. He instructed the captain to unfurl the sails and prepare the ship as if to depart. He also had a messenger confiscate the gratuities of blue baft—coarse cotton cloth—that he had advanced as a gesture to encourage swifter deliveries from the interior. Isaacs called the reluctant traders "Maraboos" (marabouts), a word which typically refers to a Muslim religious leader in West Africa. Ben described the "poor Maraboos . . . crying for their Bafts," though "Mr. I[saacs] had not the least intention of leaving the place at the time."[54] His strong-arm tactic made little difference, and trade proceeded erratically.

The delays provided Isaacs time to scribble down some observations about the locals, perhaps for a sequel to *Travels and Adventures:* "The Maraboos . . . believe in Gregorys [gris-gris] which is made in leather of various shapes." These amulets, he explained, "contain a piece of paper of Bullocks Hide with a few Arabic letters written on it—and they say supposing you take a Knife and offer to cut me, the Knife will break and cannot cut. They also say if you take a Musket or Pistol and load it with ball, and fire it at them the ball cannot enter any part of their Body." Isaacs found the custom "a very foolish one—and I do not think they are courageous enough to let any person cut or fire at them. For they would find a great disappointment in their Gregorys."[55] Whether he connected the practice of wearing gris-gris for protection to the Jewish ritual of wearing *tefillin* (phylacteries)—leather boxes strapped on the bicep and forehead containing parchment inscribed with biblical verses—he did not say.

Isaacs and Ben remained aboard the *Matchless*, eager for further commerce. "At the end of July the rains set in," Isaacs recounted, "attended by hot winds and calms then follow loud peals of thunder and terrific flashes of lightning which ends in a strong breeze, or what is commonly called a tornado that purifies the atmosphere [and] throws a smooth veil over the ocean and abates the overwhelming surf." Masts could be snapped by the violent gusts, and anchors dragged or lost. Ben endured several of these violent squalls and wrote of the sky turning black and heavy rains pelting the ship. Then the "sand blew up mountains," and the *Matchless* rocked on her beams.[56]

For a month, sultry days of boredom alternated with evenings of stormy terror. By August, Isaacs could wait no longer. The sultan had indicated that gum might be found at another spot, Little Portendick, about thirty miles south in an area under firmer Traza control, where there was an "abundance of Water and food for the Camels [and] Cattle." The British lacked a treaty with the French allowing them to trade there. Nonetheless, Isaacs had agreed to survey the area for the Gambia's governor and inform him of the possibilities of establishing a British outpost in the vicinity. He now ordered his sailors to rig the brigantine's sails and cruise down the seaboard. Along the way, Isaacs spotted the Trazas' principal minister, Abdullah, dressed in white and surveying their progress from the shore.[57] The *Matchless* anchored, and Isaacs prepared to make landfall as soon as the heavy seas allowed.

The next day, with the "surf breaking violently," Isaacs and four crewmen pulled at the oars of a longboat. Near shore a surge upset the boat and tossed the men into rough waters. Fishermen waded in to save them. Watching from aboard ship, an anxious Ben thought Isaacs "much exhausted" as he was helped to safety. But his risk paid off. Within days, Isaacs had purchased and loaded another six or seven tons of gum. Shipping conditions at Little Portendick remained challenging: Ben noted that "it is only once in a week that the weather is fine," and even then "the boat will get half filled with water."[58] Yet Isaacs and his crew managed to load several more tons during the succeeding days.

By the third week of August, Isaacs determined to visit the sultan of the Trazas in his encampment near the legendary gum forest. He had heard rumors that the French had attempted to bribe Lahabebe to end English trade at Portendick, and he wanted to assess the truth.

Meanwhile, the captain of the *Matchless* oversaw transactions on shore, while Ben superintended the loading of the cargo. Isaacs set off into the western Sahara perched atop a camel on a "tedious and unpleasant journey" through a parched and windswept landscape. Two weeks later he returned with tales of the sultan's hospitality. Isaacs confirmed that French authorities had indeed approached Lahabebe to grant them exclusive rights to Traza gum supplies in exchange for a one-time fee. Isaacs also learned that he had been the "first 'white man' that had traveled so far in the Interior." He reported to Ben that the landscape offered "nothing . . . to please the eye" and that "the whole face of the country is covered with sand." As soon as he returned, they set sail for Bathurst. Ben tallied accounts during the journey south.[59]

Perhaps during a stopover in Senegal, Isaacs had a relationship with Nanette Guey (or Gueye), who was Black or of mixed race. Nanette may have been a *signare*, a woman of the more privileged classes in French West African colonies who engaged in concubinage with wealthy Europeans. Nothing is known about her other than that she gave birth to a daughter, Hannah (or Anna) Isaacs. A late nineteenth-century photo of the adult Hannah Isaacs portrays her as an attractive woman sporting European fashions, including a brocade jacket. She wears her hair in an elaborate coiffure ornamented with bows. Hannah has a broad forehead, lustrous eyes, and a wry, close-lipped smile. Isaacs seems not to have acknowledged Hannah as his daughter, though she may have later encountered him in the region north of Sierra Leone, where she lived for a time and gave birth to a daughter.[60] Hannah, like Porter, exists in the historical record as little more than one of Isaacs' abandoned children.

Back in Bathurst, Isaacs and Ben wrapped up their accounts before boarding the *Matchless* for the return voyage to London on 4 October 1836. Aboard ship, Ben and his bunkmate were tormented by cockroaches, and both Ben and Isaacs suffered bouts of malaria or another debilitating fever. Most of the crew were stricken as well. During the voyage, Ben spotted a southbound Portuguese brig whose deck was crowded with passengers—"or, rather," Ben added, giving voice to his suspicions, "I think they were passengers."[61] The ship may have been innocently conveying cargo, crew, and travelers; equally possible, given Ben's doubts, it might have been ferrying slaves from one Portuguese territory to another.

Once Ben recovered from his illness and the crosswinds lessened the ship's ceaseless yaw and pitch, Ben amused himself by reading Isaacs' recently published *Travels and Adventures*, a copy of which had found its way to Isaacs in the Gambia. "I have passed time away by reading Mr. Isaacs' work . . . which I find both amusing and interesting," Ben recorded. "And I must further add," he continued, "that general information might be obtained from this work which would prove of great service to the public." Ben had to interrupt his reading to attend Isaacs when he fell ill. After finishing the second volume of *Travels and Adventures*, which Ben thought "much more interesting than the first," he picked up *History of the Earth*, a popular account of the natural world by Oliver Goldsmith, the eighteenth-century novelist, poet, and naturalist. With Isaacs slowly recovering his strength and Ben's bunkmate still "very weak," Ben may have paged through Goldsmith's scientific work seeking practical medical advice. Goldsmith's warnings about a tropical distemper that "thins the crew of European ships, whom gain tempts into . . . inhospitable regions," would not have made for reassuring reading. The theory that bad air—miasma—caused tropical fevers like malaria prevailed well into the nineteenth century. To ward off miasmic vapors, crews in West Africa fumigated their vessels with tobacco or rinsed the decks with vinegar or lime juice.[62] The discovery of the role of mosquitoes in transmitting malaria was decades away.

The *Matchless* was battered by wind and waves on her return voyage. Rain leaked into the cabins, making the passengers' "posteriors wet." Possibly the arid climate of the Saharan coast had dried out the brig's boards and the pitch that sealed them. Ben struggled to compile a cargo manifest by dim candlelight in advance of their arrival in England. The dawn of 7 November brought his fourteenth birthday, more rain, and colder temperatures. Ben wrapped himself in his blanket coat against the chill and was cheered slightly by his ailing bunkmate's birthday promise to buy him a dozen tarts once they docked in London. Foul weather followed the *Matchless* into the English Channel, where a thick fog enveloped the ship. They could not take their bearings until a sudden clearing revealed them to be abreast of Dover's white chalk cliffs.[63] Isaacs went ashore and from there headed to London to meet with Redman. Ben stayed on board to accompany the cargo to the dockyards.

After a delay at Gravesend, where the *Matchless* was quarantined owing to fear of disease, Ben arrived in London on 17 November 1836. "I had been expected to be out not longer than 8 months but what may happen during an African voyage . . . is impossible to foretell," he wrote. He had been away nearly two years. Yet only three days after his feet touched British soil, he expressed his eagerness to head to sea once more: "I have made the best use of my time and paid the greatest attention to the duty Mr. I[saacs] has allotted to me . . . and to tell you the truth I am almost prepared for another trip—the climate of Africa I am happy to say agrees with my constitution uncommonly well, and which is a great blessing thank God."[64] Within a few months, Ben would again ship out with his mentor.

How he and Isaacs spent their leave in and around London is unclear. Ben surely enjoyed his reunion with his father, sisters, and extended family, regaling them with tales from afar. Isaacs may have found time to visit his mother and sister in Canterbury. His sister, Hannah, was now of marriageable age, and she soon met and married Isaac Schwersensky, a Prussian-born jeweler. They were married in Canterbury's synagogue. Isaacs also kept busy hustling between Redman's offices and the dockyards preparing for another voyage and tabulating accounts. Based on the latest figures supplied by Isaacs, Redman revised his total damages due to the French blockade of Portendick upward to an astounding £51,481 11s. 4d.[65] Facing such losses, he and other affected merchants pressed the British government to present their claims to France.

During his shore leave, Isaacs would have learned that one of his allies in the effort to annex Natal was now in London promoting colonial reform. The lawyer Saxe Bannister, the intemperate critic of Cape officialdom and would-be colonizer, had returned to England. He insinuated himself into the highest echelons of science, religion, and political reform; in the 1830s these three spheres overlapped to a great extent. Bannister's likeness peers from the background of a famous group portrait of delegates to an antislavery convention. A brown forelock distinguishes him from the mass of balding, red-cheeked burghers in the foreground. A former agent for Indian affairs in Canada, the first attorney general of New South Wales (Australia), and a defender of Indigenous rights in the Cape Colony, Bannister had varied colonial experience. At the Cape, he had traveled through Xhosa lands and later charged British forces with complicity in atrocity. In 1828 and early 1829, he worked with Farewell to outline the proposal for the systematic colonization of Natal. By 1836, Bannister had authored several books and pamphlets that were grounded in his familiarity with Indigenous peoples on three continents. These works offered scathing criticisms of Great Britain's colonial policies. Even Bannister's allies considered him indiscreet—a significant breach of gentlemanly conduct.[66]

Bannister was dogged in London by his reputation as stiff-necked and high-minded. But his knowledge of Great Britain's colonies meant that his unpopular opinions on protecting the rights of aboriginal peoples could not easily be ignored. Bannister's crusading efforts admitted him to elite intellectual circles engaged in political reform, philological study, and ethnographic pursuits. He became acquainted with the pioneering physician and social reformer Dr. Thomas Hodgkin, now chiefly remembered for his identification of Hodgkin's disease. In 1836, Bannister and Hodgkin cofounded the Aborigines Protection Society, which arose as an adjunct to the Parliamentary Select Committee on Aboriginal Tribes, in session from 1835 to 1837.[67] The term *aborigines* referred to any Indigenous people colonized by Great Britain.

Violence among settlers, Boers, military forces, and Xhosa along the porous eastern borders of the Cape Colony had thrust government policies in South Africa into the political spotlight. Skirmishes, slave holding, cattle rustling, reprisal raids, and lawless slaughter proved endemic. Bannister had done much to bring the bloodletting to public attention. In response to the ongoing abuses, MP Thomas Fowell Buxton, who led the antislavery cause, launched the select committee to investigate settler and military actions and to formulate recommendations that would safeguard Indigenous rights. An evangelical with strong ties to Quaker communities, Buxton proved an adept envoy to a Colonial Office helmed by allies who professed evangelical beliefs and remained circumspect about haphazard territorial expansion, a group that included the new undersecretary for colonies, Sir George Grey.[68] The select committee's efforts and its subsequent published report remain touchstones for understanding humanitarian attitudes of the era.

Bannister contributed details to the select committee on the mistreatment of the Xhosa, again drawing attention to the massacre of 1828, in which colonial forces at the Cape claimed to have defeated the Zulu but instead "destroyed ... another tribe by mistake." Today the committee's findings seem tainted by paternalism, such as the assessment that decades of colonial experience among "the negro race" in the Crown colony of Sierra Leone had demonstrated that Africans possess a "good average intellect" and the capacity "for mental culture." The evangelicals who guided the committee's fact-finding process and who synthesized its witness statements were indeed paternalistic, even when they were women. Nonetheless, its damning conclusion that "the effect of European intercourse ... has been ... a calamity upon the native and the savage nations whom we have visited" echoes down the ages as a condemnation of British policies.[69]

The committee's report had an enduring influence on the Aborigines Protection Society. The early years of the society saw Thomas Hodgkin come under the influence of Edward Gibbon Wakefield's program for systematic colonization, proclaiming it as a "means of rendering important services to the aborigines," and one that "form[s] a contrast with every instance of modern colonisation now on record."[70] Hodgkin's enthusiasm for the protections promised by Wakefield's program, and his zeal for establishing the society in the first place, partly derived from his involvement in one of the burning scientific debates of his day: monogenesis versus polygenesis.

Monogenesis holds that there is one common origin for humankind and that variations of physical appearance, language, and culture are often the result of environmental factors related to human dispersion. Polygenesis maintains that such variations reveal evidence of multiple human origins and a lack of common descent. Hodgkin shared a Quaker background with many society members, and monogenesis accorded with the Christian belief in a universal Adamic ancestor—a belief that also informed Quaker and evangelical abolitionist agitation. Efforts to reconcile polygenesis with the biblical account of creation, by contrast, often led to the crude conclusion that some peoples and races were not fully human, and that, like animals, they might be exploited, enslaved, pushed toward extinction, or even killed. Such were the explanations for evolutionary difference and the assumed superiority of Europeans favored by later so-called scientific racists and some social Darwinists.[71]

Hodgkin championed monogenesis following the lead of his proto-anthropologist friend James Cowles Prichard, who was an honorary member of the Aborigines Protection Society. Prichard theorized an essential unity between all human beings, whatever their race. The first edition of Prichard's influential *Researches into the History of Mankind* (1813) rejected as an "absurd hypothesis" the popular view that Africans might be "the connecting link between the white man and the ape." Nonetheless, Prichard did maintain a belief in cultural hierarchies that would have comforted his white readers: "For we find that all nations who have never emerged from the savage state, are Negroes, or very similar to Negroes." Yet Prichard could also be positive in his assessment of African peoples. In the expanded 1837 edition of *Researches into the History of Mankind*, Prichard made clear that the Zulu were a "bold and warlike people, of noble carriage," an observation he based on the published reports of William Fitzwilliam Owen.[72]

Christian education and productive labor held out the promise of civilization to all people, according to advocates of monogenesis like Hodgkin or Prichard. Indigenous peoples could develop into the citizens of new Englands planted abroad with the help of the "Bible and the plow," as Buxton put it. Prichard's opponents, such as the polygenist anatomist Robert Knox, insisted on fundamental distinctions between races that were underscored by a crude environmental determinism. Whites, Knox believed, could not flourish in torrid Africa. Nor could they successfully export European culture to other races. As a consequence, precisely those individuals who did not believe in racial inferiority emerged as vociferous supporters of colonization, while those who pursued visions of racial purity and superiority often opposed imperialist expansion.[73] Monogenist abolitionists often stoked colonialism; polygenist racialists, by contrast, made good anti-imperialists.

Hodgkin, for example, interpreted Prichard's monogenist position as a humanitarian call to "civilize" Indigenous peoples under colonial sway. Culture, not biology, explained Anglo-Saxon superiority.[74] It was internal conviction, not outward appearance, that fostered Christian rectitude and industriousness. And Hodgkin's junior partner, Saxe Bannister, believed he knew just how to engineer civil and Christian African polities.

The Aborigines Protection Society went on to fight a pitched battle with the political establishment to institute a scientific and philanthropic approach to colonization. Today we may well view the society's proposals as insidious and reeking of Christian moralism; yet Bannister, Hodgkin, and their parliamentary allies, like Buxton, were considered progressive or even radical at the time. For seventy-three years the society was the leading organization in the United Kingdom devoted to the welfare of Indigenous peoples. Its members dedicated themselves to "protecting the defenceless, and promoting the advancement of uncivilized Tribes."[75] Protection was to take the form of rigorous legal frameworks that would guarantee the rights of the aboriginal peoples and deter settler violence. Advancement was synonymous with assimilating colonized peoples to Christian belief, commercial trade, and British cultural norms.

A similar goal of "ameliorating" the conditions of the impoverished Jewish immigrants in London existed among Christian missionizing societies, as Isaacs was surely aware. These proselytizers to London's poor Jews promised a domestic version of the humanitarianism practiced abroad by the Aborigines Protection Society. "Anything short of savages can hardly be more demoralised than many of the Polish Jews on their first arrival in this country," one minister mused in his survey of nineteenth-century Jewish immigrants in London. Great Britain's impoverished Jews might not be as degraded as its aboriginal subjects, but immigrant Jews were nonetheless characterized by "dirt, stupidity, and obstinacy" and as possessing "scant sentiments of integrity." Therefore they stood in need of the "amelioration" promised by British Christians.[76] At times, the rhetoric of philo-Semites who aimed to minister to London's Jews sounds uncannily similar to the language of those missionaries who collaborated with the Aborigines Protection Society in order to save the heathen souls of the Zulu and other African peoples.

Another charter member of the Society, Dr. Stephen Lushington, was a member of Parliament, a staunch abolitionist, and a supporter of full Jewish enfranchisement—a controversial cause in his day. Lushington, possibly by way of Bannister, learned of Isaacs and Redman's commercial losses at Portendick due to the French attack on the *Eliza* in 1835. In the House of Commons, Lushington pressed the foreign secretary, Lord Palmerston (Henry John Temple), to seek reparations from the French for the seizure of the *Eliza* and to redeem the stain on Britain's national honor. Palmerston replied that an answer would be forthcoming, but he provided no details. A poor orator who reportedly could not pronounce his *r*'s, Lushington agreed to wait for the outcome of negotiations with the French rather than resume his verbal sparring with Palmerston on behalf of Redman and Isaacs.[77]

Isaacs, meanwhile, had been busy looking after his own interests. He and Ben Moss again embarked for the Gambia in early 1837 on different Redman company ships. On his arrival in Bathurst, Isaacs learned that the French were newly "threatening to oppose any English vessel's trade in gum, by sending French Ships of War to drive the English from Portendic." Wary of further losses, Isaacs reconnoitered other ports of call, while Ben remained behind to see to Isaacs' dockside business. Ben spent his free hours collecting shells, two parrots, and a monkey that he hoped to sell once back in England. Such curiosities could fetch a handsome price and help supplement his family's income. London's thriving market for exotic curios was entirely in the hands of Jewish traders. Later Ben sent his father twenty-eight stuffed birds, which "cost a mere trifle," in the care of one Captain McCormack. Ben hoped his father would be able to resell the birds as objets d'art. He also reminded his father to treat Captain McCormack to brandy, "as you are aware that he is very fond of the bottle."[78]

At the end of June 1837, Ben informed his father of a "dreadful calamity that has befallen Sierra Leone through the ravages of the epidemic"—a combined outbreak of yellow fever and smallpox. But Ben reassured his father that he and Isaacs were still "500 miles from that place, and . . . we are as healthy here, as we should be perhaps in England." Within a month of Ben sending off this anxious missive, Isaacs sailed heedlessly for Sierra Leone "to settle an affair of importance"; Ben instructed his father, *This I do not wish to reach the ear of* [Isaacs'] *mother.*[79] This request suggests that the Moss family remained in contact with the Isaacs family of Canterbury and that Lenie Isaacs and her son still maintained close emotional bonds. Isaacs, it seems, didn't want his mom to worry.

With Isaacs away in plague-stricken Freetown and another clerk "on his 'Sick Bed,'" Ben, now promoted to bookkeeper, managed business in Bathurst on his own. "To tell you the truth," Ben wrote, "I am left entirely in charge of Mr. Isaacs' affairs here." With his increased responsibilities, Isaacs would be paying him £25 per year. Isaacs had also agreed to provide his charge with "board, lodging, medical attendance, and Washing," and consented to raise Ben's annual wages to £35 in the coming year. He could afford to be generous to Ben. Isaacs earned at least £600 per year as Redman's agent—about ten times what a skilled tradesman earned—and he likely supplemented that sum with lucrative dealmaking on the side as opportunities arose. "Were I in 'Dear England,' I could not enjoy better health," Ben assured his father, and he closed this optimistic letter with wishes of "health and prosperity."[80]

This letter was the last one Ben would ever send. Soon afterward, perhaps in August 1837, Ben died, possibly of malaria, and likely without Isaacs by his side. We can imagine the lonely adolescent suffering blinding headaches, dosing himself with quinine, shivering uncontrollably, and sweating out a high fever on his bed while the West African sun beat down on the slate roof of his Bathurst residence. Had he risen from his bed to peer out his window to look for a sail signaling Isaacs' return, the glare on the River Gambia would have made him blink back from its appalling whiteness.

Isaacs must have been shocked and grief-stricken by Ben's death. But no surviving letters or journal entries record his reaction. Did Isaacs bury the fourteen-year-old who had been so dear to him? Did he pay a shiva call to Samuel Moss when he returned to London? Did he give an account of the circumstances of Ben's death to his family? Perhaps they never learned the details of the boy's final days. The only documentation I have found is a laconic notation made by one of Ben's sisters decades later: "Benjamin Moss died at River Gambia."[81] That is all that we know for sure.

Isaacs returned from West Africa to find England altered once again. Queen Victoria had assumed the throne. A newly installed monarch presented fresh opportunities for influencing government policy. The energetic Bannister did not miss his chance to establish yet another organization, this one aimed at implementing a version of Wakefield's plan for systemic colonization while adhering to the humanitarian impulses of the Aborigines Protection Society. The South African Land and Emigration Association, with Bannister as its provisional secretary, presented Her Majesty's Government with a formal proposal for the colonization of Natal in the first months of 1838. Bannister explained: "Endeavours should . . . be made by Government to exercise a salutary controul over enterprising men, by laws and by institutions suited to their position, but which will not crush their activity. And the State ought to accompany them in advance, so as to make their energy tend to improve, instead of oppressing the native people. Such it is conceived will be the character of the proposed Colony at Natal." In his plan, Bannister referenced Isaacs and the lands granted to him by King Shaka. Active supporters of the association included Wakefield's brother and G. C. Redman. But like all previous plans for the colonization of Natal, the association's proposal was rejected.[82] Meanwhile, piecemeal settlement in Natal grew apace—and soon led to disaster.

A wagon train of Boers seeking to live beyond British control at the Cape pierced the heart of the Zulu Kingdom in early 1838. The leader of the party, Piet Retief, offered Isaacs' old friend Henry Francis Fynn a "handsome salary" to join them as an aide-de-camp. Fynn, ever the loner, turned him down. About one hundred of Retief's Voortrekkers, as the emigrants were known, sought out King Dingane to negotiate settling in his lands. Retief wrested a land grant from the Zulu monarch, much as Farewell, King, Isaacs, and Fynn had each received concessions from Dingane, and from Shaka before him. But Dingane, perhaps recalling Hlambamanzi's description of the British method of creeping territorial control—that "white houses" would accrue to overwhelm the Zulu nation—lured the Voortrekkers into a trap.[83] His *amabutho* massacred the Boer company on a hillside, an infamous deed that still stands out in South Africa's long and bloody cycle of aggression and revenge.

To avenge the slaughter, a motley group of Englishmen and Africans, led in part by the castaway and colonial pitchman John Cane, launched an attack on one of Dingane's regiments. The Zulu defeated them in a horrific battle at the uThukela River in April 1838. Cane was speared through his chest by assegai and tumbled dead from his horse with his pipe still clenched between his teeth. Years later, the shattered bones of the one thousand or so men killed along the uThukela still littered the landscape.[84] The Voortrekkers then supported a rival aspirant to the Zulu throne. Dingane was finally defeated in battle and assassinated by his enemies in 1840.

Despite the carnage, Isaacs continued to see promise in Natal. He picked up pen and ink while in Sierra Leone and addressed Fynn in grandiose fashion: "Since leaving you my dear Fynn, I have been variously and occasionally profitably employed, I have made and spent fortunes." He gave vent to his frustration over the many failed plans to settle Natal: "I have . . . published many things and done much to induce the British Government to colonise Natal, and fail[ed] in this from the narrow views, and stupid economy of the present administration." Nonetheless, "I shall never rest until Natal becomes a Settlement of the first power." To accomplish this goal, he had "induced a very influential and opulent mercantile body to form a company to colonize that part of Natal where my early days have been spent, and which I have always yearned for since leaving it." Isaacs here refers to his work on behalf of the South African Land and Emigration Association. Even after the blunt rejection of their schemes by the secretary of state for the colonies, Bannister, Redman, and others had continued to promote the plan.[85]

Isaacs assured Fynn that "Natal for some time past has been exciting great feeling in London." He proposed that they resume their "juvenile partnership" and exploit their land grants in the Zulu Kingdom, with financial capital to be supplied by Redman and other wealthy merchants among the association's supporters. Isaacs averred that "a company established on wise principles can govern a colony best, as has been seen by the EIC [East India Company] and the South Australia Company."[86]

Isaacs also enquired whether his old friend had received the volumes of *Travels and Adventures* he had sent by way of his brother, Benson (Benjamin) Isaacs, who was then at the Cape. Indeed he had. Fynn's copy, inscribed by Isaacs, still resides on a library shelf only a few miles from the site of what was once Fort Farewell.[87] Fynn apparently never acknowledged receipt of the books. Perhaps he was miffed that Isaacs, formerly his understudy in adventure, had been the first to publish a history that forged the mythology of King Shaka and the Zulu nation. Isaacs gave up on his Natal proposal and remained in Sierra Leone to seek another outlet for his enterprising spirit.

CHAPTER 6

Railroad Christianization, 1837–1853

In 1763 John Newton, a white Englishman, recounted that while he was living in Sierra Leone, he had "grown *black*." Newton reassured his more literal-minded readers that his complexion had not changed. Rather, he had "assimilated to the tempers, customs, and ceremonies of the natives." He regretted that he had ever adopted the heathen rituals of the "blinded Negroes" who had tempted him to believe in the heresy of necromancies, amulets, and divinations. Now a properly repentant Anglican minister, Newton lamented the former "spirit of infatuation" that had led him into "closer engagements"—presumed to include sexual liaisons—with West African inhabitants. He would have "died a wretch amongst" these people had a ship not found him wandering the coast and transported him back to his homeland. Safe in England, Newton gave thanks that he could now clearly see God's hand in all that he had endured.[1]

The language used by Newton to record his wayward life in Sierra Leone, evoking blindness and wretchedness, presages the first verse of the hymn he subsequently penned, "Amazing Grace" (ca. 1772–73). That Newton authored "Amazing Grace" decades after working as a slaver in West Africa is well known; that the hymn was composed more than a decade *before* he repudiated slavery and joined the abolitionist movement is less celebrated.[2] Such a reckoning would be bad news for gospel.

A century or so after Newton wandered the shores of Sierra Leone—a name that was used variously to refer to lands extending from today's

Liberia to Guinea-Bissau—Nathaniel Isaacs made the region his base of operations. Like Newton, Isaacs would "grow black" by forging alliances with local leaders and adopting the languages, cultures, and customs of various groups inhabiting Sierra Leone and adjacent territory.[3] Also like Newton, Isaacs would indulge in "closer engagements" with women on the coast and find himself entangled in the brutal economy of slavery.

Ports along the West African coast from today's Senegal south to Angola had long been central to the transatlantic slave trade. By the late eighteenth century, Sierra Leone and Guinea, situated in the western bulge of the African continent, became terminals for the Middle Passage. The region was known to be unhealthy for Europeans, but it was a convenient location for slavers cruising for human cargo. Bunce Island, a notorious slaving center in the Sierra Leone River, was about a thousand miles closer to the labor-hungry British sugarcane plantations in the Caribbean than the so-called Slave Coast along the Bight of Benin. In time Sierra Leone would become the focus of Great Britain's most ambitious abolitionist project: to create a model colony of freed slaves.

In the last decades of the eighteenth century, heterodox Anglicans, many of them followers of John Wesley's incipient Methodist movement, and Dissenters, mostly Quakers, began to agitate for an end to slavery in the British Empire. The Reverend John Newton fell under the influence of Wesley's evangelism and was drawn to the controversial minister's objections to slavery. Anglicans and other Protestant groups had tried in past decades to convince plantation owners in the American colonies and the Caribbean to coax their slaves to accept the Christian faith. After all, these clergy and lay leaders argued, "instruction of the slaves in the principles of religion" had encouraged "habits of regularity and industry" and thereby increased the value of their human property by one-third. An appropriately Christianized slavery could thus serve the cause of commerce. Missionary efforts were therefore aimed at ameliorating the conditions of slavery, not ending it. Political abolitionism in Great Britain later coalesced around figures who were frustrated by the meager number of African converts to Christianity in the colonies and outraged by the distant plantocracy's refusal to mitigate the cruelties of slavery.[4]

Many of these high-minded opponents of slavery considered Great Britain's virtue to be tarnished by the slave economy. Unless they stopped the trade in Black souls, their own eternal spirits would suffer damnation, and the empire would continue to endure the lashings of

Providence. King George III had finally lost the American colonies in 1783; some thinkers saw this defeat as divine retribution.[5] Early abolitionist endeavors were as much products of evangelical eschatology and nationalist sentiment as of humanist and philanthropic sympathies.

In the mid-1780s, William Wilberforce, a wealthy member of Parliament, underwent a religious awakening. Wilberforce was a popular gentleman-about-town with penetrating eyes, a friendly countenance, and wavy shoulder-length hair. His conversion moved him to use his charisma and political connections to challenge the trade in human lives. Wilberforce's acknowledged "spiritual father" in this venture was John Newton, who had personally counseled Wilberforce when the horrors of slavery first intruded upon the younger man's conscience.[6]

By 1787, Wilberforce and other evangelical notables fired by spiritual piety, social activism, and biblical study formed the Committee for the Abolition of the Slave Trade. The ubiquitous symbol of the movement was an image of a chained Black man on bent knee imploring, "Am I Not a Man, and a Brother?" This motto likely originated with Peter Peckard, an Anglican minister whose antislavery pamphlet of the same name railed against the "abominable traffick in the Human Species." Peckard's rhetorical question echoes Shylock's plea in *The Merchant of Venice* (III.1) for recognition of his essential humanity: "I am a Jew. Hath not a Jew eyes?" Peckard had previously defended British Jews against public venom and violence during the failed efforts to enfranchise them under the 1753 "Jew Bill." Perhaps the similarity between Peckard's abolitionist slogan and Shylock's protest is more than mere coincidence. The scholar David Feldman has suggested that the phrase *Jewish emancipation*, which frames the discussion of the Jewish encounter with modernity, derives from the language of abolitionist efforts to emancipate Black slaves.[7] Two despised peoples once thought to be theologically damned—Jews and Blacks—found common emancipators in the devout Peckard and others who sought to remove Jewish legal disabilities.

Peckard and John Newton moved in the same evangelical circles, and both men were financial subscribers to an effort to establish a colony of freed slaves in West Africa: the Sierra Leone Company, founded in 1792. The company centered its efforts on resettling former American slaves who had fought for Great Britain in its unsuccessful bid to suppress the American Revolution. These loyalists followed British troops to Halifax, Nova Scotia. Some journeyed on to London, where they joined other Blacks who had been taken to England as slaves by Caribbean and North American masters and later emancipated under British law, but who now endured hardship, poverty, and discrimination. Many of these "black poor," as they were known, and some of their white British spouses, emigrated to Granville Town, a settlement in Sierra Leone. This settlement lasted only from 1787 until 1789.[8] The Sierra Leone Company's subsequent founding of Freetown would prove far more enduring.

Settlers erected the first structures of Freetown in 1792, at the northern end of a mountainous and forested peninsula jutting into the Atlantic Ocean. A vast bay provided one of the coast's few deep harbors, sheltered from the unpredictable surf. Black settlers from London and the so-called Nova Scotians were joined in 1800 by deported maroons, escaped slaves who had formed their own communities and fomented rebellion against white planters in Jamaica. These groups formed the core of the early settlers, who came to be known as Krios. Later, some Krios mixed with the Fula, Mende, Temne, and other peoples, giving rise to a creolized population comprising multiple ethnic, linguistic, national, and religious identities.[9]

Freetown grew apace, attracting trade from the interior, and by 1808 the town was solidly established. That same year Sierra Leone became a Crown colony, with a governor appointed by the British government and administered under the authority of the colonial secretary. The abolitionist leader Thomas Clarkson proclaimed the Freetown experiment a model for the "civilization for Africa," and he attributed the movement's successes "to Christianity alone." In other words, he implied that civilization could not exist without Christianity. The London Missionary Society's Reverend Philips, who had promoted Native rights in South Africa with Isaacs' former editor, John Fairbairn, made the claim even more powerfully. "The missionary who labours among a savage people has no right to expect much success if he neglects their civilization," he wrote; and those who "make the attempt without the doctrines of the Cross" were doomed to fail.[10] Such an attitude encouraged the proselytization of supposedly backward peoples, such as Black Africans, and stubborn holdouts, like the Jews.

Some in London thought Jews to be virtually indistinguishable from Black Africans. The anatomist Robert Knox described being alarmed by the grating voices of Jewish peddlers on London streets. Surveying "the Jewish physiognomy" during these urban safaris, what he saw was a preponderance of "large, massive, club-shaped, hooked nose[s], three or four times larger than suits the face," and other features "which stamp the African character of the Jew." Knox was a veteran of Waterloo who began to collect human crania to launch a polygenist theory of racial difference during a posting in South Africa. Later he was disgraced when it emerged that he had purchased cadavers from a pair of murderers who killed to profit from the trade in corpses. Another, less scandal-prone anthropologist, John Beddoe, began recording observations of physical characteristics of Europeans in the late 1840s. He established a pseudoscientific scale of blackness on which Jews ranked higher in "nigrescence" than their European peers. Beddoe concluded that the Jews possessed an unusual ability to flourish in disparate climes, which related to their possessing a "double physical type": both "xanthous" (Caucasian) and "melanous" (Black).[11] The Jew was defined as a racial shapeshifter.

Similarities between Jew and African, and the equation of civilization with Christianity, left Georgian-era and early Victorian Jews characterized as neither entirely savage nor entirely civilized.[12] A largely Christian triumphalist society perceived Jews as a cursed people maintaining backward customs and an outmoded faith. At the same time, Jews went about their lives on the margins of Great Britain's developing modern polity and distinguished themselves as avatars of a global economic trade network. In short, Jews were simultaneously archaic and modern, never fully one or the other, and therefore wholly dubious.

The period that witnessed the passage of the Slave Trade Act (1807) and Sierra Leone's formalization as a Crown colony (1808) also gave rise to efforts to convert England's subaltern Jews. The historical proximity of these events suggests the scale of evangelical ferment at the time. Wilberforce and other evangelical "saints," as they were cynically called by detractors, launched the London Society for Promoting Christianity amongst the Jews in 1808. The society proselytized indigent Jews, primarily in the city's eastern districts, which were crowded with new arrivals from continental Europe. Their aim was to convert and thereby assimilate these unruly Jewish immigrants to decorous Anglo-Saxon civilization. Wilberforce donated his "affection, patronage, time, and advocacy" to the society, served a term as its vice president, and addressed it eight times at its annual meetings.[13] Many of the same tactics employed to contain the Black Other across colonial Africa—the limitation of civil rights, a presumption of cultural inferiority, paternalistic educational models, confinement in racialized spaces, and proselytization as a means for spiritual "improvement"—had been pioneered in England's metropole and directed against domestic aliens: the Jews.

The conversion and civilizing of Jews, like the conversion and civilizing of Africans, accorded with deeply held notions of British moral superiority and the evangelicals' eschatological beliefs concerning the Second Coming. Promoting Christianity and ending the slave trade emerged as twinned imperatives of British moral imperialism. London had its missions; distant Freetown emerged as a mission city. Early in its history, the colonial government even parceled out Sierra Leone into administrative parishes overseen by missionaries.[14] Proselytization and abolitionism proved to be useful tools for imperial entrenchment.

Ammunition for the fight against the slave trade also came from those with more worldly aims. "It appears from the experience of all ages and nations," Adam Smith pronounced, "that the work done by freemen comes cheaper in the end than that performed by slaves." For Smith, the labor of free individuals, whether African or European, increased the wealth of nations. When the great powers who had allied against Napoleon convened in 1815 to redraw the map of Europe at the Congress of Vienna, they reaffirmed a commitment "to induce all the Powers of Christendom to proclaim the universal and definitive Abolition of the Slave Trade." The signatories included many of the most influential diplomats in Europe, but the language of the resolution was mealy-mouthed and lacked any timeline or method of enforcement. Parallel debates over resolutions to enshrine Jewish emancipation raged in Viennese drawing rooms while congressional delegates met. But any clauses ensuring the inclusion of Jewish rights were defeated as well.[15] The congress offered no meaningful redress to Jews or Blacks.

At a sequel to the Congress of Vienna held in 1818 in Aix-la-Chapelle (now Aachen in Germany), slavery and Jewish emancipation were once again the central humanitarian concerns. At the urging of Wilberforce, Thomas Clarkson promoted the abolitionist cause to Russia's Tsar Alexander I. The Reverend Lewis Way, the man who had induced Wilberforce and other like-minded evangelicals to join his London Society for Promoting Christianity amongst the Jews, also met at least four times with the tsar to press for Jewish emancipation. Neither Clarkson's nor Way's efforts resulted in any accord protecting Jewish rights. Nor did the Congress at Aix-la-Chapelle agree to meaningful checks on European slave trading. Yet abolitionism in Great Britain continued to enjoy substantial moral capital, and the prestige of individual abolitionists helped sway British citizens and government authorities to support the cause of human liberty.[16] And it was Freetown that emerged as the showpiece of Britain's antislavery efforts.

Freetown became the naval base for the West Africa Squadron of the Royal Navy, whose sailing ships and steamers hunted slavers for nearly sixty years at enormous public expense. Captured slave vessels were brought to port as prizes and auctioned off. Their human cargo was released on Freetown's docks, giving rise to a heterogeneous group known as "recaptives" or "liberated Africans." Those freed were often emaciated and beset by gruesome ulcers due to the squalor of the slave ships. Blinking at the light, the liberated emerged from filthy holds, their legs sometimes incapable of supporting bodies smeared with urine and feces. Though rescued from the Middle Passage, many of those freed plunged into madness and found themselves confined in Freetown's overcrowded infirmary.[17]

These traumatized men, women, and children settled among the creolized populations in and around Freetown. Some became traders, forming mutual aid societies and peddling goods throughout the colony. Successful merchants were sometimes called "African Jews." A diversity of ethnicities and races participated in the colony's trade and governance under a colonial regime that, for all its paternalism, aspired to benevolence.[18] That regime's dominion, however, was limited to Freetown and its immediate surrounding enclaves. Most of modern Sierra Leone, including much of the two hundred or so square miles of its Western Peninsula, remained a patchwork of sovereign Indigenous territories. Domestic servitude and slave trading thrived just beyond the colony's porous borders.

By the mid-nineteenth century, Freetown had evolved into a town "laid out with the regularity of Philadelphia consisting of seven parallel streets kept free from grass, with thatched huts on either side around which are smaller plots of ground full of bananas and trees." This orderly outpost must have surprised Victorian travelers accustomed to London streets choked with peddlers, pedestrians, and omnibuses that left in their wake steaming, ankle-deep piles of horse dung. This mixed with straw and household waste until rain washed the city's effluvia towards the Thames River, which was rank with human sewage, animal innards, and swirling rubbish. Cholera, spread mainly through contaminated water, killed tens of thousands in London and throughout Britain.[19]

British visitors to Freetown, meanwhile, would have sailed into the sweeping estuary of the Sierra Leone River, marveling at its lush forests and clear waters. Canoes glided across the bay to Freetown's market, which offered varieties of rice, beef and mutton, dried bats and skewered rats, land snails, fresh and dried seafood—barracuda, snapper, and

FIGURE 7. "Free-Town, Sierra Leone," ca. 1850s, from J. L. Wilson, *Western Africa* (J. L. Wilson, 1856).

cockles—and produce such as ginger, bananas, yams, and mounds of raw groundnuts. Pins and needles, cotton cloth, leopard skins, baskets, straw hats, and a profusion of other items tempted consumers.[20]

Commerce was brisk, but spiritual welfare was not neglected. Visitors to Freetown could attend one of at least thirty chapels. Church attendance was not compulsory in the colony as it had been on St. Helena, but the Sunday Sabbath was scrupulously observed. Contemporary visitors were impressed by streets "thronged with well-dressed negroes, on their way to church." One recorded with wonder that "every Sunday morning thousands of free natives . . . may be seen going to those respectable places of worship, clean in their persons . . . & in their dress." Churches and missions, particularly those of the Wesleyan Methodist Missionary Society (WMMS), dominated the religious, social, and educational life of the colony. The society's directors in London dispatched preachers to Freetown and beyond. Religious schools administered by the Church Missionary Society (CMS) reinforced the Christian character of Freetown and were said to have "largely contributed in expressing the benefits of civilization to the Native population of the Colony."[21] Wilberforce and Newton had both been founders of the CMS's predecessor organization.

Yet the stubborn hold of Indigenous religions bedeviled Sierra Leone's churchmen. People worshipped thunder, lightning, and crocodiles, and they crafted wooden or clay fetishes. "The dirty idols . . . were the god and goddess of thunder," one Wesleyan missionary reported, likely referring to manifestations of the Yoruban thunder spirit Ṣàngó, "and every day they are pretended to be fed with palm oil and fufu." The outraged minister proposed purchasing these figurines, but the unnamed idolater responded with angry chants and contortions of great "savageness" that culminated when he grasped a chicken: "Its head he took in one hand and the legs in another and tore the head from the body, immediately thrusting the neck of the fowl into [his] mouth he sucked out the blood and biting off the entire neck and grinding it in his mouth, while blood streamed down on the ground." Aghast, the missionary withdrew and knelt to pray. He later recorded his conviction that "the prince of darkness reigns" in Africa in a manner "darker than I could ever have imagined." The "degraded" state of those living in Sierra Leone moved another missionary to remark that he was "not surprised at some denying [the inhabitants] the rank of man." Even abolitionists could doubt the humanity of those whose cause they promoted. Worship of the elements and animals was finally outlawed by the colonial police in 1851, though it is doubtful that the ordinance put an end to such rituals. That same year, the Great Exhibition in London's Hyde Park featured a "strange-looking fetiche" from West Africa, "a household god, with eyes composed of looking-glass."[22] Perhaps the idol on display had been purchased by another missionary in Sierra Leone.

According to the 1851 census, Wesleyan Methodists were the largest religious group in Sierra Leone (numbering 13,946), followed closely by Episcopalians (Anglican Communion, 13,863). These two groups alone made up about 63 percent of the colony's total population of 44,501. "Pagans"—that is, those who adhered to Indigenous religions—numbered 6,192, and a smaller number of Muslims (2,001) existed on the margins. There were precisely three Jews living in the colony. One of those three was Nathaniel Isaacs, who was active in and around Freetown from 1837 until 1854. This fact is in and of itself remarkable. The average annual mortality rate recorded for Europeans in Sierra Leone hovered near 50 percent in the 1830s. Yellow fever, with its telltale black vomit, malaria, cholera, smallpox, dysentery (the dreaded "bloody flux"), typhoid fever, sleeping sickness, venereal disease, guinea worm, and endemic intestinal parasites preyed on new arrivals and veterans alike. Whatever gods they prayed to, many arrivals to the colony quickly met their respective makers. Sierra Leone was referred to as "the white man's grave" or the "red grave," owing to its iron-rich soil. A wag suggested that the constant dispatch of doomed churchmen to Freetown was an effort to erode the foundations of Christianity itself.[23]

Though physically unimposing—Isaacs was described as "5 feet 5 inches in height, [with] prominent features, high forehead, brown or brownish hair"—he outlasted ten colonial governors and numerous other officials, merchants, and missionaries. He grew prosperous enough to purchase property on Gloucester Street, near Freetown's harbor and close to the seat of government. The lessons he had learned during his residence with his uncle in Jamestown were put to good use. Isaacs made himself indispensable to local commerce and governance.[24]

Nearly three thousand miles away, Saul Solomon remained at the center of power on St. Helena. He was a consular agent for France as well as for Brazil, Spain, and Austria, and he served as vice consul for Belgium and commercial agent for Holland. When the British agreed to return Napoleon's body to France in 1840, Solomon hosted the Prince of Joinville, the king's son, who had been sent to repatriate Bonaparte's remains. The prince's entourage included two generals who had served Napoleon and followed him into exile, as well as several others who had been part of the emperor's household on St. Helena. Solomon provided the lodging, meals, wine, and spirits, as well as the horses and carriages, for those attending the prince. He entertained nineteen members of the prince's retinue, though Solomon recorded that only eighteen of these men had partaken of his invigorating "warm baths." It is not clear which individual in the French party neglected to bathe. Solomon also served as on-site caterer during the painstaking exhumation of Napoleon's body from his tomb in the forested hills above Jamestown. The total bill Solomon presented to the colonial government for his exertions was £609 15s., or about eight years' wages for a skilled tradesman in England.[25]

Isaacs undertook his own lucrative commercial operations while continuing to work with G. C. Redman. While Isaacs traveled to the Gambia and elsewhere in West Africa, Redman remained in his London offices, still pleading with officials for indemnification from France for the damages he claimed to have sustained in 1835 with the blockade of Portendick. Though he was considered a respectable merchant, Redman had a series of business dealings with the infamous slaver Captain Théodore Canot (a.k.a. Théophilus Conneau). The captain recorded for posterity how he outfitted his vessels to carry human cargo to Cuba and other destinations. A twenty-two-inch-high slaving deck was installed between the structural decks, and the grating above was locked to prevent escape. The narrow space forced the enslaved to "lie down spoon fashion, one in the other's lap."[26] Redman may not have been aware of the extent of Canot's trade in human beings, though he must have suspected something of his past.

As for Isaacs, he likely knew Canot, or at least knew of him, through Redman and the network of European traders operating in West Africa. Canot did not mention Isaacs in his memoirs, although he did note other Jews active in the coastal trade, jocularly referred to as "amphibious Jews." Nonetheless, Isaacs' influence may plausibly be discerned in one of Canot and Redman's joint ventures. In 1841, while acting as Redman's agent, Canot secured a deed of possession from a king who ruled Cape Mount, a coastal headland northwest of Monrovia in today's Liberia. Canot and Redman's land grant recalls Isaacs' earlier charters for territorial control in the Zulu Kingdom. Of course, concluding treaties for land was not a novel idea. "This was always the way with white men," one West African chieftain, Namina Lahay, chided in 1841: "They first sent quiet people [missionaries] to do good; then merchants; then as their numbers increased, they built forts and brought guns; and, at last took away your country."[27]

By the time Canot joined forces with Redman, he swore he had abandoned slaving. Canot initially used his Cape Mount fort as an entrepôt for produce and other goods from the interior, but he later complained that he could not make a living as a legitimate merchant. British reports suggested that Canot, an "atrocious miscreant," simply could not resist the temptations of slaving. Though his ships escaped detection by West African Squadron interceptors, sailors discovered three blacksmith's forges for making slave irons at Canot's Cape Mount "factory," as coastal trading posts were known. Squadron men also identified an ingenious hose system that Canot had engineered to provide freshwater to slave ships as they waited beyond the surf to take on a full hold of human souls.[28]

Isaacs, by contrast, prospered through legitimate trade. From his Gloucester Street residence near the docks, he likely witnessed the arrival of the Niger Expedition, whose shallow-keeled steamers docked at Freetown in June 1841. Each of the vessels boasted a ventilation system designed to filter out the swamp-borne vapors—miasma—then thought to cause malaria. The presence of these technologically advanced steamers in the colony at a time when fewer than 5 percent of all British-flagged vessels could boast steam power was a landmark event. In Freetown, the expedition vessels took on coal and wood for fuel and recruited more than one hundred Africans from the colony to serve as laborers and interpreters. Since the captain of one of Redman's schooners had the honor of dining aboard ship with leaders of the Niger Expedition, Isaacs was surely well aware of the flotilla's presence.[29]

The Niger Expedition was launched by the abolitionist luminary Thomas Fowell Buxton and members of the Church Missionary Society in London as part of a plan to completely cleanse Africa of the slave trade. After chairing the Aborigines Select Committee, Buxton developed a proposal for fostering legitimate commerce while civilizing inhabitants under Christian tutelage. Buxton aspired to redeem an Africa he believed to be "bound in the chains of the grossest ignorance ... [and] savage superstition." Like Wakefield's earlier colonization plans, Buxton's vision promoted an informal empire based on private enterprise, in which "the merchant, the philanthropist, the patriot, and the Christian, may unite." He was no doubt heartened by the ready acceptance of the gospels of commerce and Christianity in Sierra Leone, especially among the colony's Krios.[30]

The Niger Expedition sailed under the banner of the Society for the Extinction of the Slave Trade and for the Civilization of Africa (also known as the African Civilization Society) and with the patronage of Prince Albert, members of Parliament, aristocrats, and religious leaders. One young supporter would soon emerge as the heroic embodiment of Victorian humanitarianism: the missionary Dr. David Livingstone. He proclaimed that the Niger Expedition had made clear to Black Africans the "English love of commerce and English hatred of slavery." As it turned out, the expedition mostly made clear that Englishmen died in West Africa by the score. Fever struck nearly all of its 145 white members, and about one-third of them died.[31] Among the Africans aboard there were no casualties. The much-vaunted ventilation systems did nothing to prevent malaria, of course. Nor did the expedition make any progress in uprooting slavery among the peoples they encountered.

Apparently envisaging that the expedition might discover members of the long-lost tribes of Israel bending their heads in Torah study along the Niger's malarial banks, abolitionist patrons provided the Niger Expedition's commander with twelve Hebrew Bibles. Another member of the expedition carried letters of recommendation in Hebrew, drafted by the leading Ashkenazi and Sephardi rabbis of London, Solomon Hirschell and David Meldola. Rabbi Meldola, of Bevis Marks Synagogue, requested that these imaginary Jews report back to London about the "books you possess ... and what customs you adhere to."[32] Even if

Hirschell and Meldola did not entirely believe that long-lost African Jews would be located, they were at least willing to humor their evangelical petitioner with signed Hebrew encomia.

While this notion may seem absurd today, profound belief in the Bible made it conceivable that the lost tribes might be located at the most remote points of the globe. The scholar Tudor Parfitt has characterized "the universal construction of imagined Jewish and Israelite histories for a vast number of the world's peoples" as an "endemic feature of the colonial project."[33] Any ritual practice, physical feature, or custom that echoed Jewish observance or any linguistic peculiarity thought to be a cognate for biblical Hebrew could be construed as evidence for the Israelite origins of newly encountered peoples, especially in Africa.

White colonizers and philanthropists could be quick to find resemblances between Black Africans and stiff-necked European Jewry. Henry Francis Fynn, Isaacs' one-time partner, identified similarities between the "laws, customs, and habits" of Jews and Black South Africans, such that he was "led to form the opinion that they were Jewish proselytes." Fynn solemnly informed the bishop of Cape Town that the Zulu had borrowed from the ancient Hebrews such traditions as circumcision, Levirate marriage, the marking of dwelling-place entryways, animal sacrifice, and purification after contact with the dead. Isaacs was almost certainly Fynn's source for his spotty knowledge of Jewish ritual. Fynn's remarks then became the basis for a history that invented a Jewish past for the Zulu people. Accounts of Jewish influence on the Zulu endured for a century, disseminated by churchmen, colonial functionaries, doctors, ethnographers, and historians. The former South African administrator and author H. Rider Haggard also helped to link Jews with Africa in *King Solomon's Mines* (1885), in which the ancient Israelite king's source of fabulous wealth is located just beyond the Zulu Kingdom.[34]

In Sierra Leone, the abolitionist physician Thomas Masterman Winterbottom concluded that climate determined racial appearance: Jews in England were "fair," but those in Africa were "nearly black." Such a naive environmental determinism made the supposed Jewish origin for Black African rituals seem plausible. Winterbottom also reported on local customs that resembled those of the Jews. Individuals accused of witchcraft, he observed, were made to drink a poisonous decoction made from the bark of a tree in order to prove their innocence. He believed that this "red water" ordeal was similar to the "bitter water" trials described in the Book of Numbers (5:12–31). The identification of West African rituals with those of the Bible proliferated throughout the nineteenth century. It became a means to comprehend and domesticate the alien and to find a tenuous kinship between Jews, white Protestants, and Black Africans who practiced Indigenous religions.[35] Efforts to demonstrate a correspondence between supposedly primitive African peoples and an only slightly less backward Jewry also served the prevailing Christian triumphalist discourse that considered Judaism to have been superseded and its practitioners doomed to remain culturally obsolescent.

Isaacs did not need to imagine his own close ties to Black Africans. On St. Helena, he had been served by slaves in his uncle's home, and he met many free African sailors in the bustling port of Jamestown. In South Africa, Isaacs had lived among and fought beside Zulu men, and he had fathered children with at least one African woman. In the Gambia, he had worked and socialized with African traders and fathered another child with a local woman. Along the Mauritanian coast, he had traded with and lived among the Traza. And in Sierra Leone, Isaacs labored alongside numerous Black and mixed-race individuals. Although Indigenous peoples and liberated Africans might not have had the status of Englishmen in Freetown, there was as yet little racial segregation in the spaces of everyday life. One resident of many years described how "Europeans freely mingle with persons of every dye." In January 1842 Isaacs might even have witnessed the arrival of those who had fought for their freedom aboard the *Amistad* and then again in American courtrooms. After making the Middle Passage twice, once in bondage and once in liberty, these men received an exuberant welcome in Freetown.[36]

In November 1843, Isaacs was almost clapped in irons himself after running afoul of the colony's collector of customs, who had caught him evading import duties on rum and sugar. He paid a fine of £100 and escaped prosecution. In the same month, G.C. Redman's case against the French government was finally concluded. He was awarded only a percentage of a mere £1,700 settlement, a far cry from the astronomical sum he had demanded after the hijacking of the *Eliza*.[37] The French, like Isaacs, had gotten off easy.

Earlier in 1843 the British had annexed Natal. Isaacs' dreams of obtaining a colonial freehold there, on the basis of the land grant he claimed from King Shaka and Dingane, were shattered. His ambitions to control territory in Natal had been frustrated for more than a decade, and now in Sierra Leone he feared future entanglements with pesky customs agents. Fortunately, Isaacs stumbled upon a business opportunity that would allow him to carve out his own fiefdom. He purchased the lease for Matakong (or Matacong) Island, which had been held by William Gabbidon, a trader of maroon heritage. Gabbidon had established a small factory on Matakong but had fallen on hard times. When he was unable to pay his considerable debts, his property passed to his creditors, from whom Isaacs purchased the island for the sum of £215.[38]

Matakong lies approximately fifty-three miles northwest of Freetown by sea and is about three miles in circumference. The island's location offered several advantages. Sixteen rivers, each rising into different parts of the interior, flowed nearby. Canoes could not navigate the flood-swollen rivers to reach Freetown in the rainy season, but Matakong remained accessible by water throughout the year. Freshwater springs dotted the island, oysters and mullet abounded offshore, and wild birds—curlew and snipe—could be hunted a short distance away. Furthermore, because Matakong was de facto neutral territory, Isaacs could establish a duty-free port, without interference from British or French customs inspectors.[39]

In the local Susu language, the name Matakong means something like "hidden place." Even today Matakong is hard to reach. The narrow channel that once separated it from the mainland has become an isthmus choked by swamps and vegetation. Villagers say that the jagged basalt rocks extending far offshore house spirits that protect their island.[40] These rust-colored rocks, some of massive size, are said to roll about to defend the islanders from would-be marauders. Matakong can be approached by vessels of any draft only from a northern channel. It was there, at the lip of a snug anchorage protected from winds, that Isaacs revitalized Gabbidon's ramshackle factory and set himself up as an outcast of the islands.

Just as the Jewish Isaacs was neither entirely white nor Black, neither European nor alien, but an in-between middleman, a part of polite society but apart from it, Matakong was neither French nor English, European nor African, colonial nor precolonial. Islands are spaces of contact and contest between land and sea, meeting grounds between peoples, sites isolated and solid, yet connected by the ever-shifting element of water.[41] Matakong was thus perfectly suited for a man like Isaacs.

According to Isaacs, Matakong was "almost in a primitive state" in 1844: no trade, no shipping, no cultivation, no habitable dwellings, no roads. The nine people he found living there were "in a barbarous and impoverished condition." He oversaw construction of a fortified pier, quays, and warehouses along a wharf protected by a "battery of great guns." The wharf featured a crane and tramway for efficiently loading and unloading ships. Thanks to these improvements, Isaacs became a favorite of the local people, who helped him turn the island into a robust entrepôt. His ownership of Matakong was attested to by Indigenous kings who elevated him to the status of a chief. These kings agreed to protect him according to the landlord-and-stranger compacts that prevailed on that part of the coast.[42] This relationship bound the landlord-host to protect his "stranger" and assume responsibility for his actions. In return the stranger was obliged to trade for the benefit of the landlord and to assent to his host's political decisions.

Isaacs knew how to prosper under patronage, having learned the steps to this delicate dance from his uncle on St. Helena and from his time in King Shaka's orbit. The principal chiefs in and around Matakong signed their marks to a document extolling Isaacs' "uniform and constant endeavors to promote lawful commerce among us upon fair and just principles and of friendship and good understanding which have been thereby established between us."[43] As was the custom in such compacts, Europeans provided guardianship and employment for youths and family members selected by the landlord. These extended kinship networks gave rise to a range of hybridized religious, racial, ethnic, religious, and group identities, as well as to complex class structures and mutable political alliances.

Two towering kapok trees marked the site of Matakong's modernized port. Tall trees were more than just a physical landmark, however: to many West Africans, they served as portals to an immanent spiritual world. The majestic kapoks may have signaled to the island's African residents that Isaacs' dominion enjoyed divine sanction. Four hundred yards from the kapoks, Isaacs erected his compound on a bluff. Maps of the era indicate his residence with the notation "The House." This stronghold consisted of several slate-roofed buildings, including a Wesleyan chapel and a Sunday-school classroom, arrayed around a central courtyard whose high, protective walls were "pierced at intervals with loopholes for musketry." These battlements were "strong enough to withstand any force the natives might bring against it." A "spring of the purest water" within the courtyard could be accessed during a siege.[44] Isaacs' close alliances with local chieftains meant that he had little need for these military defenses, which served mostly as a deterrent.

The chapel and Sunday-school room provided more practical camouflage that allowed Isaacs' outpost to receive the blessings of officials in zealously Christian Freetown. In December 1852, the superintendent of

Wesleyan missions in Sierra Leone received a request to send a minister to the island. He recorded that "Isaacs, Esq., the chief Magistrate" of Matakong proclaimed that he "was willing to do any thing to help the cause" of Christianity's spread. The superintendent visited Matakong twice to inspect Isaacs' operations. "Matakong is an admirable station for a Missionary," he concluded in a report to superiors in London.[45]

Isaacs may have received word around this time that Saul Solomon had died of "softening of the brain"—probably a hemorrhage—while on a visit to England. But he had extracted promises that he be laid to rest on his beloved St. Helena. At great expense, contrary to health regulations, and against all common sense, his daughter sealed her father's corpse in an airtight casket, disguised it as crated cargo, and accompanied it aboard a passenger ship to South Africa. From there she sailed to Jamestown, reaching her destination more than two months after her father's demise. Solomon's body was removed from the ship's hold, transferred to shore, and buried in the town center in a service led by an Anglican chaplain. Some of Solomon's personal properties were sold off, but his family-owned company soldiered on.[46]

Nearly two hundred years later, Solomon & Company endures as a publicly held company. The Solomon logo remains ubiquitous on the island, and the company's portfolio includes not only importing and exporting concerns but also construction works, food markets, real estate ventures, a department store, filling stations and fuel depots, shipping services, insurance agencies, financial services, agricultural production, and a near monopoly on baked goods, all of which are branded with a striking capital S. Solomon & Company's offices are located in a beautiful Georgian building on Main Street, a quick walk from the seafront. Saul Solomon's six-foot-long stone grave marker—removed when the original burial grounds were redeveloped as a children's playground—rests upstairs in the corporate conference room with this inscription:

> Sacred to the Memory of S. Solomon, Esq.
> who died in England
> on the Sixth of December 1852
> Aged 76 years[47]

From his own headquarters on Matakong, Isaacs specialized in the export of groundnuts, or what most English speakers outside West Africa refer to as peanuts. Cultivation of the legume, which is of South American origin, boomed in West Africa, and beginning in the 1830s it became the top commercial crop in the vicinity of Sierra Leone. Most of these peanuts were not destined for human consumption, though some may have found their way into the trade at London's Duke's Place market, which was described as "entirely Jewish" and where "the business done . . . in nuts is immense." The majority of Isaacs' groundnuts were exported to France, where they were pressed to extract oil. Isaacs accordingly became a French consular agent, as his late maternal uncle had once been on St. Helena.[48]

Groundnut oil was also used for other purposes, including as a lubricant for machinery and trains during a period of rapid rail expansion.[49] Thus an indigenous crop of the Americas, transplanted by the Portuguese to West Africa, came to be cultivated by many of the same people engaged in slavery—and their victims. Their toil supported industries, particularly railroads, that altered the landscape of Europe, transported goods far and wide, and connected disparate peoples. Africans were linked to Europe's capital cities in a network of global exchange. Railroads soon became essential elements of European imperialist projections of power across Africa, a power that extracted resources from colonies by force of arms in an ever-expanding military-industrial-oleaginous complex.

Railroads also became a contrary metaphor for the drive to civilize, and hence Christianize, Africa. In reviewing the legacy of the 1841 Niger Expedition, Charles Dickens belittled "the heated visions of philanthropists for the railroad Christianization of Africa and the abolition of the Slave Trade."[50] Dickens saw how industrial capitalism transmuted everything, including the spiritual, into material for production. But even he may not have realized how apt his metaphor was, given the extent to which West African peanut exports—a trade dominated by Europeans, Isaacs among them—helped fuel industrial expansion, which in turn drove colonialist fever dreams for the further extraction of resources by rail. More than a decade after the Niger Expedition, Dickens mocked charitable English efforts to educate and evangelize Africans. His fictional philanthropists in *Bleak House* (1852–53) ignore injustice at home in the hopes of establishing workshops to carve wooden piano legs along Africa's riverbanks. Isaacs too might have laughed at Dickens' satiric take on British do-gooders ignorant of traditional mores in West Africa.

Respect for Isaacs prevailed among his African neighbors, thanks to the trade he brought to Matakong, his wealth, and his willingness to engage in social relationships, kinship bonds, and political pacts. He was as much a collaborator in the ongoing construction of West African cultures and communities as he was a colonizer. The governor of the Gambia paid a visit to Matakong and afterwards praised Isaacs to the secretary for the colonies for "entering personally into communication with the natives." Indeed, eyewitnesses recorded that he learned the local Mende language, and he may have also been conversant in Susu, a related tongue. He had earlier proved himself a reasonably adept linguist by learning isiZulu fundamentals at Natal, and like young Ben Moss, he probably acquired some basic Wolof while trading in the Gambia.[51] His involvement in Indigenous networks and his facility with languages allowed him to gain his neighbors' respect. Mutual trust and indebtedness developed between Isaacs—the sole European residing on Matakong—and those living near his island redoubt.

Isaacs was not unique in forming close contacts with Africans. Europeans who established themselves in and beyond Freetown frequently intermarried or cohabited with women from local landlord families—to forge beneficial economic and social ties, to consolidate power, and to seek companionship and love. Though these relationships certainly involved imbalances in power, they could be less coercive and more mutually advantageous than might be supposed. Over time, the children born of these liaisons introduced another group—Eurafricans—into an already heterogeneous population. Several Eurafricans, women as often as men, held positions of influence among the commercial and social classes of the colony and its environs.[52] These individuals would not have viewed themselves as servants of empire, and in some instances they subverted imperial goals. They sought to enrich their families, educate their children, and ensure their futures and those of their communities. They emerged as economic and cultural mediators between white American, African American, European, Afro–Caribbean, and resident African populations, negotiating competing mercantile, ethnopolitical, and colonial interests. Isaacs had relationships with several Eurafrican women whose backgrounds and careers expose the multifarious gender and racial relations in the region. These relationships and the economic trade on which they were premised highlight the limits, rather than the omnipotence, of formal imperial power. The Crown relied on intermediaries like Isaacs and his Eurafrican partners to make sense of those peoples caught up in Britain's ever-widening grasp.

At some point between 1843 and 1846, Isaacs established a romantic relationship with Mary Ann Skelton, who was the daughter of Elizabeth Fraser, a Eurafrican woman, and William Skelton Jr., a Eurafrican man. Both came from notorious slave-trading families who operated factories on rivers north of Matakong. Both had been educated in England and hence shared similar experiences and sociocultural touchstones. Skelton went into the family business and captained slave ships to the Americas. Their union produced two daughters, Emma and Mary Ann Skelton.[53]

When the Yankee trader Enoch Ware visited the Skelton family home in 1843, he observed that Emma and Mary Ann "both write bold mercantile hands, which has been taught them by the father, who is fully competent, having passed a long time in a counting room in England after receiving an excellent education. Probably either of them [Mary Ann and Emma] are better instructed . . . than many of the young wives among us who go into very good society." About three weeks before Mary Ann's father passed away, Ware noticed Isaacs near the Skelton compound, trading for hides. After William Skelton died, Isaacs likely courted the teenaged Mary Ann in a connection perhaps based on mutual attraction and surely on economic interests. Isaacs and Mary Ann sustained their relationship for two to three years and had two children together, Emily Emma and Alfred Isaacs.[54] The documentary record reveals little about the children's lives beyond the simple facts of their existence.

By August 1846, Isaacs and Mary Ann had parted company, and she married another man, Joseph Richmond Lightburn (or Lightbourn). He was the son of a notorious white American slaver, Styles (or Stiles) Edward Lightbourn, and the equally notorious British-educated Eurafrican Elizabeth (or Isabella) Bailey Gomez, also known as Neria Bely. Neria Bely ran her husband's slaving establishment in West Africa during his frequent absences, and she became a formidable leader after his disappearance at sea. According to African oral traditions, she was endowed with magical powers that she used to ensorcell and render docile those held in her barracoons before their eventual shipment across the Atlantic. Travelers described her as imperious, resentful of European meddling, and strikingly beautiful. Some sources describe her as wearing men's clothing and strutting about her compound with two pistols at her waist while she inspected the cannons that defended her illicit trade.[55] That Mary Ann Skelton bound her fate to the scion of yet another Eurafrican slaving dynasty illustrates how these families constituted a kind of royalty sustained by chattel slavery.

The preeminence of the Lightburns and Skeltons in the region endured because they lived beyond effective British jurisdiction, possessed transatlantic kinship and commercial bonds, forged alliances with local leaders, and engaged in legitimate as well as illegitimate trade. The colonial government and missionary organizations attempted to draw a clear distinction between the two forms of trade, but in practice this was impossible. Legitimate export crops were often grown and tended by domestic slave labor. Moreover, many peoples in West Africa considered slavery a perfectly legitimate form of commerce, sanctioned by custom if not by law. The legitimacy or illegitimacy of the slave trade was itself a Eurocentric notion.[56]

The American Enoch Ware attested to the intertwining of legitimate trade and slavery. He recorded that "scarcely a hundred pounds of tobacco . . . is sold but what sooner or later is used for purchasing slaves though it may go through half a dozen hands first . . . each time being carried farther into the interior, perhaps five hundred, perhaps a thousand miles, when slaves are bought with it." The same was true of Isaacs' imports. Like other British merchants, Isaacs traded cotton cloth, beads, gunpowder, and firearms for peanuts, palm nuts, hides, and other valuable items.[57] He would have known that goods imported from an industrializing England could end up in the hands of slavers to grease the wheels of human trafficking, much as his peanuts could end up lubricating the distant French *chemins de fer*.

Isaacs profited from a widespread commercial hypocrisy that celebrated a merchant's ability to barter cheap cotton cloth for peanuts; both crops had likely been planted, tended, and harvested by slave labor, the one in the American South, the other in West Africa. Even committed Christian abolitionists like Isaacs' fellow castaway Charles Maclean (a.k.a. John Ross) carried on trade with slavers. In 1845 Maclean, now captain of a merchant vessel, arrived in Wilmington, North Carolina, with a crew of West Indian British subjects. The harbormaster gave notice that the free Black crewmen were to be jailed while in port. Maclean refused to hand over the men. "What a libel . . . it is," Maclean told Wilmington's port authorities, "when you call this a free country!" He vowed armed resistance should the police attempt to board his ship to jail his crewmen.[58] Yet despite the protection he offered his Black crew, Maclean willingly engaged in commercial transactions with a slave-holding regime.

The interpenetration of licit and illicit commerce complicated efforts to uncover clear evidence of slaving in and around Sierra Leone. On Matakong, Isaacs absorbed the lessons imparted by his Eurafrican friends and in-laws. He prospered thanks to his own networks, extending to ports across the globe, while also cultivating relationships with local chieftains to invoke the protection of the landlord-and-stranger compact. And his growing stature as an influential trader enabled him to insinuate himself into the corridors of British colonial power.

In October 1847, Sierra Leone's governor, Norman William Macdonald (in office 1845–52), dispatched Isaacs and a captain in the military to conclude a treaty outlawing slavery with King Demba of Dubréka, an area near Conakry, capital of today's Republic of Guinea. Though Demba was at first hesitant, he later welcomed Isaacs. The treaty was "read and explained" to Demba, who made "comments and replies . . . upon each clause [which] he evidently fully understood," and he agreed to sign in exchange for British gifts. Isaacs lauded the "pagan" Demba as a "straight forward [and] ingenious man," as opposed to the crafty "Mahometans" in the vicinity. Later Isaacs employed King Demba's son as a superintendent of his stores on Matakong.[59]

Isaacs reported to Macdonald on the success of the mission, detailing the extent of Demba's domain and the potential trading outlets along the Dubréka River. He also highlighted other kings with whom treaties should be negotiated in order to prevent unscrupulous Europeans from using them as "outlet[s] for the Slave Trade."[60] Grateful for Isaacs' service to the abolitionist cause, the governor appointed him to head additional diplomatic missions.

In 1848 a British trader complained to authorities that his factory had been destroyed by men under the rule of Bori Lahai, a chieftain residing along the Mellacorée River, about fifteen miles southeast and upriver of Matakong. Isaacs set off with a colonial appointee to investigate the charges. They found witnesses who testified to the truth of the complaint, and Isaacs traveled to Bori Lahai's village to seek redress. A French artist depicted the village of Mellacorée as a collection of a dozen circular thatched buildings protected by dense riverbank vegetation.[61] The village occupied a strategic spit of land where the waterway narrowed and a pair of boulders required careful navigation. A wary Bori Lahai invited Isaacs to his residence, and they entered into an hours-long negotiation involving a choreographed exchange known as a palaver, a word that derives from the Portuguese palavra ("word" or "speech"). Bori Lahai admitted that his men had indeed burned the British trader's factory to the ground, and he claimed to have offered the wronged man a cow as compensation. Isaacs protested that a cow was hardly equal to the value of the goods lost and informed Bori Lahai that he had broken a treaty with the British government for failing to protect or indemnify the trader. "Let it be so," an insulted Bori Lahai pronounced, and he immediately broke up the palaver.[62]

Isaacs returned to Matakong, probably in a narrow pirogue paddled by his servants close to the serpentine shoreline. Later that night, messengers from Bori Lahai arrived, begging Isaacs to return for another palaver. He at first refused, but then agreed to journey back to the chief's residence on the muddy shoals of the Mellacorée. There Isaacs and Bori Lahai entered into further negotiations, during which the chief insisted he and his people did not possess goods equal to the value of the trader's burned factory. Bori Lahai requested that Isaacs front the money to indemnify the British merchant, since, after all, Isaacs and the wronged man were of the same nation. Bori Lahai promised to repay the loan. Isaacs refused. He moved to break up the palaver and threatened to inform British authorities of Bori Lahai's bad faith. The intimidated chief finally agreed to pay and apologize for the attack. Isaacs reported back to Freetown of another diplomatic triumph. His patient handling of sensitive negotiations earned him a reputation for tough-minded fairness up and down the coast.

Thus, in addition to being a broker of economic commodities, Isaacs emerged as a significant culture broker.[63] Balancing his roles as protected stranger among his African hosts, merchant lord of Matakong, and emissary of imperial power, Isaacs presented British demands for ending slave dealing to resistant peoples who had long practiced and profited from the trade, while at the same time he defended the ways of Indigenous peoples to an often uncomprehending and unsympathetic colonial administration. And he always made sure to represent his own financial interests by promoting trade in the region while projecting his status as a kingmaker.

Isaacs had the opportunity to play the role of kingmaker quite literally in 1849. On Matakong, he employed the mixed-race, British-educated George Stephen Caulker, son of the prominent chieftain Thomas Stephen Caulker (a.k.a. Ba Tha). At least three other members of the same powerful clan worked for Isaacs on the island. Ba Tha's kingdom occupied lands south of the Freetown peninsula and extending westward to the nearby Plantain Islands—the same islands where John Newton had "grown black" a century earlier. Jealous rival chiefs from another faction of the Caulker lineage set their eyes on Ba Tha's territory, which was well placed for trade and rich in timber. The region descended into open "civil war," with rival Caulker bands attacking and taking slaves from one another.[64] Hostilities extended into British-held territory and threatened to engulf the colony.

FIGURE 8. Mrs. Clarke, sketch of
Canray Bah, ca.1840s, from R. Clarke,
"Sketches of the Colony of Sierra
Leone and Its Inhabitants,"
*Transactions of the Ethnological
Society of London* 2 (1863): 320–63.

An account of the conflict appears in the Caulker Manuscript, written by one of Ba Tha's descendants. Ba Tha's enemies, led by the chief
Canray Bah, invited him and his entourage for a summit to settle their
differences. Despite offering them a ceremonial welcome, Canray Bah
plotted to massacre his guests. But Ba Tha was tipped off. Fleeing by
night, Ba Tha disguised himself to elude Canray Bah's spies and slipped
into a mangrove forest to board a waiting boat. On discovering their
escape, an enraged Canray Bah slit a suspected informant's throat and
sent his soldiers to strike at his enemy's stronghold.

I have not found an image of Ba Tha, but a sketch of his antagonist,
Canray Bah, depicts him as having large, oval eyes, high cheekbones,
and a pencil-thin moustache. Canray Bah resembles a young Billy Dee
Williams dressed as a pirate; he sports a bandana, a collared shirt rak-
ishly open at the neck, and a string tie held in place by a brooch.[65]

Ba Tha escaped with a handful of loyalists to a sanctuary at Bendu,
on the Sherbro River, where he rallied supporters. He built stockades to
mount a last-ditch defense as Canray Bah's forces gathered to rout him.
He also sent a messenger to his son, George Stephen Caulker, then under
Isaacs' tutelage on Matakong. The chief's situation at Bendu looked
grim until his messenger returned bearing news that Isaacs would send
weapons and reinforcements. One of Isaacs' schooners soon arrived,
laden with guns and gunpowder, swords, and supplies, and much-
needed mercenaries. "A brave and daring warrior" named Jongah

(a.k.a. Thomas Nightingale) inspired Ba Tha's beleaguered defenders. Isaacs also sent the king a cadre of "war boys," young men hired out by rulers up and down the coast. Another experienced warrior who enlisted in Ba Tha's cause was the feared marksman Mbongoombah (a.k.a. Saw Party).[66]

Despite these reinforcements, Ba Tha's men were still outnumbered. Canray Bah's forces felt certain of victory and took their time massing beyond the palisades of Bendu. Rather than attacking at night as was typical, the overconfident troops waited until dawn to launch their assault. A legendary warrior named Kehkehgbookeh, described as a "Goliath," led Canray Bah's men against Ba Tha's stockade on land, while a flotilla of Canray Bah's canoes swept along the channel to overrun Bendu from the sea. Ba Tha surveyed the approaching forces but remained undaunted by Canray Bah's superior numbers. He stirred his soldiers: "I have done nothing; it is an unrighteous war against me and God will deliver them into your hands!"[67] Whatever Ba Tha may have lacked in eloquence on the eve of battle, his soldiers made up for in the frenzy of combat.

The giant Kehkehgbookeh launched Canray Bah's charge against Ba Tha's defenders. Kehkehgbookeh was accompanied by his personal Muslim cleric, who prayed for his patron's success. But Saw Party lay in wait. He steadied his rifle and fired, blowing a hole through the holy man's head. Enraged, Kehkehgbookeh raced at Bendu's gates and furiously swung an axe-like weapon to force his way inside and exact revenge. Again, Saw Party loaded, took aim, and fired, killing Kehkehgbookeh and scattering the invaders who were stunned by their leader's sudden death.

Jongah meanwhile led Ba Tha's counterattack against Canray Bah's 150 battle canoes. He perched in the lead canoe, which was helmed by twelve skilled men who sat back to back so that the craft could change direction without the paddlers having to shift position. Though few in number, Jongah's fleet repulsed the attackers. Dead and wounded men tumbled from Canray Bah's vessels that day until "the creeks and rivulets were filled with dead bodies which became a feast for the fishes and alligators [crocodiles]." The severed heads of those slain on land were piled in a six-foot heap. Other skulls were strung up on thorny trees "to represent the fruits" of treachery.[68]

Ba Tha's victory allowed him to reassert control over his hereditary lands and conquer territory belonging to his adversary. For years to come, Ba Tha remained a potent force in the region. Isaacs' timely assistance assured him of the protection of the victorious Caulker faction. He had also succeeded in quieting a conflict that had troubled British authorities.[69]

Between autumn 1847 and winter 1852, Isaacs negotiated treaties outlawing slavery with nine different kings, whose domains stretched in a patchwork from the Little Scarcies River to the Pongo River. Today these territories range across roughly one hundred miles north of Freetown and up the Guinean coast. But Isaacs must have traveled many hundreds of miles more along the innumerable waterways that led to the principal villages. There he would conduct tiresome palavers, present British terms, and bestow gifts such as silver snuffboxes, scarlet cloaks, and cocked hats in exchange for promises to end the slave trade.[70]

During a period of repose on Matakong, between trading journeys and diplomatic missions, he received shattering news from home. His mother, Lenie Isaacs, had died on November 13, 1849, at the age of seventy. She was interred beneath a sycamore tree at the eastern end of the old Canterbury Jewish Cemetery.

For weeks on end, and once for fifty-eight days straight, Isaacs and his fellow commissioners endured physical exhaustion and combatted the "prejudices, fears, intrigues, and suspicions" of the local peoples, while also contending with the slave dealers and hostile French traders who sought to undermine their abolitionist embassy. Ultimately Isaacs and his fellow commissioners foiled "the wily tricks" of the "pro Slavery party." But while humans could be reasoned with or outsmarted, intractable nature could not be placated. Isaacs' pirogues sat grounded on sandbars for hours between tides. His colleagues fell to sunstroke and fever. Insects plagued them, and crocodiles were a constant threat. Yet Isaacs had moments of ease and reflection, too. He met the king of Kambia, who impressed Isaacs with his intelligence and the efforts he made on behalf of his people. He floated down stretches of calm river and admired the abundant vegetation and lofty palms. And he praised the African spirit of trade that sent produce and provisions to Sierra Leone and far beyond.[71]

The colony's board of council, including the most senior members of Sierra Leone's administration—the governor, the chief justice, the queen's advocate, and the police magistrate—conveyed their gratitude to Isaacs and his fellow commissioners for their sacrifices to "secure the onward progress of civilization." Governor Macdonald formally commended Isaacs and the other commissioners to the secretary for the colonies for persisting in their antislavery diplomacy despite myriad

MAP 8. Sierra Leone.

"difficulties of no ordinary character." And he personally thanked Isaacs for the "highly creditable and satisfactory manner in which the very arduous duties entrusted to you have been performed . . . and it is highly gratifying to me to . . . [extend] my high appreciation of the very able manner in which you have carried out the instructions furnished to you. . . . I beg you will accept . . . my best acknowledgments for the zeal, ability, and patience exercised by you in overcoming the difficulties . . . encountered." Even by the standards of the day, Macdonald's gratitude stands out as effusive. The governor made a habit of flattery: he took great pains to send Queen Victoria a living specimen of an "electric fish."[72]

Macdonald also informed his superiors in London about more pertinent matters. He warned that the slave dynasties of the Lightburn and Skelton families would attempt to scuttle or circumvent the treaties Isaacs had negotiated.[73] The governor did not mention, and was perhaps unaware of, Isaacs' intimate ties to these families. Isaacs may have been a kind of double agent, sometimes pursuing abolitionist aims, sometimes subverting them, but always acting to advance his own prosperity and political influence.

He had a great deal of practice in the art of concealment: eluding shipboard abuse as an adolescent, learning subterfuge from his uncle on St. Helena, masking his fear of King Shaka and transforming himself into a white Zulu chieftain, covering up the criminal machinations of James King and Fynn, obscuring his Jewish identity as an author, and ensconcing himself far from the prying eyes of colonial administrators on Matakong. Concealment may have served as a means for Isaacs to outpace a vulnerability he despised. Experience had taught him to trust himself first and best. Many people dream of living on a private island as a sanctuary from the incessant demands or manipulations of others. Isaacs achieved a measure of this fantasy even as his trade expanded.

In 1852 Isaacs claimed to have loaded seventeen ships at Matakong bound for Marseille alone. A year later he recorded twenty. The gross tonnage of Isaacs' vessels bound for Marseille was in excess of 3,100 imperial tons (about 3,500 US tons) per annum. Marseille had long been the headquarters of French soapmaking, a craft that in the late 1840s came to use groundnut oil because of the rising cost of olive oil. Almost all West African groundnuts found their way to Marseille, where numerous brokerage houses resold groundnuts to the many oil-pressing facilities nearby. The resulting oil was used for soapmaking and increasingly for industrial applications, such as lubrication. The demand for peanut oil had pushed the price of groundnuts in Marseille to about 35 francs per kilogram in the early 1850s. Isaacs' vessels surely also contained other cargo, such as palm nuts, which were also pressed for oil, but the Gambia's colonial governor estimated that Isaacs imported about £60,000 worth of British goods a year and exported £70,000 worth of groundnuts, palm nuts, cayenne pepper, and other commodities.[74] His income from trade in high-demand, high-value com-modities was hardly, shall we say, peanuts.

An ocean away, British and American children of the 1850s were drilled in a "moral geography" that would have flattered Isaacs. This pedagogy asserted a pseudoscientific basis for elevating European civili-zation over the cultures of the "western and southern part[s] of Africa." One textbook lectured that those "living in cities are called civilised," describing these fortunate few as people who built sturdy homes, made clothes, read books, and valued the arts. Another defining feature of civilization was the use of "ships, which go to various parts of the world, to exchange the fruits and manufactures of one country for those of another." By this measure, with each vessel Isaacs dispatched, he fulfilled his obligation as an agent of civilization. Perhaps he was reas-sured by this vision as he traded peanuts sown by domestic slaves in West Africa for cotton picked by slave labor in southern American plan-tations, cotton which was then processed into cloth by young children and desperate men and women toiling in Britain's "dark satanic mills"— factories where the color of textiles was fixed with gum arabic har-vested by slaves at the edge of the Sahara. And so by another measure— the geography textbook's axiom that "people are expected to be habitually humane and kind in proportion as they are civilised"— Isaacs' commercial reach had made him a key player in a brutal pag-eant, and his kingdom on Matakong a travesty of the values endorsed by civilized people.[75]

One can also view Matakong as a logical extension of the British belief that trade transformed raw nature and premodern societies. At the Great Exhibition of 1851, held in London's spectacular Crystal Pal-ace, empire's reach was clearly visible. While the exhibition's full name gave a polite nod to publicizing the "Works of Industry of All Nations," the event enthroned the United Kingdom as the preeminent world power. A tongue-in-cheek illustration by George Cruikshank depicted the exhibition as the crown of all the globe. People from around the world rushed toward to the British flag that fluttered high above Crystal Palace. From the east, Arabs rode camels, and Asians sat atop elephants; from the west, steamers churned the waters and a locomotive belched, while from the south, Black figures poured from reed huts in a headlong race to embrace civilization. Goods from Britain's colonies made their own journey to the exhibition, including a number of West African items on display: gum arabic, hides, ivory, nuts, palm oil, pepper, stuffed birds, and timber.[76]

Isaacs traded in many of these same commodities. His empire functioned as a scaled-down version of the East India Company, which had inspired Saul Solomon's own operations on St. Helena and in turn had been Isaacs' blueprint for his unrealized settlement in Port Natal. The structures of free-trade imperialism replicated themselves with each trading post and missionary station established in Africa. With its entangled legacies of abolitionism and colonialism, Sierra Leone itself served as a model for Isaacs' island headquarters on Matakong. Both Freetown and Matakong were militarized beachheads for a territorial and commercial acquisitiveness that professed the humanitarian aim of abolishing slavery in a West Africa whose promised "civilization" and Christian character were constantly deferred.[77]

Sometime after Mary Ann Skelton married Joseph Richmond Lightburn, Isaacs formed a relationship with another Eurafrican woman, Hannah Hayes. She too possessed valuable trade and political connections in Sierra Leone and beyond. Her mother, Philippa, was the daughter of an American trader and an African woman. Philippa married the white Englishman William Hayes. After he abandoned her, she assumed control of his holdings and played a role in slave trading on Bali Island, today part of Liberia's capital, Monrovia. The upwardly mobile Philippa sent Hannah to be educated in Freetown, where she found herself favored by Governor Charles MacCarthy (in office 1814, 1816–20). Hannah assumed the role of MacCarthy's "housekeeper": in Freetown housekeepers were often involved in long-term, intimate relationships with their employers. These men typically acknowledged their mixed-race children and provided for their education.[78] We have no record of whether Hayes entered into this arrangement with MacCarthy out of affection or merely as a means to an end.

Following MacCarthy's departure from Sierra Leone, Hannah Hayes embarked on a relationship with the Freetown judge Walter William Lewis, an Englishman. This romantic bond with another member of the colonial establishment indicates that Hayes sought out and was accepted by the genteel classes. Hayes likely wielded the Victorian virtues of modesty, discretion, and what the poet Coventry Patmore called "subtle sweetness" to pursue her romantic interests along with social and financial stability, even if that meant compromising her personal autonomy. Modern marriage frequently involves a similar tradeoff.[79]

Hayes had three sons by Lewis: Walter, John, and Thomas. The younger Walter Lewis made his way into government service like his father, rising to the position of second writer in the colonial secretary's office in Freetown. Writers were junior clerks in the administrative hierarchy, often selected from among educated Native people. They would have been required to take dictation and summarize documents in a clear and bold hand. Walter's brothers would have honed other skills working for their boss, Nathaniel Isaacs. Hannah Hayes became Isaacs' consort and consul on Matakong. While Isaacs was negotiating treaties to end the slave trade, Hayes played hostess and nurse to other commissioners when they became ill and were forced to remain on the island to recuperate. Governor Macdonald thanked Hayes in his florid style "for the unremitting kindness which has characterized your conduct . . . and your skill in treating the fevers of this Colony."[80] Hayes's courtliness afforded Isaacs further entrée into the good graces of the colonial administration. And now Isaacs had a stepson of sorts, Walter Lewis, as his eyes and ears in the colonial government.

Isaacs' compacts with Indigenous leaders, his willingness to countenance local cultural practices, his eagerness to abet regional conflicts by providing mercenaries and European matériel, and his officially sanctioned government roles all enabled him to consolidate his rule as "king" of Matakong. His efforts to improve the spiritual life of his subjects also gained him respect. "The powerful influence of the Gospel," a missionary reported to London, may be seen "from the state of things at the island of Mata-Cong, which are increasingly cheering." Isaacs' Sunday school had enrolled nine pupils, seven boys and two girls. His factory employed and trained local men as shipwrights, carpenters, masons, blacksmiths, quarry men, gold workers, and general laborers. Isaacs added a dry dock at his port to "repair and copper vessels," a necessity in a climate wracked by scorching sun, sand-blasting winds, and oppressive humidity. Approximately three hundred people now lived and worked on the island. Matakong had once been "a headquarters of the Slave Trade," the Gambia's governor wrote, but thanks to Isaacs' exertions, the island had been transformed, making it "neither unreasonable or visionary to hope that Religion and Civilization will advance hand in hand with the Commercial reciprocity of the Western Coast of Africa."[81] To the governor, Isaacs was little short of a miracle worker.

Isaacs' renown was such that an article extolling his "skill, energy, and enterprising spirit" appeared in the *Illustrated London News*. It noted that Isaacs "holds the island upon a secure tenure" and has "all the rights and privileges of a chieftain conferred upon him." Given Isaacs' penchant for self-promotion, he may have guided the pen of the

FIGURE 9. "Matakong on the West Coast of Africa," *Illustrated London News*, 2 December 1854.

unnamed reporter. The article also described the buildings of red sand-stone Isaacs had erected and the nearly one hundred handpicked men, trained in the use of firearms, and three mounted cannon that defended the island.[82]

Shipwrights, carpenters, coopers, and blacksmiths worked in well-appointed workshops along the noisy harbor front. As many as eighty vessels under various flags—American, British, French—arrived in Matakong in 1853. The *Illustrated London News* correspondent admired the scale of trade as well as the chapel and Sunday-school room as signs of Isaacs' success in "promoting the beneficent views of Her Majesty's Government, as well as of the Christian philanthropist." Yet, unbeknownst to the writer, Isaacs had hedged his spiritual bets by returning to the Jewish fold, at least nominally. Isaacs remitted annual dues of one guinea to the Canterbury Hebrew Congregation in 1853.[83] Even in matters of the soul, Isaacs played both sides.

The reporter noted that Queen Victoria's birthday had been declared a holiday by Matakong's monarch. Visiting dignitaries from Sierra Leone enjoyed a cannonade and royal salute, a musical program that included "God Save the Queen," and a lavish meal. British visitors and Africans danced together late into the night. An accompanying illustration depicts top-hatted gentlemen, a bonneted white woman, and men in African garb arrayed along the sturdy stone pier Isaacs had built. In the background, his private militia stands at attention in the shadow of warehouses dwarfed by enormous kapok trees. Isaacs had at last found his private sceptered isle. In addition, he was making money—literally. He and his junior partner, Thomas Reader, minted coins stamped with their name and circulated them as currency in and around Matakong.[84] Isaacs doubtless modeled his minting operation on that of his uncle on St. Helena.

Given this evidence of prosperity, it is small wonder that the bankrupt former owner of Matakong, William Gabbidon, tried to wrest back control of the island through legal and other means. Gabbidon charged Isaacs with the illegal occupation of Matakong in autumn 1852 and provoked a handful of chieftains to oppose him. Governor Macdonald rebuked Gabbidon for "exciting the Native Chiefs to rise up against Mr. Isaacs in order to drive him by force" from the island. He invited Gabbidon to submit evidence that he was indeed the rightful owner and warned him that if he had none and continued "disturbing the peace of the Country and injuring the commerce of the Colony," he would "answer for such conduct at [his] peril." In tandem, the governor dispatched a letter to King Tombo Mahomadoo, one of Gabbidon's allies, advising him that Gabbidon had no right or title to Matakong and warning the king not to interfere with Isaacs' trade or property.[85]

In a meandering response to the governor, Gabbidon grumbled that an unstated "youthful indiscretion" had unfairly prejudiced Macdonald against him, that Isaacs had forged his deed to the island, and that the colonial administration had no jurisdiction over Matakong in the first place. Clearly Gabbidon did not know how to entreat authority. King Tombo Mahomadoo's response was also inept. He complained to the governor that his domestic slaves had fled to Matakong, where Isaacs provided the runaways with "Guns, Powder, and Ball." Isaacs then sent these armed men back against their masters with instructions that "they must fire at them." Macdonald would not have been favorably impressed by Tombo's admission that he held slaves in contravention of existing treaties. Neither would he have been much alarmed to learn that Isaacs was arming Tombo's people against him in the cause of freedom—or at least that of free trade. By the 1850s the notions of free trade and freedom had in any case become intertwined, even indistinguishable, moral imperatives in the British political order.[86]

King Tombo noted that he had "always thought Mr. Isaacs was a Jew," but now charged that Isaacs had "turned Mahomedan and got a maraboo [marabout] at Matacong making gregories [gris-gris]" to help him maintain his hold on the island through necromancy. To those in power, whether in Freetown or London, Isaacs' Jewish identity might weigh against him. The claim that Isaacs had embraced Islam and associated folk practices would have been scandalous, even perhaps to the laissez-faire Macdonald, had it been credible. West African Muslims were notorious for holding domestic slaves, a fact that vexed colonial policy makers. Vestiges of child slave dealing in the "*Christian* Colony" by "petty Mahomedan traders" were protested by more than five hundred of Freetown's citizens in a public letter to the secretary of the colonies.[87] Isaacs was one of the signatories, suggesting that he did not want to give the appearance of being an ally to Muslim domestic slave holding. In any case, Gabbidon's and King Tombo's intrigues gained little traction with Macdonald as his term as governor drew to a close.

But Gabbidon stepped up his campaign to discredit Isaacs and reassert control over Matakong with the arrival of the new, reformist governor, Arthur Edward Kennedy (first term in office 1852–54). Kennedy had grown up under the influence of evangelical abolitionists and sought to root out any remnants of slavery. Possessed of a "harsh ruthlessness" and characterized by "imperiousness," Kennedy landed in Freetown in October 1852 and promptly began cleaning house. If the claim that Isaacs had converted to Islam did not arouse the new governor's antipathy, then it was certainly kindled in 1853, when the colony's surveyor fled to Matakong ahead of a tightening police dragnet. An audit uncovered evidence that the surveyor had defrauded the colonial government for years by clandestinely importing items and selling them to "the public service by charging exorbitant prices ... upon inferior goods." Kennedy believed that in exchange for a "large sum," Isaacs had helped the surveyor abscond to Cabo Verde "in one of [Isaacs'] coasting boats." Seeing an advantage against a now-compromised Isaacs, Gabbidon informed Kennedy that Matakong was a covert slave-trading factory.[88]

One of those willing to testify against Isaacs was Momodu Yeli, a Mandinka headman and Arabic teacher resident in Freetown. Though respected for his status and learning, Momodu Yeli outraged the colony's Muslim community by denouncing slave traders and siding with British authority. As a sign of his hereditary standing and education, Yeli wore a turban-like head-covering pinned in place by a frontlet of twinned "horns," which in illustrations look like plump cocktail shrimp. His ritual attire moved one contemporary observer to compare these ornaments to the regalia believed to have been worn by ancient Jews as a mark of distinction.[89] Thus the Muslim Yeli betrayed traces of being a crypto-Jew. Gabbidon, of course, believed that Isaacs was a crypto-Muslim. He hoped that on the strength of Yeli's evidence, Kennedy would arrest Isaacs, but the governor suspected that Gabbidon's intelligence was a self-serving means to repossess Matakong.

Kennedy nonetheless resolved to send a West African Squadron steamer to investigate the allegations, but "cautiously without offending Mr. Isaacs." The desire to avoid offense indicates the prestige Isaacs had acquired in and around the colony. Privately, however, Kennedy disdained Isaacs' embrace of African folkways, dismissing the ritual negotiations of the palaver as "a meeting of naked drunken savages."[90] Isaacs had made considerable efforts to adhere both to local folkways and to colonial Christian expectations, yet he never quite escaped the suspicion of his social betters, at least in part because of his Jewish origins.

Kennedy, for example, assumed that the Wesleyan schoolmaster Isaacs had secured for his Sunday school had been "inveigled to Matacong by Mr. Isaacs . . . (he himself being a Jew)." In nineteenth-century British parlance, this identification was virtual proof of Isaacs' corruption. Kennedy referred to Isaacs as a "Jew or of the Jewish extraction" in other official correspondence to his superiors. Kennedy also outed another Sierra Leone merchant as "a Jew" and charged him with being a smuggler. Thus he managed to impugn two of the colony's three Jewish residents in a single letter. He likewise disparaged what was arguably the largest and most successful of the liberated African groups he presided over, the Aku (a Yoruban identity), who, he concluded, "may be termed the Jews of Africa (bearing all the strong characters of that race in Europe)."[91] For Kennedy, Jews and Africans were interchangeable objects of derision, at least when they achieved commercial success.

Early Victorians would have shared the governor's reflexive distaste. Jews were conspicuous in the criminal underworld in London, including the fence Isaac "Ikey" Solomon, who was the subject of lurid press reports. Just a few years before Isaacs established Matakong as the center for his trading empire, Charles Dickens partly based one of his most memorable characters on Ikey Solomon: Fagin, who appeared in the serialized novel *Oliver Twist* (1837–39). Fagin is referred to as "the Jew" well over one hundred times in the novel, in a manner that renders his religion synonymous with his criminality. Fagin presides over a gang of pickpocket street toughs who pilfer handkerchiefs, watches, and jewelry to enrich their miserly impresario. Readers first meet him as he roasts sausages on the end of a "toasting-fork"—a three-pronged devil's trident—over a flaming fireplace. [92] His red hair further highlights the association between his diabolical character and his particular Jewish cupidity.

Much as Isaacs' *Travels and Adventures* escorted readers through unfamiliar landscapes and detailed the customs of the "savage" Zulu, *Oliver Twist* offered readers a journey through a London underworld peopled by villains, like the "savage" Bill Sikes and the demonic Fagin. Illustrations by Cruikshank rendered Fagin as a deformed wretch. One of these images features the condemned Fagin on his stone bench in his jail cell, wrapped in a caftan, biting his nails, his terrified eyes peering out from above his enormous nose. In the novel, the doomed Fagin drives away the rabbis who come to comfort him, and for good measure he also rejects the church bells' message "tell[ing] of life." The jailer, disgusted at Fagin's indifference to the spiritual, wonders, "Are you a man?" Here, Fagin lacks the basic human dignity of Shylock or even of an African slave in chains. When Oliver begs Fagin to bend his knee and pray for his condemned soul, "the Jew" refuses and can only release "cry upon cry" of desperation.[93] Fagin's subverbal moans liken him to a brute, while his dismissal of Oliver's Christian charity marks him as renouncing civilization itself.

Isaacs, by contrast, was willing to bend his knee in imitation of Anglican piety to advance his career, even if Kennedy could never forget or forgive Isaacs for being a "Jew or of the Jewish extraction." For his part, Isaacs grew to loathe Governor Kennedy, arguing that "it is the misfortune of the Colonial Service that men utterly ignorant of the habits customs and disposition of the Natives are generally employed to confer and treat with them."[94] Here Isaacs highlights the shortcomings of colonial administration while promoting his own efficacy as a culture broker. Isaacs' adventures afforded him a worldly familiarity with civilizations and their discontents, and he adopted a hard-hearted cynicism toward the measures of power valued by others: hierarchies, religion, money, decorum, legal authority, race, and class. Kennedy embodied the officiousness and pedigree that Isaacs despised. The scene was set for a clash of wills.

Before the West African Squadron vessel sailed to Matakong to investigate the rumors about Isaacs, the embittered Gabbidon and his supporters, including the slave-holding King Tombo, launched an attack on 22 March 1853. "9 a.m. . . . . A battle has been fought," Isaacs informed Kennedy by messenger. "My people have succeeded in driving the enemy, taken 4 guns, 1 prisoner. It is not yet ascertained how many are hurt. The enemy are in ambush waiting fresh forces." A statement issued by chiefs allied with Isaacs told a similar story of what they referred to as Gabbidon's War: "People went to Matacong to plunder Mr. Isaacs and . . . Allah helped him against his Enemies. . . . Mr. Isaacs killed one Man and took several Prisoners." Isaacs assured Kennedy in a dispatch that "the party of marauders which attacked Matacong . . . were induced to come over by William Gabbidon." Word of the attack made American newspapers.[95] Now that events had turned violent, Kennedy had no choice but to launch a formal investigation into the unsettled state of Matakong.

One week later, Lieutenant Commander Henry Christian of HMS *Bloodhound* arrived at Isaacs' island "without previous notice being given" to investigate Gabbidon's claims that slave trading persisted on Matakong. Commander Christian interviewed eight men, women, and children he found on the island and reported that none of those questioned wished to leave. All maintained they were free, and one headman went so far as to "express astonishment" at being asked "if Mr Isaacs ever bought Slaves."[96] Kennedy shelved his investigation; Gabbidon's integrity was in tatters. So things rested until Isaacs sought revenge.

CHAPTER 7

The Queen versus the King, 1853–1856

By the early 1850s, the Crown colony of Sierra Leone had expanded to become an orderly outpost of about forty-five thousand souls. Wooden houses with brightly painted shutters fronted wide dirt roads in Freetown. The airy interiors of these "board houses" were decorated with illustrations torn from European journals and pasted to the walls. Whitewashed picket fences bordered gardens planted with fruit trees and thriving vegetable plots. Residents chatted with passers-by, sometimes inviting them to share pinches of snuff or smoke pipes of home-grown cannabis.[1] Women carried heavy baskets of produce—yams, shallots, guavas—on their heads as they strode through the heat, carrying their infants on their backs wrapped in yards of colorful cloth. Liberated men wearing straw hats sang along the docks. West African Kroomen, famed for their maritime skills, ferried puncheons ashore from anchored ships. Beneath the lofty kapok tree near Government House, where Kennedy presided, uniformed Black policemen exchanged greetings with civil servants.

Church bells clanged from the belfry of St. George's Cathedral to remind townspeople of the miracle of their freedom and increasing prosperity. Missionary students tugged at the collars of their cotton jackets to catch the faint breeze stirring across the broad pan of the bay. These young African men were trained in the faith of the colonial government at the imposing brick and wrought-iron facade of Fourah Bay College, the first modern institute of higher learning in sub-Saharan

Africa. The college was said to have been built on the site of a former slave factory and the rafters of its roof milled from the reclaimed masts of slave ships—fitting symbols of civilization's advance. A nearby race-track drew the less reverent to wager on horses that thundered around the oval, kicking up a mist of red soil that hung in the still, moist air. In the stands, well-dressed Europeanized ladies shielded themselves from the sun with silk parasols and cooled their perspiring flesh with folding fans. Others ornamented themselves with silver bangles around their wrists or with delicate belled anklets that tinkled with every footfall, jewelry that one visitor believed signaled a connection between African women and the Israelite women described by Isaiah (3:18).[2]

An official report concluded that the colony was "steadily and perceptibly advancing—morally, socially, commercially, and financially." Freetown attracted a diverse mix of liberated Africans, merchants, traders, and migrants from the interior. One Fourah Bay College scholar recorded that more than one hundred languages were spoken along Freetown's byway. The college took special pride in instructing students to read biblical Hebrew.

Yet for all its growth, commerce, and straight-backed churchliness, the colony was also a site of decay, poverty, and desperation. Poorer people lived in hovels that were rudely plastered and thatched with vegetation that harbored vermin. Downpours rendered roads impassable during the rainy season. Bridges alternately swamped by floodwaters and desiccated by harmattan winds collapsed beneath the hooves of horses. Tornadoes flattened homes. Swarms of cockroaches ate through the hose of Freetown's fire engine and destroyed the bellows of a church organ. A young boy was sold into slavery for a heap of tobacco. A man sold his friend for a quantity of rice. One missionary looked out over the colony and despaired at the ravages of disease, warfare, and slave trading: "O Africa! To what advantage hast thou been robbed, and spoiled of thy children?"[3]

Governor Kennedy was driven to distraction by the persistence of slave trading and the courts' inability to secure convictions. His review of cases brought before Freetown's grand jury in June 1853 revealed that only three of a paltry eleven men arraigned for slave dealing had been convicted. He abolished the grand jury on learning that suspects escaped prosecution through influential business or family connections within the colony. Even liberated Africans indulged in the commerce of slaving.[4]

The endurance of slaving was in part due to entrenched patterns of "Mohamedan domestic servitude," as well as to patterns of licit com-

FIGURE 10. Remaining facade of original Fourah Bay College building, Freetown, Sierra Leone. Photo by author.

merce. Many of the peoples who inhabited lands adjacent to the colony, whether they were Muslim or not, held to the age-old practice of enslaving criminals, the impoverished, debtors, prisoners of war, or those who strayed too far from their home villages. Agricultural laborers were in high demand among chieftains for cultivating palm nuts and peanuts to sell to European traders. White merchants embedded in the fabric of coastal societies north of Freetown also held domestic slaves, though their Eurafrican wives and family provided camouflage for the practice. Domestic slaves were often treated as dependent and inferior kin, rather than as chattel: that is, they possessed some minimal rights, though this did not prevent their being mistreated, abused, or sold into the transatlantic slave trade when opportunity allowed. British jurists concluded in the 1830s that no legal distinction existed between domestic and chattel slavery. However, they were "aware indeed of the mitigated form or quality of Slavery among Mahomedans, as contrasted with the nature of that Slavery which it has been the great object of British Policy to extirpate."[5] This sort of mixed messaging regarding slave holding was refined into an art in Sierra Leone.

The secretary of state for the colonies issued a directive in September 1855 that the British Government had no "intention or policy . . . to suppress the internal and domestic Slave Trade of the African Tribes." But that same month, Sierra Leone's governor proclaimed that any individual held as a "pawn"—that is, as a domestic slave—would henceforth be considered to be enslaved, and anyone buying, selling, or dealing in pawns in the colony would be treated as a felon. Nonetheless, the Royal Navy's West Africa Squadron was twice instructed (in 1844 and 1866) not to interdict canoes carrying domestic slaves. These promulgations, reversals, and subsequent qualifications gave the colonial administration leeway in its approach to domestic slavery. By temperament and conviction, Governor Kennedy was more vigorous in pursuing suspected slavers than his predecessors had been, and he was encouraged in his efforts by praise from his superior, the secretary of state for war and the colonies.[6] Kennedy was different, and Isaacs knew it.

In May 1854 Isaacs learned that his rival, Gabbidon, was trading near Freetown and had met with Kennedy. Isaacs addressed a letter to the governor seeking to head off Gabbidon's machinations: "The countenance Mr. Gabbidon receives must arise from non-acquaintance with the circumstances in his history well known to everybody else, and which your Excellency certainly ought not to be unacquainted with. . . . [Y]our Excellency cannot be aware that there is a warrant outstanding for Mr. Gabbidon's apprehension on a charge of Slave dealing."

No such warrant against Gabbidon appears in the archive, though records indicate that he was accused of slave trading by a merchant northwest of the colony in 1850.[7] Isaacs may have sincerely believed there was a warrant out for Gabbidon. Or perhaps he made his allegation secure in the knowledge that Gabbidon, like every other legitimate trader in the region, carried on suspect dealings. He must have wagered that Kennedy's probity would cause him to expel Gabbidon from the colony, or at least lose regard for him.

But Isaacs' intemperate letter did not set Kennedy against Gabbidon as he had hoped. Rather, Kennedy read it as implying that the governor himself was consorting with known slave dealers. The outraged Kennedy wrote to his superiors: "The imputations on my self contained in [Isaacs'] letter are simply without a shadow of grounds. . . . I ordered [Isaacs'] letter be submitted to Mr. Gabbidon who called upon me, and said he was prepared to meet any charge Mr. Isaacs had to make against him, and in turn accused Mr. Isaacs of Slave dealing."[8] Isaacs had over-played his hand; now Gabbidon held the moral high ground. He in turn pressed Kennedy to interrogate Seeray Moodoo, the son of a local chief, who claimed to have firsthand knowledge of Isaacs' slave dealing.

Seeray Moodoo described under oath how he had "conveyed a quan-tity of ground nuts" to Matakong. On delivery of the cargo, Isaacs had approached him and requested that he find slaves for "Mammy Hannah Hayes," Isaacs' live-in lover. Hayes offered Seeray Moodoo "4 pieces of St. Iago cloth"—coveted textiles from the island of Sao Tiago, Cabo Verde—for "the purchase of two slaves." Seeray Moodoo then pur-chased a young boy, Yerah, and returned to Matakong to deliver him to Hayes. About a month later, he arrived with a load of groundnuts and a slave girl, Nah Watah. Another witness reported that Nah Watah "was healthy and good looking and Mr. Isaacs who was present shook hands with Seeray Moodoo, thanked him and said he wanted such strong girls." Hannah Hayes presented Nah Watah "to the head man on her farm" as a gift.[9] Though the statements incriminated Seeray Moodoo as active in slave trafficking and principally concerned Isaacs' consort, Kennedy was nonetheless scandalized.

The governor now believed that Lieutenant Commander Christian of the *Bloodhound* had been duped by Isaacs and those he had interviewed on Matakong a year prior. Kennedy was certain that Isaacs was involved with the slave trade, but the colonial secretary remained unconvinced. He placed more faith in the governor of the Gambia's glowing report on Isaacs from the spring of 1854. The secretary even ordered that report be copied and sent on to Kennedy, noting that he was "curious to see what [Kennedy] will make upon it." Kennedy must have been incensed at the imputation that the governor of the Gambia was better informed than he was. With no incontrovertible evidence against Isaacs—now referred to in London as "the sole Monarch" of Matakong—Kennedy was powerless to proceed.[10]

Then, on June 30, 1854, two women, Bangah and Langoh, and Barracah, Langoh's husband, arrived in Freetown by fishing boat. They claimed they had been held as slaves on Matakong. None of them explained how they had managed to escape the island fortress. Possibly they had been assisted by individuals working for Gabbidon. Kennedy ordered the interrogations of the three refugees under oath. Their testimony provides a tragic chorus indicting the systems of human bondage prevalent in the region:

> Bangah: I was taken prisoner in a war between my country people about four years ago.
>
> Langoh: Some years ago and when I was a little girl a war commenced . . . and I was taken prisoner and carried to a village.
>
> Barracah: A war took place and I became a slave.
>
> Bangah: I was purchased . . . [and] transferred . . . over to a Sierra Leone man . . . for whom I worked as a farm labourer without wages for about a year when a Mr Isaacs of Matacong came on a visit to my master. . . . On [the] next day a large boat with mast and [a] deck house below . . . came . . . in charge of one Caulker a Clerk in the service of Mr. Isaacs. Caulker . . . carried me . . . in the same vessel to Matacong. . . . Caulker told us to jump into the boat and that we would be free people at Matacong.
>
> Langoh: I was sold to . . . a native of the Foulah [Fula] Country. I was conveyed by this man to the Island of Matacong where I remained for some time.
>
> Barracah: I was then sold to a Jolliffe [Wolof] man . . . who on several occasions carried me as his slave and labourer to the Island of Matacong.
>
> Bangah: On the morning after our arrival there [Matakong] Caulker took us to the house of Mr. Isaacs who talked with him in English—I did not understand what was said by them. After this Mr. Isaacs told me in the [Mende] language to go into his yard where I was placed at work as a general servant in beating rice, washing clothes & c. for about six months during which period I did not receive any payment for my labour. . . . I did not receive any victuals or clothing from Mr. Isaacs.
>
> Langoh: I was then placed to work as a labourer [on Matakong] without receiving any wages for my services. When I refused to work I was driven to it with a stick. After some time I managed to escape and in company with one Bangah and Barracah my husband made my way to Freetown, Sierra Leone where we claimed our freedom.
>
> Barracah: [My master] sent me to work for Mr. Isaacs from whom I did not receive any wages.[11]

The evidence prompted Kennedy to investigate further. He took advantage of the presence in Freetown of one of Isaacs' allies, Mahmodoo Touray (a.k.a. Mamodoo Turee), minister for King Bamba Lahai, whose "royal residence" was in Malageah, on the Mellacorée River. That king had previously defended Isaacs' right to occupy Matakong against rival claims by Gabbidon. According to Touray's testimony, Kennedy summoned him for a discussion during which he urged Touray to incriminate Isaacs. But Touray refused to provide suborned testimony. Kennedy reportedly threatened Touray that "if he could get anyone to give evidence that Mr. Isaacs had taken a single [slave], he would have [Touray] in irons working in the Streets. He (Governor Kennedy) would show him . . . that there were not two Governors in the Country." Here Mahmodoo Touray plausibly presents Kennedy as pursuing a vendetta against Isaacs and as jealous of his influence. Then again, as minister to King Bamba Lahai, Touray would have had reason to protect Isaacs under the landlord-and-stranger compact. When word reached Isaacs of the governor's meddling, he had his associates in Freetown print up and post anonymous placards attacking Kennedy.[12]

Even without Touray's assistance, the governor believed he had amassed enough evidence to order Isaacs' immediate apprehension "wherever he may be found in this colony." He dispatched the superintendent of police aboard the eight-gun *Britomart* to execute the warrant on 17 August 1854. The brig set sail from Freetown at 2 p.m., but because of "boisterous weather" it did not arrive at Matakong until twenty-three hours later. August is the height of the rainy season in Sierra Leone, when black clouds mass along the coast before thunderclaps shatter the quiet and forks of lightning rend the sky. Storms can last for hours.[13] By the time the *Britomart* reached Matakong and the police trudged uphill to Isaacs' citadel, their quarry was nowhere to be found.

Hannah Hayes' eldest son, Walter Lewis, the second writer in the governor's office, had gotten wind of the planned storming of Matakong and tipped off Isaacs' Freetown agent, Emmanuel Lyons. Lyons raced to the docks and "hired a boat at an exorbitant price." By hugging the coast he arrived at Matakong five hours before the *Britomart*. Kennedy learned that Lyons "took Mr. Isaacs off the Island after a few minutes delay there, & conveyed him on Board a French Merchant vessel," possibly one loaded with groundnuts bound for Marseille. Isaacs had escaped Kennedy's grasp. The superintendent of police could only leave a note with Isaacs' senior clerk voicing the hollow demand that his boss "immediately repair to Freetown" and turn himself in. That Isaacs' clerk was another of Hayes's sons, Thomas Lewis, made it unlikely that he would comply. Instead, Thomas Lewis registered a formal protest with the *Britomart*'s captain, accusing him of exceeding his command.[14]

Frustrated officials disembarked from the *Britomart* to interview those in and around Isaacs' compound. They found nineteen men, women, and children who claimed to have been slaves for Hannah Hayes and Isaacs. These individuals begged for passage to Sierra Leone and were accordingly taken to the colony as liberated Africans. Several of these men and women later testified under oath to their sufferings on Matakong. Their tales of misery describe the universe of captivity that thrived just beyond Freetown:

> Ballah (a man): Many years ago I was taken prisoner of war and sold for Bullocks. . . . I was then sent in a canoe to Matacong and sold. . . . This man gave some wrought gold to Dembah [Bambayah] the master of the canoe as payment for me. . . . Mr. Isaacs told me that I was his slave but if I worked well for him he would give me freedom. He sent me to his yard where I worked as a labourer for about two years. During this time I did not receive any wages. Mr. Isaacs gave me some coarse clothes. After working two years Mr. Isaacs agreed to pay me . . . in goods. . . . On such terms I worked for about six months when I obtained leave from him to go to a town . . . to place a child belonging to me with a nurse. Its mother had died. . . . Some time afterwards I returned . . . and found that my child had died.

> Beahmah (a woman): I was taken in war and sold to a Foulah [Fula] man. . . . I remained for two years when I was removed to the island of Matacong where I worked for . . . Mr. Isaacs. I did not receive any wages but I was frequently flogged.

> Jeremiah: When I was . . . young I was . . . sold . . . to a . . . man called Dembah Bambayah. [. . . A] Jolliffe man . . . came . . . to Dembah Bambayah from Mr. Isaacs at Matacong to tell him that he must send up . . . boys that (Dembah) had promised to Mr. Isaacs. I was then sent up to Matacong . . . I worked for Mr. Isaacs for seven years without receiving any pay. . . . I was allowed a quart of Rice a day and four pounds of Beef per week. Mr. Isaacs ordered my wife to cook for his laborers. . . . I went to tell Mr. Isaacs that I could not agree to let my wife cook for his men . . . Mr. Isaacs sent a man . . . to take away my wife, who he claimed as his property. I went and begged Mr. Isaacs and Hannah Hayes not to take my wife away. She was given back to me. My wife is now at Matacong. . . . My wife is a slave to Mr. Isaacs and . . . Hannah Hayes.

> Daniel: Some years ago I was taken in a war . . . and sold to a . . . man named Dembah Bambayah . . . where I remained his slave for about a year. . . . About 8½ years ago [ca. 1845] Dembah Bambayah took me to Matacong . . . [and] told me that Mr. Isaacs wanted me to stay with

him. . . . I worked for four years for Mr. Isaacs without receiving any pay. I worked as a slave and when I got stubborn, Mr. Isaacs would ask me if I did not know that he had bought me from Dembah. . . . Mr. Isaacs allowed me a quart of rice daily and six pounds of beef weekly.

Leah: When I was young I was caught in a war . . . by a man whose name I do not know. By him I was brought down . . . [and] sold to one Dembah Bambayah. . . . I worked as a slave to Dembah Bambayah. I was then sent to Matacong . . . and he gave me to Hannah Hayes to learn to sew. . . . Mr. Isaacs gave me to wife to Daniel . . . I was married after white man's fashion by a white man from Sierra Leone. . . . Mr. Isaacs did not give me any food after I was married [and] my husband had to feed me. . . . When I was sick, Hannah Hayes looked after me. I recollect a man-of-war officer coming to Matacong. He came in a steamer. This officer asked me if I was a slave and I told him I was not. Hannah Hayes told me to say I was not a slave. . . . I was brought down to Sierra Leone. . . . I was sorry to leave Matacong because my child was so very young—not a month old. I am very sad that I am now at Freetown, and [not with her on] Matacong. I am now free.

Eliza: I was a free woman. My cousin [owed] a debt in the country . . . and I was given in part of payment of his debt. . . . I worked as a slave for Mr. Isaacs. I beat rice for market and carried water. Mr. Isaacs never gave me any payment. He allowed me a quart of rice per day and no meat. Hannah Hayes gave me to live as concubine to . . . a slave man belonging to Mr. Isaacs. . . . Mr. Isaacs never gave me anything but rice—neither money nor clothes.

Ephraim: When young but without any recollection I was sold by my family. . . . [A man] gave me to Mr. Isaacs to pay for his son's debt. I remained with Mr. Isaacs and worked for him as cow driver for four years. I never received any payment from him whatever. He allowed me a quart of rice a day.

Nathan: When I was young I was caught in a war and sold to a . . . man whose name I do not know. That man took me in a canoe to Matacong. . . . Mr. Isaacs then told me that I belonged to him. . . . I believe I have lived about four years with Mr. Isaacs & during that time I have worked for him without pay of any kind. I was allowed half a tub of rice a week. I believe myself to have been a slave to Mr. Isaacs, because if I was free I would not be obliged to work without pay.

Leah's testimony that she had been told to lie to the British officer who visited on the steamer—HMS *Bloodhound*—accords with the statement of another witness, William Spence, a Mende man who had enlisted as a soldier in the 2nd West India Regiment and who visited Matakong as part of a military detachment. He had been told by one of his fellow countrymen on Matakong that when government vessels visited the island, Isaacs would dress his slaves "in good clothes" and intimidate them into stating that they were free, should anyone enquire. Even more damningly, Isaacs went to the trouble of providing his slaves with "papers describing them as free labourers," which the illiterate men and women would then present to any inquisitive authority. These depositions were all recorded and entered into evidence in the case, archived under the title of "The Queen vs. Isaacs for Slave Dealing."[15]

The star witness for the prosecution's case was Dembah Bambayah, the slave trader mentioned by many of those freed from Matakong, who reported that he had carried on a trade with Isaacs for years. According to Bambayah, he met Isaacs at his ally Ba Tha's outpost at Bendu and there bartered tobacco, rum, gunpowder, and textiles for rice, palm nuts, and slaves. As part of routine trade, Isaacs contracted with Bambayah to convey a cargo of thirty-seven slaves to a man named Louis, who operated from a slaving port called Seabar, or Shebar, at the southern mouth of the Sherbro River, not far from Bendu. A contemporary map records Shebar as having once housed a slaving factory run by the notorious Captain Théodore Canot.[16]

Bambayah brought the slaves to Louis, whom he identified as "a Spaniard," but the man's schooner was full, and so the slaves were instead distributed among Bambayah's own brothers and retainers. Because the slaves were never loaded onto Louis's ship, Bambayah owed Isaacs for the goods he had received (as noted in Jeremiah's testimony above). Over the years, Bambayah chipped away at his debt by delivering "five boys" to Matakong, slaves who were "part of the lot I had before purchased and carried to Seabar" for Isaacs.[17] The connection between Isaacs, Bambayah, and the mysterious Louis was not pursued by Governor Kennedy or others. Had this supply chain been investigated, Isaacs might have faced greater legal peril.

Louis was likely M. Louis Lamaignère, a.k.a. Don Luis, not a Spaniard but a Frenchman. Another man who had worked on Matakong also reported that Isaacs had traded slaves at Shebar with Don Luis, a longtime agent for the Cuban slaver Pedro Blanco, who had employed Canot at Shebar and elsewhere along the coast. Six of the thirty-six survivors of the *Amistad* rebellion testified that they had been purchased by a Spaniard named Luis south of Freetown—likely the same individual.[18] Isaacs' contemporaries therefore place him within the orbit of Don Luis and Pedro Blanco and in the vicinity of their known slaving terminals.

Don Pedro, as Blanco was known, was considered an educated, well-bred man who posed as a grandee, generous with his wealth and feared for his murderous rages. Informants claimed that the cigar-chomping Don Pedro owned a network of coastal factories hidden in thickets of vegetation and mangrove swamps, which could hold hundreds or even thousands of slaves in well-guarded barracoons prior to their making the Middle Passage. He may have shipped upwards of ten thousand slaves a year across the Atlantic, in alliance with a powerful regional king. One visitor to his slaving stations claimed that Don Pedro's influence extended to respectable society in the colony of Sierra Leone. Indeed, a copy of an IOU in Freetown's archives indicates that Havana's Don Pedro Blanco & Co. owed Isaacs a large sum of money in 1846—possibly related to the unconsummated slave-trading deal with Dembah Bambayah.[19]

The profitability of slave dealing with Cuba increased dramatically after the British government repealed prohibitive taxes on slave-grown sugar in 1848. Likewise, domestic slavery in West Africa increased during this period because of the high demand for groundnuts and palm nuts.[20] Despite the dangers of interdiction by the West Africa Squadron, Isaacs may have realized that large profits could be made on small-scale slaving. The often-incomplete documentary record connects Isaacs with Don Luis and the business interests of Pedro Blanco, and the economic data suggest that a keen businessman like Isaacs may have involved himself not only in traditionally countenanced domestic slavery but also in transatlantic chattel slavery.

The evidence linking Isaacs to Pedro Blanco's infamy may not have amounted to the sort of smoking gun required by the courts, but testaments to Isaacs' conduct are nonetheless damning:

> Ballah: I was taken to Matacong and delivered to Mr. Isaacs who placed me in irons in a gaol where I was kept for a month and a half. I was then tied up and very severely flogged in the presence of Mr. Isaacs who told the persons who were flogging me that if they did not flog me heavily he would flog them. I was flogged with a cat . . . which was every minute steeped in a pan of rum lying close to me. This increased the pain to such extent that I fainted away. . . . After the flogging the irons were kept on my arms and legs for about six weeks.

> Jeremiah: About seven months ago I was accused along with Daniel . . . by Miss Hannah Hayes of stealing goods from Mr. Isaacs' store. . . . Mr. Isaacs and Hannah Hayes were present at the trial. I was found guilty and Mr. Isaacs himself sentenced me to 3 months imprisonment with hard labor and to receive 100 lashes. I was tied up . . . and flogged to the extent of 50 lashes. . . . I felt I wanted to die. . . . I was flogged right and left with two cats of nine tails each. Mr. Isaacs was present with a drawn sword

and threatened to run [the jailer] through if he did not flog properly. After I was taken down my back was rubbed with salt and water, and I was sent to work in chains.

Daniel: Mr. Isaacs . . . was always present during the flogging. About seven months [ago] I was accused by Miss Hannah Hayes of stealing cloth and good[s] from Mr. Isaacs' store. She ordered me to be caught [and] chained hand and foot and locked up in a store. . . . [W]e were . . . put into irons by Mr. Isaacs who called the Blacksmith to make them fit tight. I was left in irons for a week when I was brought before Tom Lewis, Hannah Hayes' son, . . . Hannah Hayes herself and Mr. Isaacs. . . . I was sentenced by Mr. Isaacs to . . . imprisonment with hard labor for six months, and to be three times flogged. I was sentenced to receive 100 lashes at each flogging. I underwent a flogging of 100 lashes . . . I was flogged with a cat of nine tails and after receiving the 100 lashes my back was rubbed with salt. . . . After having been flogged I was sent to work in chains.

Ephraim: I was a slave to Mr. Isaacs. He has flogged me frequently with a cat of nine tails. I saw Daniel . . . flogged. I was never flogged so bad as he was. I saw a slave called Ballah flogged. . . . Mr. Isaacs was present then and at all other times. He gave the orders to flog. Hannah Hayes told me I was slave to Isaacs.

Kennedy was outraged by these reports and by the revelation that Matakong's Wesleyan chapel, once consecrated by the missionary Thomas Reader, had been "converted by Mr. Isaacs into a cowshed." Reader may have been a self-proclaimed missionary, or perhaps a lay leader affiliated with the Methodist movement. In any case, Reader himself sank into drunken debauchery, luring African women with invitations to Christian study before throwing himself on them.[21]

On 24 August 1854, a week after the raid on Matakong, Kennedy sent a sworn copy of preliminary testimony against Isaacs to the secretary of state for the colonies. Kennedy hoped that the evidence would be "sufficient for Mr. Isaacs' capture should H.M. Govt think it advisable to proceed further."[22] Meanwhile, Isaacs remained at liberty, making his way back to England one step ahead of the law. When he arrived on English soil, he would have found his country changed yet again.

The population of England had doubled in his lifetime. Railway construction had altered the topography of London and scarred the countryside. With rapid industrialization, England became the first nation in the world with a majority of its people living in urban centers. Abolitionist societies, unions, women's groups, and religious leaders united to preach free trade as a panacea for all that ailed the world. The use of commerce to displace the African slave trade reigned as government policy and drove a marked expansion of imperialist projects. Free-trade policies prevailed as economic and moral doctrine and gave rise to an empire of private enterprise. The libertarian spirit extended to matters of faith. Propelled by evangelical revivals, more Christians now worshipped outside the Church of England than within it. And the final step in Jewish emancipation, the repeal of the requirement for Jews to swear a Christian oath to take elected office, was on the brink of enactment. The "Jewish Question" of the 1840s would soon be resolved.[23] After 1858, English Jews would no longer be second-class citizens.

Yet Isaacs—the brave young adventurer, admirer of Zulu culture, the soldier-chieftain under King Shaka, the public critic of slavery, the ambitious trader, the acculturated stranger in West Africa, and the man who negotiated nine treaties outlawing slavery for Great Britain—had transformed himself into a slaver recalling the nastiest anti-Semitic stereotypes and cruelest villains of abolitionist tracts. What had turned Isaacs into a sadist who flogged men for petty theft and then rubbed salt into their wounds? How had the merchant of Matakong become a merchant of Venice, exacting his pound of flesh?[24]

This transformation may not be all that surprising given the historical context. In the wake of the Xhosa Wars, with their violent clashes between Europeans and Indigenous Africans, South Africa's leading liberal humanitarians, including Isaacs' editor friend John Fairbairn, recanted their universalism and retreated into a conservative politics defined by opposition to Black equality. Meanwhile, influential voices in the metropole grew more circumspect about creating little Englands in West Africa. Dickens' disparagement of the Niger Expedition and the myopic philanthropists he portrayed in *Bleak House* helped disillusion British readers regarding such ventures.[25] Likewise, despite the emergence of evolutionary theory, later to be best articulated by Darwin, polygenesis and racial pseudoscience were widely accepted and mutually reinforcing.

Blacks, Jews, and Irish Catholics were frequent targets of the newer scientific racism from midcentury onward. These groups had long been the objects of missionizing fervor, which was often frustrated. Isaac D'Israeli, for example, mocked the London Society for Promoting Christianity among the Jews for having converted only six Jews, despite a yearly expenditure of £10,000. His baptized son Benjamin Disraeli similarly poked fun at efforts to convert Irish Catholics in his popular novel *Tancred* (1847). Perhaps a frustration over the inability to convert and "civilize" all three groups—heathen, Hebrew, and Irish Catholic—

was at the root of the turn toward a more strident racism. Isaacs' inhumane actions at the edge of empire therefore serve to crystallize a familiar but by no means universal Victorian phenomenon: the abandonment of Georgian and early-nineteenth-century humanitarianism in favor of authoritarian imperialism.[26]

Perhaps Isaacs' penchant for chicanery and force of arms indicated a criminal seed that needed only corrupt soil to flourish. He had grown up poor in Canterbury, but from Saul Solomon, James King, Henry Francis Fynn, George Clavering Redman, Don Luis, and possibly Don Pedro he had learned how to transmute the demand for commodities such as imported pasta and Champagne, ivory, gum Arabic, groundnuts, palm oil, and African bodies into huge profits. Scarcity in one place, he understood, could yield abundance in another. He himself had undergone just such a change of place and witnessed a change of fortune. Isaacs' own triumph over deprivation was predicated on the increasing mobility of precious resources. Like the ships his trade depended on, Isaacs sailed into the prevailing economic storms and tacked between crosswinds of competing interests.

Because Isaacs was facilitating slavery more than twenty years after the passage of the Abolition Act of 1833, his actions cannot be dismissed as mere products of his time. By the 1850s, abolitionist efforts were a source of pride in Great Britain, and the persistence of slavery was seen as repugnant. That Sierra Leone was intended as the showpiece of Britain's antislavery policies provided Governor Kennedy with ample reason to view Isaacs as an infuriating embarrassment. Isaacs "is in my opinion the worst description of slave dealer," he railed, "inasmuch as he was reputed respectable and he so far overreached or deceived this Government as to have been employed in negociating Treaties with Chiefs to the Northward of this Colony and thus acquired an influence over the Chiefs which he has turned to mischievous account."[27]

Perhaps what really rankled with Kennedy was the fact that he had been outwitted. He might have taken comfort in knowing he was in a company of dupes. The governor of the Gambia, Sierra Leone's superintendent of Wesleyan Missions, and Governor Macdonald, Kennedy's predecessor, had all been misled. Macdonald had praised Isaacs' efforts, characterizing the antislaving treaties he negotiated as having extended Great Britain's "moral influence." But Isaacs was never much interested in moral accounting. He was concerned with the balance sheet. Whether he benefited or harmed others on the way to earning profits does not appear to have been much of a consideration. We might see Isaacs as an early "savage capitalist" who leveraged power in all its guises for gain; the racial, ethnic, or national identities of those he manipulated mattered little to him.[28] His sole surviving principle was the principle of survival.

Colonialism all but required Europeans to demonstrate or project their cultural superiority over Indigenous peoples; with circular logic, Europeans could thereby justify their claim to being fully civilized, a civility synonymous with being white. Perhaps, as a Jewish man, Isaacs' anxiety about whiteness led him to abuse his power. His experiences would have taught him that power accrues only through its exertion. Moreover, European paternalism presumed Africans to be developmentally akin to children and therefore in need of cultivation, education, and discipline. Colonialists like Isaacs may have felt a perverse duty to employ corporal punishment. A master who spared the rod, like a bad parent, spoiled his childlike African servants. The inculcation of British values—including fortitude and self-control—paradoxically required the colonialist to lash out.

Poetry allows for a clear expression of this sensibility. Isaacs probably did not much fancy verse himself; Kennedy described him as "deficient in education," though he was certainly not unlettered.[29] Still, Isaacs could plausibly have read Alfred Tennyson's popular poem "Locksley Hall" (1842) in a moment of leisure. He might have identified himself with the speaker, "a trampled orphan, and a selfish uncle's ward," who longs "to wander far away, / On from island unto island," and there "take some savage woman" to rear his "dusky race." We can imagine Isaacs priding himself on serving as an agent of progress thanks to his role in literally greasing the wheels of the railway that Tennyson's poem praises. Should any midnight pang of conscience disturb him, Isaacs could reassure himself that he was on the vanguard of a glorious culture—"heir of all the ages"—performing the noble task of introducing the benighted—Tennyson's "squalid savage"—to the ways of the modern world. From where Isaacs sat in his island compound, high on the bluff overlooking a modern wharf teeming with African workers and private soldiers eager to do his bidding, prosperity confirmed his preeminence.

The turning point for Isaacs may have been his "marriage" to Mary Ann Lightbourn, whose Eurafrican family's trade flourished at the law's margins. Until about 1845, Isaacs' character was not yet irredeemably stained by his commercial manipulations and sometimes cavalier approach to the truth. By virtue of his relationship with Mary Ann, he would have

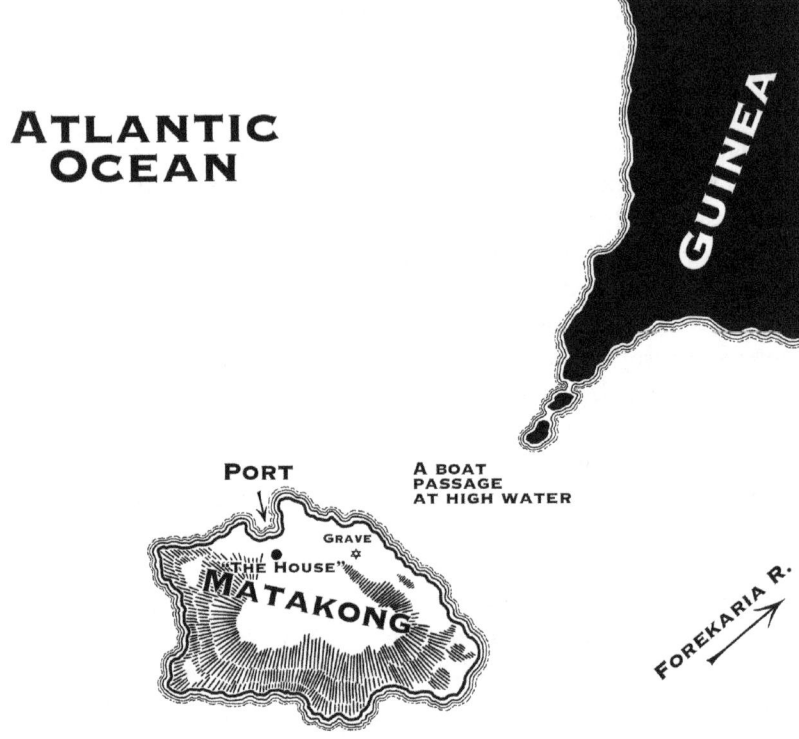

MAP 9. Detail of Matakong Island.

earned the trust of the closed circle of slaving dynasties that operated north of the colony. He would also have benefited from passing as a proper Englishman rather than being identified as a Jew among her mixed-race family and business associates. And he would have been initiated into the shadowy networks that shipped human cargo from West Africa to the sugarcane plantations of the Caribbean and South America. Even before Kennedy arrived in Freetown, colonial authorities had been alerted to "the large collection of Slaves in the Barracoons . . . belonging to the notorious Mrs. Lightburne." These barracoons were "situated about two miles in the Bush . . . [and] the road to them was only known to a few of Mrs. Lightburne's old and most confidential Servants."[30] The British government remained unable or unwilling to root out the Lightbourn family's slave trading, facts which Isaacs would have been intimately familiar with.

After breaking with Mary Ann and abandoning Freetown, Isaacs placed himself entirely beyond the colony's social mores and legal control. On Matakong, Hannah Hayes and Isaacs joined in a brutal duet pitched to depravity. At the same time, Isaacs shrewdly co-opted the colony's approach to ending the slave trade by adopting the trappings of British policy. Isaacs duplicated the government's gradualist program of abolition by negotiating treaties with regional leaders over a period of years. These efforts granted him considerable influence as a culture broker mediating between Indigenous and colonial interests and expanded the sphere of British influence, at least to the extent that Isaacs represented it, to include informal acquisitions of territory and resources. Moreover, he militarized his efforts through his private army and arms dealing, and by encouraging Indigenous leaders to believe that the colony would back up his threats with force. Isaacs' son-in-law later recollected that he had designed his Matakong freehold in such a way as to demonstrate his Englishness and thus his alliance with imperial power. Once Isaacs became "King of the island," he dotted the landscape with grazing cattle, carved a lime-tree-bordered avenue down to the beach, and even built a racecourse.[31] Supported by influential chieftaincies and Eurafrican alliances, vetted by the governor, privy to insider information, and well armed, Isaacs felt confident he could manipulate the will of empire from his stronghold.

Perhaps his was a descent by degrees: first Isaacs dabbled in the domestic slavery that surrounded him, then avarice tempted him to act as a middleman with African slavers like Dembah Bambayah, until finally the serpent's egg of unrestrained power hatched within his soul, and he brutalized the bodies of those who sought little more than scraps of clothing, a bowl of rice, a morsel of meat The more prosperous and influential he became, the greedier he grew. The very fact of Isaacs' successes may have made it easy for him to despise the unfortunate who were held in his thrall. He was strong where they were weak, and therefore he deserved all that he had taken by force of will. While Isaacs was overseeing the floggings on Matakong, Theodore Parker, an American abolitionist, was preaching the now often-paraphrased belief that the moral arc of the universe bends toward justice.[32] Isaacs' moral career, by contrast, traces the arc of a flail biting into a Black man's flesh.

To Englishmen who belittled, caricatured, demonized, or pitied Jews—artists like Isaac and George Cruikshank and William Hogarth, authors like Charles Dickens and Frederick Marryat, men of science and pseudoscience like John Beddoe and Robert Knox, imperial servants like Hudson Lowe and Arthur Edward Kennedy, the misguided churchmen of the London Society for Promoting Christianity amongst the Jews—Isaacs' alleged crimes revealed his true nature. Today we (hopefully) recognize that it is anti-Semitic to impute Isaacs' misdeeds to his ethnocultural and religious identity. While there is little doubt that Isaacs acted reprehensibly, he did so as an individual pursuing rational aims within social and discursive systems that marginalized him, a political system that curtailed his civil rights, financial systems that enforced scarcity, and commercial systems that rewarded the exploitation of those deemed inferior.

Isaacs may have rationalized his slave holding as merely a version of the domestic slavery that was the norm for many of the West African peoples among whom he lived. Possibly he likened his exploitation of African peoples to the colony's official system of drafting bewildered and recently liberated Africans into the ranks of the military, or forcing them to labor as indentured apprentices.[33] Even in England, as Isaacs well knew, a boy's apprenticeship could be a cruel sentence. Isaacs would also have known firsthand that young men in the maritime trade endured numerous dangers at sea, not to mention sexual abuse from older crewmen. He would have been familiar with the British penal system that forced debtors and their families to labor in workhouses and pay for their upkeep. Such debtors' prisons endure in our imagination thanks to the novels of Charles Dickens, who, like Isaacs, had suffered expulsion from a childhood Eden (in Kent). And as an admirer of American industriousness, Isaacs might have looked on the continuation of slavery in the southern United States as a tacit validation of his actions on faraway Matakong. Wherever Isaacs cast his eyes, whether along the mangrove-choked shores of West Africa, across the Atlantic to the Americas, or north to his home in England, he would have seen Black men, women, and children—but also poor, unfortunate whites—yoked in bondage. Better, far better, to be wealthy and powerful for as long as luck would hold.

With the disruption of Isaacs' Matakong operations and his disappearance from the colony, the urgency of apprehending him waned. Moreover, Kennedy, the driving force behind the criminal proceedings, was preoccupied. He had been appointed governor of Western Australia. His recall to London in advance of his new posting kept him busy throughout September 1854. But Kennedy had not altogether forgotten Isaacs. When he embarked from Freetown on October 13, 1854, aboard the steamer *Forerunner*, Kennedy carried with him the cat-o'-nine-tails that had been used to flog Matakong's slaves, and he took along a complete dossier of original testimonies documenting Isaacs' crimes. He intended to ensure that his nemesis did not evade justice.

FIGURE 11. "Wreck of the 'Forerunner' African Mail Steamer," *Illustrated London News*, 18 November 1854.

Twelve days later, after a stop at Madeira, the *Forerunner* departed on a clear day on the last leg of the homeward journey. She was churning the waves with a full head of steam near evening when the captain steered her too close to a well-charted rock. The ship's keel splintered "with a grating sound . . . tear[ing] exactly like a piece of brown paper." Chaos erupted, with passengers and crew fighting to reach the lifeboats, some of which took on considerable water. The captain went below deck to rescue his chronometers and a money box. Kennedy, aghast at such a dereliction of duty toward the passengers and crew, swiftly assumed control of the evacuation. Brandishing a thick wooden staff, Kennedy threatened to beat the "herd of panic-stricken greasy-jackets," as he referred to the crew, should they not follow his orders. When a powerful roller swamped the ship, the *Forerunner* lurched and began to sink bow down, her screw propeller high above the water. Kennedy was sucked into the vortex of roaring, hissing steam. He was dragged below and held fast underwater by the "despairing clutches" of those flailing around him. But being a strong swimmer, he struggled free and managed to tread water until he was picked up by a lifeboat. All that remained of the *Forerunner* was a slick of palm oil floating on the waves.[34]

The evidence against Isaacs—including the cat used to flog Ballah, Daniel, Jeremiah, and Ephraim—now lay at the bottom of the sea. The wreck of the *Mary* at Port Natal had set Isaacs on course for unrivaled freedom; the wreck of the *Forerunner* saved him from certain imprisonment.

The disaster captured headlines in England. Kennedy's first order of business once safely landed in London was to bring the *Forerunner*'s captain and crew to justice. A bitter exchange of claims and counter-claims soon appeared in the pages of the *Times*. Kennedy asserted that the ship had been unseaworthy and her crew's conduct abysmal: he depicted himself as having coolly taken charge of the evacuation. An indignant sailor, however, insisted that Kennedy's "arrogant, and med-dling interference" had in fact cost the lives of fourteen people. Moreo-ver, the sailor hinted that Kennedy's vaunted poise during the ship's last minutes had been due only to the good fortune of his having secured a lifebelt. Those without flotation devices knew they would perish in the cold, rough waters. The public scandal forced the Royal Navy to launch an official investigation. Testimony from the crew and captain tended toward self-exoneration, at Kennedy's expense. Statements by the gov-ernor, of course, emphasized his own heroism. While the court of inquiry deliberated, Isaacs must have delighted in Kennedy's distress. No evidence was found to support Kennedy's contention that the *Fore-runner* had been unseaworthy. However, a year after the wreck, the vessel's captain was indeed found negligent.[35]

Though distracted by the naval inquiry, Kennedy refused to concede the matter of "The Queen vs. Isaacs" even after the evidence had sunk into the depths. He pressed his superiors to have Isaacs arrested on the basis of the documents he had earlier dispatched to London. He knew that Isaacs was "in England and his address . . . known." But legal coun-sel insisted that the documents they possessed were insufficient to remand someone for crimes alleged to have been committed in another jurisdiction. Authenticated copies of the depositions taken in Freetown would be required now that the originals had been lost at sea—either that or the victims and witnesses themselves would need to be sent from Freetown to London to testify.[36] The officiousness of British bureaucrats that had frustrated Isaacs so many times before now proved a boon.

Sierra Leone's new acting governor, Robert Dougan, was described as an ambitious, mixed-race man. He filed a clear-headed report to Lon-don on the way Walter Lewis, Hannah Hayes's son, had learned of the police raid on Matakong and tipped off Emmanuel Lyons, Isaacs' loyal agent. Dougan assured his superiors that he would investigate Lewis's "participation in the nefarious practices" of Isaacs. A scrawled adden-dum to Dougan's letter, possibly by the colonial secretary himself, noted that Isaacs' case "will be a remarkable & important one, & will involve considerable expense."[37] The full force of the government would be marshaled to make an example of Isaacs—if only the legal authorities in London would permit his remand.

From offices on Fenchurch Street, at the heart of London's financial center and not far from the wharves his African trade depended on, Isaacs resumed his paper war against Kennedy. He fired his first volley to the colonial secretary, accusing Kennedy of acting vindictively against him. Isaacs charged that his opposition to certain of Kennedy's fiats—including abolishing Freetown's grand jury—had earned him the governor's enmity. He further asserted that reliable sources informed him that powerful chiefs north of the colony, along the Mellacorée River, were "greatly incensed . . . [and] have demanded an explanation as to the seizure of their Subjects, my Servants, at Matacong" on Kennedy's orders.[38] His informant may have been Benson (Benjamin) Isaacs, who had been dispatched in late 1854 or early 1855 to look after his brother's business empire during his enforced absence.

Isaacs insisted that those freed from his island headquarters were lawfully employed by him at their own kings' command. Those who said otherwise were liars and opportunists who had sought to throw off their customary responsibilities for the excitement of Freetown. He warned that the regional kings would not allow their people, who represented valuable agricultural labor, to be whisked away to the colony. The implication was that Kennedy had upset delicate African ethnopolitical structures and compromised British interests. "War is imminent," Isaacs prophesied, "the Chiefs having formed a comprehensive league similar to that of the Caffres preliminary to the War at the Cape."[39] Here Isaacs refers to the bloody battles of the recent Eighth Xhosa War (1850–52) between British forces and unions of dispossessed Xhosa who resented white expansion across South Africa. Isaacs' defense strategy was therefore multipronged: he would play the role of victim, dissemble, attack Kennedy's reputation, bluster, and frighten the colonial hierarchy in any way possible. He would also bury the government in an awful lot of paper.

In December 1854, Isaacs' solicitor filed a deftly argued forty-seven-page brief denying all charges against his client. The document described Isaacs' selfless service to the colony while ascribing all manner of villainies to Governor Kennedy. A further 208 pages of appendixes, including correspondence, deeds, bills of sale, shipping registers, witness statements, and supporting documents in Mughrabi Arabic script with notarized

English translations were attached to Isaacs' "humble memorial." Isaacs contended that his rule on Matakong was a British idyll in a backward land. His coastal trade had improved the local peoples. In fact, he claimed, he had initiated a system of trial by jury that meted out only mild punishments. And he invoked his collaboration with Wesleyan missionaries and the establishment of the chapel on Matakong as evidence of his charity.[40] The plaster saints of Victorian liberalism were all on display in his memorial: free trade, progress, Christian philanthropy.

Isaacs recalled the numerous "diplomatic missions" he had undertaken for the colonial government, proclaiming that he was "proud to be the means of extending British influence, of opening new outlets for English commerce, of extending the benign influences of civilization, and of checking the inhuman traffic in slaves." Yet his exertions and triumphs had earned him nothing but Kennedy's "dislike, jealousy, and rancour." Isaacs appended a document in Arabic drawn up by "Chiefs and Headmen" defending his business practices at Matakong against Kennedy's aggression. These leaders were ready to "call upon all the good Men of the Country to arm and be prepared to assert their rights against the unlawful proceedings" of Governor Kennedy.[41] Whether or not the document accurately records the landlords' sentiments, it appears that Isaacs was prepared to support—in words if not deeds—a proxy war against Her Majesty's appointed representatives.

The governor, Isaacs charged, had harassed his ships, detained his clerk (Caulker) without cause, and encroached on customary sovereignty rights by storming Matakong and removing residents by force. Kennedy's high-handedness had led Isaacs' landlord-protectors to form "an offensive league of the Chiefs against the British Government." The governor, Isaacs explained, envied his "greater influence with the chiefs" and "would not tolerate an independent man." He further claimed that Kennedy had conspired with Gabbidon, "a person of notoriously bad life and character," into fabricating the charge of slave dealing against him, presumably so that Matakong would pass into Gabbidon's hands and Kennedy could earn praise in London for his actions against an imaginary slave depot. Though Isaacs had opened "new, extensive, and profitable channels for the disposal of British manufactures," Kennedy had paid him back by "deeply and cruelly, willfully and irreparably" wronging him.[42] Here, played out in miniature, was the conflict between the City—London's financial center, which bankrolled the free trade of merchants active in the field—and Whitehall, Britain's administrative regime, which claimed expertise from behind its expansive desks.

Officials pored over Isaacs' long-winded, self-pitying memorial and requested that Kennedy respond point by point. The governor must have sputtered in disbelief. Nonetheless, he put pen to paper in January 1855 and submitted more than fifty pages refuting each of Isaacs' charges. Kennedy's prose is initially measured but grows outraged in the face of Isaacs' "organized system to impose misstatements." At times he sounds paranoid, charging that Isaacs and his "characterless agents" had found subscribers in the colony to fund a coordinated campaign to smear him. He names lawyers, merchants (including one he identifies as a "Jew"), Methodist missionaries, a doctor, a member of the vice admiralty court, two magistrates, a former police superintendent, an army surgeon "worn out by drink and debauchery," and various suspected slave dealers as members of a vast cabal against him.[43] Kennedy makes Isaacs out to be a kind of mob boss. It is of course possible that Isaacs had corrupted all these individuals, or hoodwinked them into believing in his innocence. However, it is equally plausible that Kennedy had made plenty of enemies during his autocratic tenure who were more than happy to slander him and back the influential Isaacs, who was always good for business.

In a passage revealing at one stroke his disdain of the merchant class and his low opinion of Black Africans, Kennedy asserts that British traders do not in any case deserve Her Majesty's protection, as they are guilty of "oppressing and cheating the unsophisticated savages." He dismisses as forgeries documents tending to support Isaacs, accuses him of possessing Matakong by fraud and force, questions the veracity of his shipping records, maintains that Isaacs' service as a commissioner to negotiate treaties was solely for his "personal aggrandisement," suggests that Isaacs' attorney is certifiably insane, and charges that Isaacs was alluding to empty threats of armed conflict in order to destabilize Kennedy's untested successor, Acting Governor Dougan. Kennedy flatly rejects Isaacs' intelligence that a call to arms would upset the region. The peoples who lived in the area, he assures his superiors, are "timid and unwarlike." He advises Whitehall to ignore any such dark prophecy, though he fears that Isaacs has "armed and encouraged those in his neighborhood to maraud and 'make war' for his own purposes." Kennedy further insists that Isaacs lived "a notoriously loose life, and in open adultery, flogging slaves most inhumanly." The former governor closes his screed with an offer to "cheerfully proceed to the Coast of Africa" to participate in any inquiry the government might deem necessary, even if it should mean postponing his departure for Australia.[44]

Meanwhile, Isaacs enjoined highly placed officials to plead his case and malign Kennedy as a crony of Gabbidon, whom he characterized as a petty slave trader and miscreant. Isaacs asserted that regional headmen had held a palaver and concluded that Gabbidon had conspired with Kennedy to dispossess him of Matakong. These headmen, under the direction of King Bamba Lahai, had earlier sent deputies to Freetown to present their case against Gabbidon, but Kennedy had rebuffed them, much as British officials had toyed with King Shaka's diplomats at the Cape. This lack of respect, Isaacs wrote, "greatly incensed" King Lahai, who "ordered all the European Traders to remove with their property" from his territory as a result.[45]

Dougan responded to Lahai's expulsion order by ordering three ships to cruise up the Mellacorée River and put an end to the disruption of trade. They coerced Lahai into paying compensation to the affected merchants within three months. When Isaacs learned of these actions, he protested to the colonial secretary that "in all probability the Chiefs will refuse payment, hostilities will recommence, the interchange of goods will be arrested, and the Trade of the Season will thus be annihilated to the serious detriment both of Europeans and of Natives." An official in London reviewed Isaacs' letter of foreboding and lazily annotated it with the remark that the colonial secretary had already learned of the "favorable termination of the differences with the Chiefs" of the Mellacorée.[46] The matter was considered settled, and Isaacs' warnings went unheeded.

Isaacs remained defiant. "I am innocent," he declared, "and if innocent I am wronged." To clear his name, he announced that he, like Kennedy, was prepared to travel to Freetown to "face this ordeal, trusting in God and in the protection of the Press which I will take care shall watch every movement of my accusers and note every word of my trial, to set me right with the world." Isaacs noted that he had filed a civil action against Kennedy for being his "traducer and oppressor." He further reminded the colonial secretary that he had negotiated numerous treaties to extend commerce and suppress slavery at "large cost both of time and money" and felt confident that he would be exonerated in Freetown. "Give me but a fair trial," Isaacs wrote, "and I have nothing to fear."[47]

One adviser in the colonial secretary's office suggested that Isaacs' eagerness to sail for Freetown was a pretext for "keeping out of the witness box" in London: should his civil suit against Kennedy proceed, he would be subject to a potentially self-incriminating cross-examination.

"The scheme does credit to the ingenuity of Mr. Isaacs' legal adviser," the unnamed bureaucrat concluded. Furthermore, he wrote, Isaacs would likely be acquitted in Sierra Leone, owing to the reluctance "of juries of 'liberated Africans'" to convict slave dealers out of fear, and "the influence such a man as Mr. Isaacs" must exert among them. Only Kennedy's presence in Freetown might tip the scales of justice in the right direction. But in the opinion of the Colonial Office, it was not worthwhile to return Kennedy to Sierra Leone at government expense. Several further letters from Isaacs in winter and spring 1855 were ignored, or else responses were delayed by legal wrangling and cabinet reshuffles. Between June 1854 and September 1855, five different men filled the office of colonial secretary.

Isaacs must have been preoccupied himself during the spring of 1855, when he learned that his brother Benson had taken ill and died on Matakong on 25 April.[48] But Isaacs continued to run out the clock until a distinguished civil servant and colonial expert took a keener interest in his case. Herman Merivale, a mutton-chopped veteran of the Colonial Office, had no doubt as to Isaacs' guilt. By May 1855, Merivale was orchestrating his own clever plan for Isaacs' arrest and prosecution in London, rather than in Freetown.

Merivale directed Acting Governor Dougan to select witnesses from among those who had testified against Isaacs. These men and women "should be put on board the steamer [to England] with the mail bags *at the very last moment* so that no communication from the shore may subsequently be made." Only Dougan and the vessel's captain were to know of the plan, and once the steamer docked in Plymouth, the captain was to send a coded telegraph message to London, where authorities would remand Isaacs on a felony charge. In this way Dougan could elude Isaacs' mole in Freetown, Walter Lewis, and his spy on the waterfront, Emmanuel Lyons, and anyone else he might have corrupted. "The secrecy of this operation will depend on the judgement & caution" of the acting governor, Merivale wrote on 21 May 1855.[49] On that same day, however, Dougan's recklessness exploded in spectacular fashion in a military misadventure—one set in motion by Isaacs from nearly four thousand miles away.

When payment came due from King Bamba Lahai for having expelled British traders along the Mellacorée, Dougan sought to exact compensation by force. He announced to military authorities that King Lahai and "the Chiefs must be chastised, that the only way of punishing them [is] to destroy their Town, and if possible to bring the King down to

Sierra Leone."[50] Dougan dispatched 150 soldiers of the West India Regiments on HMS *Teazer*, a wooden gunboat, on 21 May. Many men who served in the West India Regiments were liberated Africans whose only means of support, once freed from the Middle Passage, was enlistment. These regimental forces steamed into sight of Lahai's seat of power, Malageah, and fired shells over his residence. The king hoisted a flag of truce that was answered by the *Teazer*'s Captain Fletcher.

Fletcher's troops landed unopposed and advanced through a silent village toward Lahai's residence. When they arrived, they found that the king and his inner circle had fled into the bush, leaving his minister, Mahmodoo Touray, to negotiate. This was the minister who earlier had claimed that Kennedy had tried, and failed, to suborn testimony against Isaacs. Rather than respect the flag of truce or enter into negotiations, Fletcher acted on the rash orders Dougan had given him. Touray was marched off to the *Teazer* as a prisoner of war until Lahai agreed to show himself. Soldiers then fanned out to set fire to the king's residence, the town mosque, and several other buildings using signal flares and Lucifer matches. They encountered no resistance. Structures in the region were typically built of wood wattled with sun-hardened mud and thatched with dried palm fronds and other vegetation, and hence highly flammable. The mosque was likely more substantial, with thicker walls and pillars supporting a roof. The troops were soon forced to retreat from the inferno. As they waited jubilantly to embark with Touray as their hostage, they were ambushed by small arms fire. Five soldiers were wounded before they could push their landing craft beyond range of Malageah's well-camouflaged defenders. Once aboard the *Teazer*, Captain Fletcher and fellow officers watched the blaze consume the town center but were disappointed to see that much of the settlement remained intact. Though they steamed downriver, they were determined to carry out their mission "to reduce Malagea[h] to ashes."[51]

The *Teazer* returned early the next morning. Troops on deck saw flames still licking up from the half-burned town. Officers directed fire from the gunboat into the town and raked grapeshot along the thick vegetation that had served as cover for the previous day's ambuscade. With no sign of further opposition, troops poured from the *Teazer* in rowboats. The soldiers beached their boats, slogged their way up the bank, and marched along the footpath to torch what was left of the smoldering town. As they wound their way up the narrow track, invisible defenders fired at them from the brush. Several men were cut down and bled to death in the muddy ooze. Captain Fletcher took cover behind a bank of earth and watched as his lieutenant was shot clean through. Dozens of panicked men clambered into a boat and pushed off, but the overloaded craft took on water and soon capsized. Those who did not drown were shot by Lahai's warriors, or butchered and bludgeoned to death when they crawled to shore. Lahai did not observe the flag of truce when the *Teazer* raised it. Of the 150 members of the West Indian Regiments who departed Freetown, seventy-two were killed in action, and another twelve were wounded or missing.[52]

News of the debacle reached London two months later. On 20 July an official informed Merivale of the incident and directly blamed Isaacs: "It is remarkable how very literally the prophecies and forebodings of Mr. N. Isaacs . . . have been verified by the events which have lately taken place. . . . I think there can be little doubt but that the influence which he boasts to possess with the Native Chiefs have since then been actively at work in producing the result which he then professed to foresee." It was equally remarkable how mistaken Kennedy had been in declaring the docility of the "Native Chiefs" and dismissing Isaacs' warnings. Merivale responded with his own damning indictment: "I think Mr. Isaacs' fulfilled 'prophecies' look very serious, & have little doubt he or his agents are at the bottom of some of the fatal resistance" encountered at Malageah.[53] While Isaacs could not have foreseen the rout of British forces, he likely did embolden the king to resist Dougan's demands. He may even have instructed his agents in Matakong and Freetown to supply Lahai's defenders with weapons.

Isaacs encouraged such speculations. Months after the lopsided battle, he reminded the secretary of state for the colonies that he had "foretold the deplorable consequences which must inevitably ensue" should authorities in Freetown press their demands for restitution on King Lahai. "The disregard of this warning and advice," he wrote, "has cost nearly 100 lives, has disgraced our arms with defeat, and destroyed amongst the Natives the prestige of our superiority. My prediction, based upon an extensive experience of the African character, and a thorough knowledge of the views and feelings of the Chiefs with respect to the grievances . . . has been fulfilled to the letter." Isaacs did not miss the opportunity to pin blame on his nemesis: "The whole difficulty has arisen out of the arbitrary, unconstitutional, and oppressive conduct of the late Governor Kennedy, in seizing and carrying off [King Lahai's] countrymen, my free Servants, from Matacong." He warned that unless a skilled and knowledgeable man was sent to negotiate with the local chieftains, more disturbances would occur, with further loss of life and disruption of trade. Isaacs offered to "cheerfully undertake that duty." In a postscript, he added that "a considerable quantity of my property has been destroyed by the firing of Malageah." [54] Uncharacteristically, he did not request indemnification for his losses. Nor did he mention the legal proceedings that continued to cast a shadow over his reputation.

In a remarkable act of chutzpah, Isaacs announced that he would set sail for Sierra Leone in the expectation that the government would require his services to conciliate King Lahai. The information set off alarms in Whitehall. An official noted that Isaacs might arrive in Sierra Leone just as the secretly impaneled witnesses from Freetown arrived in Plymouth. Merivale concluded that "Isaacs is quite well aware of all that is taking place . . . and has timed his departure for SL [Sierra Leone] so as to avoid the arrival of the witnesses" in England. Merivale's clandestine coordination with the embattled Governor Dougan was on the verge of collapse. If the witnesses were to embark for Plymouth, and unless Dougan had retained enough evidence in Sierra Leone to issue a warrant for Isaacs' arrest when he arrived in Freetown, he would be a free man, able to return to business as usual on Matakong. A hasty letter urged Dougan that if he had not already done so, he should wait to embark the witnesses.[55] Isaacs exploited the delays in communication between London and Freetown to stay one step ahead of prosecution, and he hoped to keep the legal authorities off balance with his movements—real or threatened—between England and the Sierra Leone. He was a pioneer of a strategy used by the wealthy and powerful today to avoid prosecution: jurisdictional arbitrage.

Unbeknownst to Merivale, the beleaguered Dougan had yet to even assemble the required witnesses. He had selected three individuals to be sent to England: the cruelly beaten Ballah, the boy Yerah, and the girl Nah Watah. Dougan believed these three would provide the strongest testimony against Isaacs. His decision remains puzzling. According to their own testimonies, both Yerah and Nah Watah had been purchased by Hannah Hayes, not by Isaacs himself, though he had acted as a middleman. Yerah had disappeared into the interior, and all traces of Nah Watah had evaporated. Only Ballah's whereabouts were known. Dougan had Ballah deposed and the documents sent off to London. As for Hannah Hayes, she disappears from the archival record: perhaps she abandoned Isaacs, as he had abandoned her. Another key witness, Isaacs' perpetual rival, William Gabbidon, had lost his footing while crossing a river and drowned within sight of helpless onlookers.[56] Dougan compiled what evidence he could to use against Isaacs should he ever return to the colony.

The acting governor had other problems to contend with. He was squarely blamed for the military disaster at Malageah. Colonial Secretary William Molesworth rebuked him for exceeding his authority. Dougan "had no authority nor sufficient justification for commencing the Expedition," he wrote. "It was planned with insufficient foresight and mismanaged in the execution and its termination was dishonorable." Molesworth, a disciple of Jeremy Bentham and a supporter of Wakefield's systematic colonization proposals, accused Dougan of betraying British ideals: "By such conduct we shall never lay the foundations of civilization in Africa, and the Native races there may justly accuse us of imitating their uncivilized example." He asserted that Dougan should "be regarded as unfit to hold any position of confidence under Her Majesty's Government."[57] Isaacs' machinations had claimed another victim.

Dougan was replaced by Governor Stephen Hill, who determined that even were Isaacs to voluntarily return to Freetown to face trial, "any jury would acquit him, and the ends of justice would be defeated, so strong is the sympathy of the jurors with persons accused of such crimes." A conviction was more likely in London, but Hill made no move to send Ballah to the capital. Though the prime minister, Lord Palmerston, was "very anxious to have justice done upon Mr. Isaacs," the prosecution stalled. The appointment of a sixth colonial secretary in less than eighteen months meant that by the end of 1855 Isaacs had mostly been forgotten. In late February 1856, Hill received instructions that witnesses were to be discharged.[58] Thereafter the clamorous archival record of "The Queen vs. Isaacs" goes silent.

Isaacs did not abandon Matakong, however. In 1856 he returned for another lengthy period of residence. Perhaps he brought with him the Hebrew and English tombstone that still marks his brother's grave:

To the Memory of Benson Isaacs,
Born: 4th February 1808,
Died: 25th April 1855
Aged 47 years,
Much lamented by his numerous Family and Friends.

The skillfully carved banana-plant motif over the lettering suits the riotous tropical flora that threatens to devour the lonely headstone. The death of his brother may have prompted Isaacs to think of his own

FIGURE 12. Benson (Benjamin) Isaacs' headstone, Matakong
Island. Photo by author.

mortality: he again became a regular contributor to the Canterbury
Hebrew Congregation.[59]

A commercial agent for a French importer joined the resurgent Isaacs
on Matakong for about eight months and penned a brief description of
his stay. He described Isaacs as a "Jew" who claimed English nationality. "But if Mr. N. Isaac[s]" was of English nationality, the Frenchman
wrote, "he loudly insisted that his government had no right to his island,
which was his personal property, and over which flew his private flag."
Isaacs may indeed have hoisted a private flag, but its design is not clear.
An undated watercolor of Matakong depicts a flagpole flying a banner
with a white blotch against a reddish background. The image remains
frustratingly fuzzy. In any case, this pastoral view of Matakong from

FIGURE 13. Watercolor of Matakong Island, ca. 1850–1880s. Artist unknown. Courtesy of the Cadbury Research Library, University of Birmingham.

the sea presents an idealized and much anglicized view of the island. The foliage, except for a few distant palm trees, would not be out of place in a John Constable landscape. From the viewer's perspective—adrift on the open ocean—the geography of this fanciful Matakong does not quite match reality. The kapok trees that mark the makeshift port are still unmistakable landmarks today, but they barely register in the watercolor, and the bluff on which Isaacs built his fortress rises less dramatically than as depicted.[60]

Isaacs impressed the visiting French agent with his contempt for British colonial authority. His boasts that Matakong was his private dominion brought a challenge from the incredulous Frenchman. Isaacs coaxed the agent to travel with him up the Forekaria River, where Isaacs would substantiate his claims with the regional king, possibly Quia Foday, with whom he had negotiated two treaties in 1851 (on 2 August and 7 December).[61] No other European had ever been received at the king's residence, Isaacs maintained. His companion would be only the second white man to make the journey. Perhaps Isaacs thought that his visitor would wilt under the challenge and bow out. Despite his misgivings, however, the Frenchman accompanied Isaacs on a sweltering journey upriver by pirogue that took more than twenty-four hours.

Upon arrival, the king's guards met them armed with bow and arrows. The king wore a variety of amulets and arrayed himself with European weapons. Gratified by the gifts the merchants had brought, he gave Isaacs a warm welcome. Isaacs purchased a load of groundnuts and then prevailed on the king to proclaim in front of his friend that he was the sole, legitimate owner of Matakong. The king did so through an interpreter. That accomplished, the two Europeans were served palm wine by the king's slave girls before their return. The Frenchman did not remain overawed by Isaacs' royal pretensions for long, however. Once back in France, he reported that Matakong was unfertile and unhealthy. A ship captain who had docked at the island went mad after witnessing his entire crew sicken and die of fever. They were buried one after another in the red soil of Matakong.[62]

CHAPTER 8

Savage Invasions, 1856–1872

A little more than a year before Isaacs became notorious in memoranda circulating in the corridors of Whitehall, his name flowed from the busy pens of London's literati. *Bentley's Miscellany*, at the time owned by Richard Bentley, printed a long article titled "Chaka—King of the Zulus" in January 1853. In the first paragraph, the author noted that his account of Shaka's life relied on Isaacs' *Travels and Adventures* and emphasized that official documents substantiated the travelogue. The essay thus claimed to offer an authoritative summary of Zulu history, with eyewitness descriptions. King Shaka's desire for Macassar oil featured prominently. This focus on Shaka and the Zulu kingdom was timely: English readers renewed their interest in the "fierce character of the general southern races of Africa" during the costly Eighth Xhosa War, which concluded in February 1853.[1] Isaacs' old shipmate and fellow castaway in the Zulu Kingdom, Charles Maclean, also began publishing his recollections in the *Nautical Magazine* that year, spurred by "the recent events that have drawn public attention."

Maclean was now captain of a merchant vessel and a committed abolitionist. He fondly recalled his "three years' residence amongst those savages" and expressed gratitude for their hospitality. He reserved his opprobrium for the "rude and lawless boors," the Dutch settlers whose "destroying hand" had waged a cruel war against the Indigenous peoples, "those once happy families of the creation."[2] Maclean's pious reminiscences and the *Miscellany* article were just the first of several publications and events that year to draw British attention to the fate of South African peoples. For several months writers would invoke Isaacs' name or indirectly rely on his authority in order to disseminate information, and sometimes misinformation, about the Zulu nation and others lumped together under the moniker "the Kaffirs."

In May 1853, more than thirty years after Isaacs, Maclean, and a handful of other desperate *abelungu* heaved their way onto the shores of Port Natal, eleven Zulu men, one woman, and her infant child landed in London. They had been brought to the metropolis to perform in an exhibition of "Native Zulu Kafirs" produced by the Natal merchant A. T. Caldecott and his son Charles Henry Caldecott.[3] The Caldecotts were among the most successful ethnographic entrepreneurs of the decade. Exhibitions featuring "primitive" or "savage" peoples—Bushmen (San), Laplanders (Sami), Native Americans—had been popular entertainments in London since the arrival in 1810 of Saartjie (Sarah Baartman), known as the Hottentot Venus. Baartman, a Khoekhoe woman, created a sensation when she was exhibited in London and Paris. She was famed for her steatopygic figure—the trait of storing fat in the buttocks. In today's era, saturated in images that offer a window onto distant places and cultures—photojournalism, documentaries, internet access at our fingertips—it is easy to malign the curiosity of past generations. Yet as the British Empire coalesced during the Victorian era, and as its global territories grew interconnected thanks to emerging technologies like steamships and the telegraph, it was natural that interest in its far-flung subjects should increase. Just as most readers of *National Geographic* in the late twentieth century did not peruse its pages merely to ogle bare-breasted women, not all mid-nineteenth-century exhibition goers were motivated by prurience.

There were few options for the burgeoning middle classes to learn about their expanding world at first hand. For a significant number of the men and women who flocked to these exhibits, creation in all its infinite splendor testified to the hand of God in nature and suggested means to better serve Him. The establishment in 1843 of the Ethnological Society of London, an outgrowth of the reformist Aborigines Protection Society founded by Saxe Bannister and Thomas Hodgkin, was one expression of this religious and scientific fascination with the varied forms of life on earth. The popularity of Robert Chambers' *Vestiges of the Natural History of Creation* (1844), which set the stage for acceptance of Darwin's *On the Origin of Species* (1859), was another. So too was the opening to the public of the London Zoological Gardens in 1847. Ethnological performances offered audiences a kind of human zoo, one that promised knowledge of the world beyond England. At the same time, such displays catered to the racial and cultural biases of nineteenth-century audiences. While exhibitions sometimes humanized their performers, they also dehumanized them in ways that confirmed the audiences' assumptions of their own superiority.[4]

The Caldecotts were aware of the humanitarian quandaries posed by their exhibition, and also of British pride in the eradication of the slave trade. They emphasized the legality of their contractual arrangements with the Zulu performers, the pleasant conditions they enjoyed in England, and the scientific value of the attraction itself. The exhibition program reprinted a certificate issued by Cape Colony authorities attesting to the fact that A. T. Caldecott had tendered a £500 surety for the "due care, proper treatment, and payment" of the Zulu and for their safe return to Natal.[5]

Caldecott rented the well-known St. George's Gallery in Knightsbridge for the performances and hired a celebrated painter to create scenic backdrops. Advertisements appeared in London newspapers, and posters and handbills were distributed throughout the exhibition's three-month run. For as little as one shilling, a sum all but the poorest could afford, audience members were treated to a dramatic representation of scenes from Zulu life: a communal meal among beehive huts, ritual songs and dances, a sorcerer's ceremony, a bridal negotiation and marriage, a pantomimed hunt, a war song, and choreographed combat.[6] The goal was edification through spectacle, an early form of docudrama.

The younger Caldecott served as master of ceremonies for the ninety-minute performance. Spellbound attendees who wanted to learn more or retain a souvenir could purchase a thirty-page illustrated program that described Zulu customs in more detail. This program bore a note on its title page testifying that the information it contained had been "compiled chiefly from Authentic documents" —principally Isaacs' *Travels and Adventures*—by A. T. Caldecott and revised by his son. The Caldecotts slavishly modeled their descriptions of Zulu life and culture on Isaacs' volumes. And like the earlier article in *Bentley's Miscellany*, the Caldecotts connected their exhibition with the "long and disastrous war" with the Xhosa recently concluded at the Cape.[7]

The program listed the names of all the Zulu performers, other than the fourteen-month-old girl, and noted the warriors' proficiency with various weapons. Despite their martial prowess, the Caldecotts wrote, "they are very pleasant people, delighted with the new world which has opened upon them in this country." London impressed the Zulu, and the Zulu impressed London. The hereditary Zulu chieftain Manyos (or Manyosi) was among those summoned for a command performance at Buckingham Palace. He concluded his performance with a direct address to Queen Victoria: "Now the hearts of the Zulu will be gladdened. When Manyos returns to his country, it will be in joy . . . [and] all that we have seen and heard verifies all that has been said of . . . the English. . . . Great has been our satisfaction in this day having been received by the great mother of the whites."[8] The exhibition was the sensation of the 1853 season.

A front-page item in the *Standard* two days after the show's opening described the Zulu as "finely-formed men, of a dark copper hue" and praised the lecture delivered by Charles Caldecott. The same day, the *Times* cheered the exhibition's "wild artists," while drolly commenting on the lack of verve exhibited by London's home-grown actors. The *Standard*'s review appeared beside an article heralding a future in which electric light and power would replace gas and coal. The review in the *Times* was likewise nestled between notice of a forthcoming lecture on "Voltaic Electricity" and a discussion of an expedition to the Arctic Sea. The message implied by the juxtaposition of these columns was threefold: first, that the Zulu Exhibition in London was as fascinating to the public as were the latest technological marvels; second, that the appearance of the Zulu on stage contributed to scientific knowledge; and third, that the Zulu themselves lacked the benefits of such knowledge and therefore remained in a state of arrested cultural development.[9]

Further reviews followed. The *Illustrated London News* treated the performance as unintentionally comic, if affecting, and reproduced a drawing of Zulu warriors. The *Spectator*'s reviewer singled out the Zulu for being "fine well-formed men" but lamented that their musical ceremonies appeared to "disregard . . . all economy of time and exertion" and thus provided evidence of their self-indulgent nature. The implication was that a steady dose of Anglican decorum would turn the childlike Zulu into a mature nation of shopkeepers. An anonymous reviewer for the *Athenaeum* dwelled on the cranial shape and physical features of the Zulu, though even he managed to escape his racialism to compliment the performers on their fine figures and stage presence. The article's invocation of a crude physical anthropology would have signaled scientific sophistication in the mid-nineteenth century, not prejudicial backwardness. News coverage of the exhibition throughout the summer praised the performance and referred to Isaacs, "a trading adventurer," as the primary source for knowledge of the Zulu people.[10]

FIGURE 14. "Zulu Warriors at St. George's Gallery," *Illustrated London News*, 28 May 1853.

The exhibition program mentions Isaacs several times, as well as Farewell and Fynn, and treats as established fact the grant of three and a half million acres to the white settlers by King Shaka. The Caldecotts' booklet borrowed Isaacs' depiction of Zulu governance as "Zulucratic" and noted the battles for succession that raged when a Zulu king "begins to exhibit the signs of age," a reference to Shaka's graying hair. The program also informed readers that many Zulu customs were "probably of Jewish origin." And like Isaacs' travelogue, the Caldecotts praised the Zulu as handsome, bold, fearless, robust, athletic, cheerful, kind, grateful, honest, and intelligent: in a word, they were noble.[11] At the same time, and again following Isaacs, the Caldecotts cautioned that the Zulu were impulsive, emotional, and cruel, and lacked all notion of (Christian) morality; hence, they were uncivilized. In this supposed intemperance audiences could recognize the need for imperial paternalism. The Caldecotts concluded their program with an exhortation Isaacs would have been proud of: "Go to Natal, the field is open to your industry, the reward is certain; . . . and perchance [you] will have it in your power to assist in the enlightenment of the poor savage, and lend your aid to the civilization of Africa."[12]

But not everyone was inspired by this display of Zulu nobility and British benevolence. The leading voice to heap scorn on the exhibition and to mock Victorian pretensions in Africa was Charles Dickens, who attended a performance ten days after the exhibition opened. Dickens was then completing *Bleak House*, and the character of Mrs. Jellyby—the self-aggrandizing philanthropist who sought to Christianize peoples along the Niger River—was likely on his mind. He soon published a darkly comic, and now deeply offensive, review of the Zulu Exhibition in his periodical *Household Words*. There is no such thing as a noble savage, Dickens declared. Savages were only "cruel, false, thievish, [and] murderous," and they existed in a state little better than that of "wild animal[s]." A savage, Dickens wrote, was "something highly desirable to be civilised off the face of the earth."[13]

The tone of the piece does not improve from there. The Zulu are described as "extremely ugly," and their songs and dances as consisting of monotonous stamping and raving. One of the performers dressed in animal hides looked to Dickens as if he had "come express on his hind legs from the Zoological Gardens." Not only is the spectacle of the exhibition here likened to one of the natural wonders on view at the London Zoo, but the Zulu performer is explicitly denigrated as bestial. Dickens reports to his readers that the Zulu sovereign's "whole life is passed chin deep in a lake of blood" and that the king is in turn butchered "the moment a grey hair appears on his head."[14] The reference to information found in Isaacs' travelogue could hardly be clearer.

Three times in his review Dickens complains of the smell of the Zulu, describing them as "odoriferous to the nose" and noting that the men had "a very strong flavour" and were "high-flavoured." Jews, like Africans, were long accused by white Europeans of having a peculiar identifying odor. Dickens presumably believed that Anglo-Saxon bodies remained odor-free even in summertime as they traveled along London's streets, thick with horse excrement and other effluvia. Of course, those of a certain class would have perfumed themselves with imported Marseille soap, whose pollution by the foul essence of slavery for peanuts would have gone undetected. Dickens further damned the Zulu for their absence of "moral feelings" and considered them "simply diabolical." If cleanliness, as John Wesley put it, is next to godliness, then lapses of hygiene are by implication demonic. In *Oliver Twist*, Dickens described Fagin's backroom lodging as "perfectly black with age and dirt" and treated the slovenly Jew as a stand-in for the devil.[15] More than a decade later, he replaced this infernal alien with the pungent Zulu. By midcentury Jews had become less exotic, and less threatening, than Black visitors from Africa.

Dickens also used his review of the Zulu exhibition to take aim at another group scorned as inferior invaders of Protestant England: Irish

Catholics. Like the Jews, the Irish were often depicted as primitive and bestial. Dickens suggested that the "scenes of savage life" enacted by the Zulu would be "understood at Cork" even without Caldecott's explanations. His misanthropic review is only partially redeemed by a tongue-in-cheek tone. In the antepenultimate paragraph of his review, Dickens winks at the reader to acknowledge that his British contemporaries are not as civilized as they like to think. Still, Dickens' concluding sentiment is chilling: "the world will be all the better" when savages, Zulu or otherwise, disappear from the face of the earth.[16]

Dickens based his invective on information supplied in Caldecott's lecture and program, documents that were in turn based on Isaacs' *Travels and Adventures*, itself a Dickensian tale: it describes a penniless waif whose peregrinations among surrogate family supply one happy coincidence after another to lead the young hero to material good fortune. That both Dickens and Isaacs were men of Kent who had weathered family reversals, endured adversity in adolescence, and then triumphed over poverty thanks to talent and resourcefulness—and in Isaacs' case, exploitation—makes the influence of Isaacs' work on Dickens's review an apt coincidence. Sadly, no documents have come to light suggesting that Isaacs was aware of the Zulu exhibition, the Caldecotts' incorporation of *Travels and Adventures* into their lectures and program, or Dickens's review.

And what did Africans of the time think of British culture? A remarkable account of the Zulu performers' impressions of London offers just such a perspective. The *Natal Journal* published a transcript of a meeting convened by curious elders following the Zulu performers' return to South Africa. The young man questioned about his travels—possibly Manyos—was described as being about twenty-two years old. The report came from an anonymous witness to the meeting said to be fluent in isiZulu, so it was likely mediated by a colonialist whose assumptions about English superiority color the piece.[17]

The speaker described to the assembled Zulu council his initial terror when he scanned the open ocean from the Caldecotts' ship and his seasickness when the vessel pitched in the waves. He and his fellow Zulu were much relieved when they reached "some island in the sea"—possibly St. Helena—before they continued their long voyage. After months aboard ship, they were told they had arrived in London, but all the Zulu could make out was "a cloud of smoke" and "poles standing out of the water, like reeds in a marsh." Soon they identified this odd forest as a collection of ships' masts. The young man described having been awestruck by the size of the city. He could see no end to the vista of buildings "so tall that they shade the streets," the crowds of people day and night, and the innumerable watercraft choking the Thames.[18] Despite the immense population, he reported, there was no scarcity of food. He and his compatriots ate plentifully, enjoying beef, bread, and beer.

The visitor continued by describing his audience with the queen after passing through high walls guarded by soldiers. He thought Victoria "not tall, but good looking . . . like any other English lady in her house." She asked him pointed questions about Zululand, King Shaka, and other "old Zulus, who had died long ago." After their audience, the Zulu were taken to see the "Queen's carriage, fastened and covered with gold" and the "soldier guards, with metal covers to their heads."[19]

The traveler was less impressed by this pomp and circumstance, however, than by the scientific and engineering marvels he had seen. He related to his incredulous audience the span of the bridges he had crossed, the journey he had made aboard a steam train—a "hot-water wagon," and the great factories where "clothes and iron things are made." He witnessed men climbing into the wicker basket of a hot air balloon that looked "like a large calabash, with the mouth downwards" and watched in amazement as the aeronauts ascended into the heavens from which the Zulu believed they originated. He had been fascinated by a visit to the London Zoo, where he saw elephants, hippopotamus, and crocodiles "living in houses." But he was horrified by "doctoring houses" where anatomists cut open the dead to "learn to cure the sick," a sight he and his listeners feared as a kind of witchcraft. He prophetically warned his elders about "the number and power of the English," who if they arrived in Zulu lands would "dig down the mountains and build up the valleys, and we should be like dogs on a flat, howling for their homes."[20] Indeed, within a few decades the eruption of the Anglo-Zulu War would shatter the Zulu social and military structure and destroy the unrivaled sovereignty they had enjoyed since the days of King Shaka.

The king's one-time *induna*, Isaacs, seems to have abandoned any designs he may have harbored to return to territories granted to him in Zulu lands. However, he continued to protect his Matakong holdings throughout the 1850s and 1860s. He traveled back and forth to West Africa unmolested by the Crown but grew concerned that spiteful authorities might pursue claims against his property. Isaacs duly acted to protect Matakong "against the designs of Governor Kennedy" through a scheme to mortgage the property to a French firm. In 1855, probably following the death of his brother Benson, Isaacs installed the ex-missionary Thomas Reader as a full partner on Matakong and established the firm of Mssrs. Isaacs, Reader & Company. Their trade increased to over £100,000 annually.[21] The new partnership was registered in Liverpool, not only because the city was then one of the busiest ports in the world but also because Isaacs' younger sister, Hannah, had relocated there with her husband, the Russian-born merchant Isaac Schwersensky.

Isaacs' brother-in-law was active in Liverpool's Jewish community. By 1856, he had been elected president of the New Hebrew Congregation. He presided over fundraising for a new synagogue and spoke at the laying of its cornerstone in July 1856. London's *Jewish Chronicle* reported on the event, and a plaque listed Nathaniel Isaacs as one of the synagogue trustees. Little more than a year later, the Hope Place Synagogue was consecrated at a ceremony attended by more than one hundred people. Isaacs was present and offered a toast.[22]

It is likely that Isaacs renewed his bonds to the Jewish community less out of conviction than for reasons of circumstance and family ties. Though he was not a man much preoccupied by the past, perhaps grief over his brother's death had propelled him back into the fold. The choice would have come at a cost, since any public identification as a Jew carried risk. Henry Mayhew's 1861 description of London noted that Jews were treated as a "misers, usurers, extortioners, receivers of stolen goods, cheats, brothel-keepers . . . [and] clippers and sweaters of the coin of the realm." Even Benjamin Disraeli, who identified with the aristocracy and wielded political power, suffered from anti-Semitism. In his novel *Tancred* (1847), Disraeli articulates the treatment meted out to Jews of his day: "Born to hereditary insult, without education . . . never treated with kindness, seldom with justice, occupied with the meanest, if not the vilest, toil, . . . an object . . . of prejudice, dislike, disgust, perhaps hatred."[23] Isaacs, always shrewd, surely knew that engagement with the organized Jewish community could have harmed his interests. Perhaps he embraced Judaism in his later years out of sheer defiance, an expression of his combative streak.

Yet the years 1860–63 mark an abrupt cessation of Isaacs' participation in the Jewish community in Liverpool and a gap in his contributions to the Canterbury Hebrew Congregation. He returned to Matakong in 1860 to attempt to salvage his business operations. Left in charge, Thomas Reader had driven Isaacs' firm to ruin. A handful of surviving ledgers from 1856–59 indicate that Reader doled out credit recklessly, though whether this was the cause of the failure is not clear.

In any case, by 1859 Isaacs had removed Reader from his island outpost and forced him back to England to dissolve their partnership. Reader seems to have returned to the vicinity of Sierra Leone sometime later: a trader named Thomas Reader was reported murdered by a "savage horde" led by Kondo, a rebel Temne warrior, during a period of resistance to British encroachment. Isaacs' relative Phillip Lemberg joined him on Matakong in 1861.[24] But Lemberg's tenure on the island was brief, and he soon struck out on his own. He remained in Sierra Leone for half a century, married into a prestigious Krio family, and twice served as mayor of Freetown.

In debt to his French trading partner thanks to Reader's imprudent management, Isaacs struggled to pay off his debts and revive his business. He complained of deception and of having been detained in West Africa beyond the limits of his patience. Isaacs was a captive to his flagging business interests on his own island realm. Perhaps he thought back to his years on St. Helena and to the fate of its most famous resident, Napoleon. If he received the news that Henry Francis Fynn had died in 1861 in Durban, the city founded on the site of Port Natal, Isaacs may have felt himself to be the last of a vanishing breed.[25]

In 1864, the year Isaacs turned fifty-six, he returned to English domesticity, living with his sister and brother-in-law. As his world shrank, age revealed a more introspective man. In a letter to a cousin to whom he had once been close on St. Helena, he lamented: "I have been such a rover, and fortune, or change, led me to such out of the way places that I appear, now, like an outcast from that part of the family who were most dear to me. . . . I am still very happy to have this opportunity of reviving feelings that have for so long lain dormant." He recalled old friends and relations whom he assumed to be dead. One was John Fairbairn, the editor of the *South African Commercial Advertiser*, who had published Isaacs' early impressions of King Shaka and life among the Zulu. "But I must not dwell upon this subject," he wrote. He then observed that years of incessant travel had "blunted my feelings in regard to death."[26]

Perhaps Isaacs did not fear his own demise after confronting so many dangers in far-off locales, but he did begin to ensure his personal legacy at home. When he returned to England around 1864, he brought with him one of his adult daughters, Phoebe Anne, who had been born in Sierra Leone in 1845 to an unknown woman.[27] She had a brother, Benson Hilton, likely born after Isaacs' own brother's death in 1855. Young Benson seems to have remained behind in Freetown, and nothing more is known of him. These children's nameless mother may have been an industrious Krio woman in Freetown or a "housekeeper" brought to Matakong to serve Isaacs' domestic needs and sexual desires. Perhaps she was a member of the Eurafrican elite of the colony, like Isaacs' former paramours Mary Ann Skelton and Hannah Hayes. Such a union would likely have enabled Phoebe to pass as white in Liverpool.

By 1866 the twenty-year-old Phoebe had met and married the Irish-born merchant Peter Manning, who was about ten years her senior. Manning dealt in all sorts of goods: cutlery, tea trays, carpet whisks, guns and gunpowder, soaps and oils, looking glasses, jewelry, music boxes, and writing desks. For a time, the couple lived together with Isaacs, his sister Hannah, and her husband Isaac Schwersensky in Liverpool. Around this time Isaacs may have partnered with his brother-in-law, who is listed as part owner of a ninety-one-ton ironclad sailing vessel suggestively named *Uncle Nat*. The schooner had once been registered to Thomas Reader.[28] Isaacs continued to conduct business in West Africa from this crowded home, though he now sought to wrap up his affairs in Matakong and rent out his distant freehold.

In June 1867, a business associate in Marseille who attempted to broker a lease for Isaacs reported that "Matacong has suffered much" and that intertribal warfare had desolated the island. Nonetheless, Isaacs' correspondent assured him, a lucrative rental agreement was imminent. A few months later Isaacs read an item in Freetown's *African Interpreter and Advocate* that threatened his real-estate deal. The publication reported that Matakong had been conveyed by local chiefs to the British government as early as 1826, well before Isaacs' purchase of the property from the indebted Gabbidon family in 1844. An alarmed Isaacs dispatched a hasty summary of his legal claim on the island to the secretary for the colonies, noting his clear title to Matakong, his long residence there, and the fact that the local king had made him a chief. He further reminded the secretary that his predecessor in the position had definitively pronounced that "'Matacong is no part of British territory.'"[29]

The Freetown newspaper's faulty intelligence did not thwart Isaacs' deal. He concluded negotiations to lease the island to a Manchester-based trading company in 1869.[30] He must have been relieved. His old antagonist, Arthur Edward Kennedy, had returned as governor of Sierra Leone in 1868, after having been knighted by Queen Victoria. Sir Arthur held the governorship until 1872 and was afterward rewarded for his service with more prestigious colonial postings.

FIGURE 15. Remains of Isaacs' pier on Matakong Island. Photo by author.

Relieved of worries about Matakong, Isaacs moved with Phoebe and Peter Manning just across the River Mersey to the town of Egremont, which attracted middle-class merchants seeking respite from the noise and pollution of Liverpool's industrial port. The Mannings' home on 8 Church Street was a few hundred meters from the ferry that steamed across the Mersey. In 1871, Phoebe Anne bore a daughter, Isaacs' first acknowledged grandchild. His last will and testament, updated before his granddaughter's birth, stipulated that Phoebe Anne, her husband Peter Manning, and Walter Lewis—Hannah Hayes's son and Isaacs' mole in the Freetown colonial government—were to equally inherit "the Island of Matacong with the buildings and everything belonging thereto." He divided his remaining estate among his sister Hannah and three of his children: Emily Emma and Alfred Isaacs (his issue by Mary Ann Lightbourn), and the mysterious Benson Hilton Isaacs. Though the value of this estate was less than £1,000, providing for his heirs may have eased his conscience. On 26 June 1872, he died with Phoebe at his side in the house they shared. The cause of death was "softening of the brain," likely a hemorrhage, followed by "convulsions and coma."[31] His uncle Saul Solomon had died in a similar manner.

FIGURE 16. Nathaniel Isaacs' headstone in Canterbury Jewish Cemetery. Credit. Photo by author.

Isaacs outlived nearly all of the kings, colonial administrators, business associates, friends, co-conspirators, and antagonists he had known. Walter Lewis died two years later en route to England from Lagos, perhaps on a voyage to claim his share of Isaacs' estate. Nathaniel Isaacs' island trading post was not sold by his heirs until 1882. Nearly a decade later, an official notice distributed by the colonial government in Free-town proclaimed that "the said Island of Matacong no longer forms a portion of Her Majesty's Dominions but is French Territory."[32] Isaacs' kingdom was formally defunct.

Nathaniel Isaacs was buried in the Canterbury Jewish Cemetery. Though his will did not stipulate his burial place, the fact that his family had his body transported nearly 250 miles for burial, at considerable expense, suggests that it was his explicit wish. Isaacs' plot lies next to the grave of his mother and along the eastern wall of the cemetery. An old sycamore tree with gnarled roots crooks the headstones and shades the graves. Ivy covers his epitaph:

Sacred to the memory of
Nathaniel Isaacs
who departed this life
at Egremont, near Liverpool, June 26th 1872
Aged 65 years
Deeply regretted by his relatives and friends
May his soul rest in peace.[33]

Postscript

Perhaps Isaacs found the peace in death that had eluded him throughout his tumultuous life. He lived his early years in the shadow of the
Napoleonic Wars, served as a mercenary in the Zulu military machine,
dealt in arms in West Africa, and clashed with various official—and offi-
cious—imperial minds in a far-flung colonial bureaucracy. Once he was
elevated by regional kings to positions of power in South Africa and
again on Matakong, he continued to do battle with his enemies, while
exploiting, manipulating, and attacking those who opposed his will.

Yet he was not only a man of the sword; he wielded his pen with
equal effect. Isaacs' *Travels and Adventures* shaped popular views on
the Zulu for decades, and its echoes still reverberate in historiography,
literature, and film. Together with other accounts of African exploration, his travelogue exerted a profound influence on British culture and
the unfolding of imperial history in South Africa. *Travels and Adventures* contributed to an ongoing discourse of "Africanism" that, like the
Orientalism examined by Edward Said, collapses a distinction between
geography and myth to project a Western fantasy—or nightmare—on
other peoples.[1] My own work on Isaacs is inevitably caught in the web
of Africanist assumptions, though I have tried to subvert these assumptions in order to displace the romance of imperialism and complicate
the imperial romance genre.

Isaacs' life charted a course parallel to that of the imperial romance
genre. The naive optimism of Marryat gave way to the assertive

jingoism of Haggard and Kipling, all finally yielding to the world-weary nihilism of Conrad. Of course, Isaacs was never as innocent as Marryat's Peter Simple, as ludicrously virtuous as Haggard's Allan Quatermain, or as murderous as Conrad's Kurtz. Nor, despite this book's title, did Isaacs conclude his career like Kipling's antiheroes in "The Man Who Would Be King." Kipling's vagabond kings die ignobly and brutally, one poisoned by drink and sun, the other beheaded in a frenzy of rough justice.

Kipling and Haggard influenced generations of boys with their tales, and Haggard in particular held outsized importance for colonial South Africa. His first full-length book (1882), a nonfiction account of British policies in Zululand, mixed historical reflections and policy recommendations with firsthand observations of the Zulu. The central aim of that work was an examination of what he called the "Native Question." A subject of considerable public and parliamentary debate in the aftermath of the 1879 Anglo-Zulu War, this question turned on how the British should treat the Indigenous peoples who had come under imperial control. Haggard sought to deal justly with the "troublesome remnant . . . [of the land's] original and natural masters: shattered fragments of the Zulu power in Natal, men who had once swept over the country in the army of Chaka the Terrible, Chaka of the Short Spear." This "remnant" naturally resented British control. And the arch-colonialist Haggard resented the corrupting influence of British culture on the Zulu. He thundered against the "power of civilization" that had "set to work to sap" the Zulu of their nobility, and he blamed this state of affairs on two helpmeets of empire: "the missionary and the trader."[2] For Haggard, the assimilation of the Zulu to European norms drained them of their most admirable qualities, qualities that precisely accorded with Haggard's own notion of masculine British virtues. As in Isaacs' *Travels and Adventures*, white and Black, African and European, ambivalently mirrored one another.

Haggard's phrase "the Native Question" evokes that other nagging question current in Europe in the 1880s: the "Jewish question" (*Judenfrage*). Emerging in public consciousness as early as the 1840s in England, this referred to a complex of concerns, including how Jews might modernize and integrate themselves into European culture, whether Jews should be accepted as full citizens and hold seats in Parliament, and what Christendom was to do with those stiff-necked Jews who could not or would not ingratiate themselves with their host communities. One answer to the Jewish question was Zionism, a program for modern political Jewish reterritorialization. In a state of their own, Jews would be masters of their political destiny.

Zionism's visionary leader, Theodor Herzl, believed that the movement would become a model for a renascent African nationalism and consequent Black reterritorialization. In 1902, Herzl published his utopian novel *Altneuland* (Old New Land), which featured a technocratic vision of a state arising in the biblical land of Israel. One character, a scientist dedicated to eradicating malaria, compares the Jewish question to the "Black question":

> There is still one problem of racial misfortune unsolved. The depths of that problem, in all their horror, only a Jew can fathom. I mean the Negro Question [*Negerfrage*]. . . . Think of the hair-raising horrors of the slave trade. Human beings, because their skins are black, are stolen, carried off, and sold. Their descendants grow up in alien surroundings despised and hated. . . . [N]ow that I have lived to see the restoration of the Jews, I should like to pave the way for the restoration of the Negroes.

To the extent that this passage reveals Herzl's own thinking, he linked Jewish fate to the fate of Africans and saw the two "races" as allies. Not surprisingly, he approved of a plan to create a Jewish polity under British oversight in the East African Protectorate, today's Kenya. Herzl's leading supporter for that project was the most famous Anglo-Jewish author of the day, Israel Zangwill. Herzl described Zangwill as being of "the long-nosed Negroid type, with very woolly deep-black hair"; in other words, Herzl saw his most vigorous Anglo-Jewish ally as somehow Black.[3]

During Zangwill's years of Zionist activity he was friendly with Sir Harry Johnston, the African explorer, author, and colonial administrator. Zangwill likely heard of Nathaniel Isaacs from Johnston, though it might have been from Haggard, with whom Zangwill corresponded and who similarly approved of Jewish territorial self-determination. In any case, Zangwill referenced "a Jew named Nathaniel Isaacs[, who] having fought for a Zulu king, was granted a large territory, with the title 'Chief of Natal.'" For Zangwill, the historical Isaacs functioned as proof that the Jewish people "produce men . . . who can win territories and men who can govern them."[4] Zangwill thus transmuted Isaacs into a symbol of labor and physical prowess that stood in contrast to the British stereotype of Jews as corrupt urban savages. It was Isaacs—not the Zulu—who corresponded to his ideal of the masculine virtues necessary to hold sovereignty. A foe of slavery and the dispossession of Indigenous peoples, Zangwill did not know of Isaacs' unsavory later career.

Isaacs was both hardy and reckless, audacious and greedy, courageous and cruel. Like most men now or then, his sails filled with the winds of self-interest, and he navigated accordingly. Since I began this project, numerous people have asked me some variation of an anxious question: "How Jewish was Isaacs?" The question seems to be a permutation of the Jewish question of earlier centuries. No one wonders "how Protestant" Francis Farewell and Arthur Kennedy were, or "how Catholic" Théodore Canot and Pedro Blanco were. Of course, those who wonder about Isaacs' Jewishness may really be asking to what degree he reflected, maintained, or abandoned so-called Jewish values.

I am skeptical of most claims for the existence of unique Jewish values. More often than not, well-meaning individuals refer not to anything singular that might be identified from the millennia-long development of Jewish religion, philosophy, and ethics, but rather to a blander sense of how twenty-first century diasporic Jewish values are conditioned by the prevailing discourses of their host societies. In other words, those who seek to determine and champion the influence of Jewish values on historical agents merely identify and promote the conservative or liberal—but always foundationally European and Christian—values of the wider culture in which they are embedded. Whatever Jewish values Isaacs retained, they were indistinguishable from the values of his nominally Christian nineteenth-century peers. Paradoxically, that may make him very much like contemporary Jewish readers.

In America and Britain today, Jewish individuals can choose to identify as Jewish (or not) and to affiliate with their ethnoreligious communities (or not). Isaacs did not have that option, at least not among Englishmen. To the extent that he could erase the stamp of his "race," he did so in Africa among peoples who did not care, did not fully comprehend his religious otherness, or were already rendered subordinate by the emergent pseudoscientific hierarchies of race. So how Jewish was Isaacs? My unsatisfying answer is that he was more Jewish than he wanted to be in England, and exactly as Jewish as he wanted to be in Africa—which is to say, hardly at all. Still, his affiliation with two synagogues, his insistence on marking his brother's grave with a Hebrew headstone, and his own burial in Canterbury's Jewish Cemetery attest to some degree of Jewish identification no matter where he sojourned.

No one I met in Canterbury, on St. Helena, across lands that once constituted the Zulu Kingdom, along the riotous streets of Freetown, or on Matakong Island's shores had heard of Isaacs. The only testament to his existence in Canterbury is his ivy-smothered headstone. In James-

town, his uncle Saul Solomon lacks even a grave, though his stone ceno-
taph rests in the conference room of the Solomon & Company headquarters. A side street in Durban's center bears Isaacs' name, but there
is no other memorial to him, despite the once-flourishing Jewish community of that city. Not a trace of Isaacs remains in Freetown. Weather-
worn Krio houses, the stone wharves where liberated Africans once
landed, and a few public buildings are all that survive from his period
of residency there.

But on Matakong, the signs of Isaacs' tenure endure. The ruins of his
stone wharf jut out into the shallow harbor. Crabs scuttle around the
massive square-cut stone blocks lapped by ocean currents. Tumbledown
storefronts and homes line what was once the busy dockside. Uphill
from the sentinel kapok trees stands a squat, one-story stone and con-
crete building. The exterior walls are painted a handsome blue that
would blend with the color of the skies overhead but for the corrugated
rusty metal roof that breaks the plane. This site, affording the best van-
tage of the coast, would have been the center of Isaacs' compound. A
rough-hewn stone foundation suggests that this might be exactly where
his home once stood. Thick, crumbling walls, likely the remains of
storehouses and outbuildings, edge the red dirt expanse nearby. A few
hundred meters distant lies Benson Isaacs' forlorn jungle grave.

Isaacs' descendants—the distant progeny of Eurafrican lovers and
unnamed African women—may yet be out there somewhere in Matakong, in Sierra Leone, in Zululand, in England, or far beyond. I hope
that, like Isaacs, they make the most of their freedom to choose who
they wish to become, and I hope that, unlike their forebear, they do so
without violating the freedom of others.

Acknowledgments

Thanks are owed to Nick Arvin, Dr. Milo Gough, Dalia Rosenfeld, Matt Rovner, and Agur Schiff for reading and commenting on my manuscript; Andrew Stuart of the Stuart Agency for taking on this project and University of California Press editors Eric Schmidt and Jyoti Arvey for believing in it and shepherding it through production; Erika Bűky for expert copyediting and more; Dr. Daniel McIntosh, Dr. Ingrid Tague, and the Office of the Dean of the College of Arts, Humanities, and Social Sciences at the University of Denver for research support; my colleagues in the Department of English and Literary Arts, and the Center for Judaic Studies—especially to Ingrid Weyher and Selena Naumoff for putting up with my deadlines, disappearances, and devilishly difficult expense reports; Dr. Michael Jolles for his pioneering work on Isaacs and for access to private materials; Dr. Judith Suissa, Hanan Suissa, and family for hosting me during multiple trips to London, and especially to Lia Suissa for being convinced to do stupid things with me in Zululand; Natalie Gordon for hosting me in Johannesburg; Dr. Adam Mendelsohn and family for allowing me to crash their Hannukah party in Cape Town; Mandy Peters of Solomon & Company in Jamestown for assistance and access; Alan Barnett for helping me find a legal way into the old Canterbury Jewish Cemetery; Billy Rovner for this book's title; Albert Moore and Mustapha Kallon at Fourah Bay College for archival assistance; Femi Boyle-Hebron and family for hospitality and conversation in Freetown; Anne Marie Perry and her team from Dore Education and Culture Exchange (Dore, Bato, Ibrahim, Jules, and Annette) for getting me to and from Matakong; Matakong's mayor, Ibrahim Kamara, vice mayor, Fantamadi Kamara, and the people of Matakong (especially my fast friend Bingo!) who welcomed me, took me out on a pirogue, and have preserved Benson Isaacs' grave for more than a century; to journalist Mamadouba Bangoura for organizing a rescue party to extract me from a brief captivity on the island;

Dr. Mbongeni Z. Malaba, Dr. Gavriel Rosenfeld, Dr. Jean Rossman, Dr. Tuvia Shlonsky, Dr. Adam Sutcliffe, Dr. Nadia Valman, and Dr. Dror Wahrman for their conversation and recommendations; Dr. Susan Grundy and Xavier Ricou for documents and references; archivists and librarians at all of the various repositories I visited—thank you for your assistance and labor; participants of the 2021 Atlantic Jewish Worlds Conference at the McNeil Center for Early American Studies and Herbert Katz Center at the University of Pennsylvania for feedback. Closer to home: Geraldine Toste for asking good questions, Abe Brennan for map illustrations, William Untiedt for literary domination, Mike Kollins and Josh Sohn for always being interested. Extra special thanks: Jessica Cohen for her patience during my travels to remote islands and for her unfailing editorial eye, and to Talia for listening to endless stories about nineteenth-century sea voyages, Napoleon, and King Shaka Zulu.

Notes

PROLOGUE

1. Based on deposition of Bangah, 5 July 1854; deposition of Barracah, 19 October, 1854; deposition of Langoh, 8 December 1854, all from TNA, CO 267/243.

INTRODUCTION

1. Wainaina, "How to Write about Africa."
2. Arendt, *The Origins of Totalitarianism*, 205.

CHAPTER 1. JEWS AND OTHER SAVAGES

1. Jolles, *Isaacs*, 255, 274–79; Lenie Isaacs genealogy in Jolles, *Isaacs*, 274–77. Jewish socioeconomics: Endelman, *The Jews of Georgian England*, 172. Jews and Royal Navy: Green, *The Royal Navy*, 28–29. Descriptions of Canterbury: Hasted, *The History of the Ancient and Metropolitical City*, 99; Gostling, *A Walk in and about the City*, 1, 64; King, *Handbook for Travellers*, xiv–xv.

2. Natte Solomon: Anderson, *A Practical Essay*, 34. See also Roth, "The Membership of the Great Synagogue," 178. Descriptions of Canterbury: Hasted, *The History of the Ancient and Metropolitical City*, 101; Gostling, *A Walk in and about the City*, 8–9.

3. Descriptions of Canterbury: Hasted, *The History of the Ancient and Metropolitical City*, 102, 127; Gostling, *A Walk in and about the City*, 74. Georgian attitudes: Endelman, *The Jews of Georgian England*, 87; Felsenstein, *Anti-Semitic Stereotypes*, 153–54. Quotation: Chaucer, *The Canterbury Tales*, 841.

4. Jews as savages: Penslar, *Shylock's Children*, 38–39; Eilberg-Schwartz, *The Savage in Judaism*, 37–39. Quotation: Penslar, *Shylock's Children*, 41. Quotation: *Encyclopaedia Britannica*, 4th ed. (1810), q.v. "savage," 511.

5. Kuper, *The Reinvention of Primitive Society*, 11.

6. Pepys, *Diary*, 14 October 1663, 46.

7. Endelman, *The Jews of Georgian England*, 45, 276. Cruikshank caricature: British Museum, www.britishmuseum.org/collection/object/P_1868-0808-7097.

8. Names: *The Navy List*, 135–37. Naval language: Mendelsohn, *The Rag Trade*, 93.

9. Green, *The Royal Navy and Anglo-Jewry*, 7.

10. Quotations: *Tales of the Wars*, 199–200. The same anecdote appears in Green, *The Royal Navy and Anglo Jewry*, 33–34, and derives from *The Naval and Military Sketchbook* (1845). The events reportedly occurred aboard the *Royal William* sometime after 1782.

11. Marryat, *Peter Simple*, 63–64.

12. Parascandola, "Introduction," xviii.

13. Further details: Schwarzfuchs, *Napoleon, the Jews and the Sanhedrin*. Quotations: Reid, *Sandhedrin Hadasha*, 6; D'Israeli, "On the late Installation of a Grand Sanhedrim," 38.

14. Bicheno, *The Restoration of the Jews*, 59; Schwarzfuchs, *Napoleon, the Jews and the Sanhedrin*, 194.

15. Quotation: Smith, *An Inquiry into the Nature and Causes of the Wealth of Nations*, 199. Jewish accommodationism: Endelman, *The Jews of Georgian England*, 3–11.

16. Governor A. E. Kennedy to Sir G. Grey, 24 August 1854, TNA CO 267/241.

17. Evangelical belief: Hilton, *A Mad, Bad, and Dangerous People?*, 175–77; Darwin, *Unfinished Empire*, 280.

18. Maclean, "The Loss of the Brig *Mary*" (hereafter "Loss"), June 1853, 298.

19. Isaacs notes that he was fourteen in 1822. His partially illegible tombstone indicates that he was either sixty-three or sixty-five when he died on 26 June 1872. If he was born in the second half of 1808, the dates of his birth and death are consistent. For dates: Jolles, *Isaacs*, 277.

20. Endelman, *The Jews of Georgian England*, 166.

21. Apprenticeship: Endelman, *The Jews of Georgian England*, 188–89. Quotations: Isaacs, *Travels*, 1: xiv–xv.

22. Parascandola, "Introduction," xiii.

23. Quotations: Isaacs, *Travels*, 1:xv. Motto: Barrow, *The East India Company*, 41. Observer quoted in Philips, *The East India Company*, 2.

24. Details of Solomon's birth: Hearl, "Saul Solomon of St. Helena," 165. Service as solider: Register of Baptisms 1680–1807, 73, SHGA. Details of route and voyage: Barrow, *The East India Company*, 12; Royle, *The Company's Island*, 19. Shipping frequency: St. Helena and Dependencies Statistical Year Book 1989, section 9.6, SHGA, notes that 197 ships arrived in 1823.

25. Baptismal records: Register of Births, n.p., SHGA. Sunday sabbath observance: St. Helena Records 34, 18 February–30 December 1822. vol. 123. n.p.; St. Helena Records 34, 8 December 1817–22 October 1818, 119:324,

SHGA. East India Company organization of St. Helena life: Royle, *The Company's Island*, 40.

26. Quotation: "Napoleon at St. Helena," 536. Pasta reference: O' Meara, *Napoleon in Exile*, 2:68. Lodging expenses: Jackson, *Notes and Reminiscences of a Staff Officer*, 126. I have used the online currency converter tool of the National Archives (UK) to provide rough estimations of value throughout: www.nationalarchives.gov.uk/currency-converter.

27. Lowe's career: Unwin, *Terrible Exile*, 88. British source: "Letters . . . Received by the Mary." Napoleon's view of Lowe: Las Cases, *Memoirs of the Life*, 2:58.

28. Number of soldiers: Unwin, *Terrible Exile*, 60. Census details: St. Helena Records 34, 8 December 1817–22 October 1818, 119:433, SHGA. Napoleon's dinners and diversions: Unwin, *Terrible Exile*, 85, 118, 74–75, 140–41.

29. Lowe's censorship: Unwin, *Terrible Exile*, 119. Quotation from proclamation: St. Helena Records 34, 8 December 1817–22 October 1818, 119:380, SHGA. Records of escaped slaves: East India Company Letter Book H (1822–25), SHGA. Complaint to governor, 5 August 1824, East India Company Letter Book [no series] (1825–26), 223, SHGA; record of slave escaped to Ascension Island, 26 September 1825, East India Company Letter Book [no series] (1825–26), 157–58, SHGA. See also IOR/G/32/143, 11 January 1823–30 September 1824, EIC. Notice of escaped slave: East India Company Letter Book H, 571–73.

30. Clandestine communications: "Napoleon at St. Helena, Part 3," 539. Lowe on Solomon name quoted in St. M. Watson, *A Polish Exile with Napoleon*, 70. See Grose, *Classical Dictionary of the Vulgar Tongue*, q.v. "Jew Bail," "Reader Merchant," "Sweating."

31. Postans, "The Emperor's Grave," 215, 217.

32. Escape plot: Solomon, *Saul Solomon*, 4. See also Solomon File/Folder E, SHGA; Hearl, "Saul Solomon of St. Helena 1776–1852," 4; Solomon, *The Solomons*, 2. Threat and response: S. Solomon to H. Lowe, n.d., Add MS 20233, 17 January 1816–24 March 1829, f.151–52, EIC.

33. Quotations on clandestine correspondence and missing letters: entry 211, East India Company Letters from England 17, 8 March 1820–18 December 1822, SHGA. See also 28 July 1821, IOR/G/32/142, 11 November 1820–16 December 1822, 318–19, EIC.

34. Quotation: IOR/B/174, 19 March 1822, 530, EIC. Relationship between L. Isaacs and N. Isaacs, possibly his paternal uncle: Jolles, *Isaacs*, 3–4, 274–75. Levy as slopseller: IOR/Z/O/1/9, no. 3319, EIC.

35. Details from this paragraph and next inspired by *Life on Board a Man of War*, 56–70; Nordhoff, *Sailor Life;* Volo and Volo, *Daily Life in the Age of Sail.*

36. Details of shipbound life: Pappalardo, *How to Survive in the Georgian Navy*, 44–46, 53, 99. Quotation: Isaacs, *Travels*, 1:xv.

37. Quotations: Isaacs, *Travels*, 1:xv–xvi. Euphemistic language of complaints: Burg, *Boys at Sea*, 124.

38. Burg, *Boys at Sea*, 128–29.

39. Quotation: Isaacs, *Travels*, 1:xvi. On wharf: St. Helena Calendar and Directory for 1830, 44, SHGA.

40. EIC quote on Solomon: 11 November 1822, St Helena Records 34, 18 February–30 December 1822, 123:399, SHGA. Quotations: Advertisement in *South African Commercial Advertiser* (hereafter *SA Commercial Advertiser*). 19 September 1832; Jolles, *Isaacs*, 225.

41. Darwin, *Unfinished Empire*, 155–56, 170, 180–81.

42. Appointment of Solomon's son: 11 November 1822, St Helena Records 34, 18 February–30 December 1822, 123:399, SHGA; see also 16 December 1822, IOR/G/32/142, 11 November 1820–16 December 1822, 778, EIC. Return of Phoebe: IOR/B/172, 4 October 1820, 37, 29 November 1820, 214; n.d., 227, EIC. Baptism of Nathaniel Lee Solomon: Register of Births, 5 June 1822, SHGA.

43. See Mendelsohn, *The Rag Trade*, 115.

44. Portrait by Walter Barshai in Solomon & Co. headquarters. Solomon's role as consul, Isaac as apprentice, and quotation: Isaacs, *Travels*, 1:xv, xvii.

45. Quotations: Isaacs, *Travels*, 1:xvii. Isaacs' cousins: Letter from A.M. [Amelia Marks], *Jewish Chronicle*, 3 May 1895, reproduced in Jolles, *Isaacs*, 65–66. The boarding school was run by Rabbi H. N. Solomon, who founded the Jewish Free School.

46. *Cape Town Gazette and African Advertiser*, 5 August 1825, KCC, file 5, extracts 7, 252. King's claim of serving ten years: King to H. Bathurst [3rd Earl Bathurst, secretary of state for war and colonies], 10 July 1824, in Leverton, *Records*, 1:40–42. King's later claim of serving twelve years: King to H. Bathurst, 7 November 1824, in Theal, *Records of the Cape Colony*, 19:105–6. King and Farewell had a habit of inflating numbers: see "Return of the Gaols, & c.—Robben Island," 1828, TNA, CO 53/65. King's command of the *Salisbury:* Leverton, "James Saunders King," 18.

47. Quotations: Isaacs, *Travels*, 1:xviii; "From the Journal of Mr. N. Isaacs," *SA Commercial Advertiser*, 9 June 1832, KCC, Stuart Papers, file 3, extracts 2, 126, Isaacs.

48. Sailing timeline and cargo: *Cape Town Gazette and African Advertiser*, 5 August 1825, KCC, file 5, extracts 7, 252. Census figures: Thompson, *Travels and Adventures in Southern Africa*, 489; estimate of the slave population in TNA CO 714/36, 2 February 1831. British South Africa as a slave colony: Pringle, "On the Demoralizing Influence of Slavery," 161–74.

49. Request for wallpaper ordered by Lord Somerset: TNA, CO 714/36, 24 May 1826.

50. King quoted in Isaacs, *Travels*, 1:xix–xx. Resupply of Farewell: Lieutenant Edward Hawes of *York* reported that Farewell and eight others were "living on the best of terms of friendship with the natives and under the protection of the king . . . Chaka" at Port Natal. See E. Hawes to Commodore Moorsom, 16 May 1825, in Leverton, *Records*, 1:49–50. Farewell and party were known to be in want of supplies but otherwise fine: *Cape Town Gazette and African Advertiser*, 4 June 1825. Farewell and party again reported to be in want of supplies: *Cape Town Gazette and African Advertiser*, 30 September 1825.

51. Farewell's injuries: F. G. Farewell to Governor Somerset, 6 September 1824, in Leverton, *Records*, 1:37–38. See also James, *Naval History of Britain*, 5:520. For marriage (17 August 1822): "Marriages," *Cape Town Gazette and*

African Advertiser, 24 August 1822, KCC, Stuart Papers, file 18. See also Morris, *Washing of the Spears*, 73. Morris is a controversial source and has been used with extreme caution. For dowry: notarial deed, 21 June 1823, in Leverton, *Records*, 1: 8–9. Farewell's partners were William Dickson, James Nisbet, and John Robert Thomson: notarial deed, 21 June 1823, in Leverton, *Records*, 1:9–11. I have not ascertained whether this is the same Dickson who later partnered with Saul Solomon. Ship registration: McLean, "Loss," February 1853, 79. Tonnage of ship: notarial deed, 21 June 1823, in Leverton, *Records*, 1:9–11.

52. Owen, *Narrative of Voyages*, 1:253–54.

53. Description of *Salisbury*: notarial deed, 21 June 1823, in Leverton, *Records*, 1:9–11. Msimbithi's background: Fynn, *Diary*, 180–81. Msimibithi's rearrest and transport: Isaacs, *Travels*, 2:252. See also Maclean, "Loss," June 1853, 299; Fynn, *Diary*, 181. Quotation: "Return of the Gaols, & c.—Robben Island," 1828, TNA, CO 53/65.

54. King and Msimbithi: Isaacs, *Travels*, 2:252. Msimbithi as interpreter: Maclean, "Loss," June 1853, 299. Msimbithi joining King and Farewell: Isaacs, *Travels*, 2:253. Owen on Msimbithi: Owen, *Narrative of Voyages*, 1:82.

55. Cargo: notarial deeds, 4 December 1823 and 29 December 1823, in Leverton, *Records*, 1:19–21, 1:24–25. Reports that three men were drowned: notarial deed, 4 December 1823. Farewell later claimed that four were drowned: F. G. Farewell to Governor Somerset, 1 May 1824, in Leverton, *Records*, 1:35–36. King claimed that six were drowned: GHA, 1/39 1824, and J. King to H. Bathurst, 10 July 1824, in Leverton, *Records*, 1:40–42. *Salisbury* cutting anchor cable: notarial deed, 19 December 1823, in Leverton, *Records*, 1:21–23. Quotations on native peoples: GHA, 1/39 1824; King to Bathurst, 10 July 1824 reproduced in Leverton, *Records*, 1:40–42. Other details from Isaacs, *Travels*, 2:253. See also King's report on Farewell's near drowning: Bannister, *Humane Policy*, appendix 1, lv; Mackeurtan, *Cradle Days of Natal*, 100; notarial deed, 19 December 1823, in Leverton, *Records*, 1:21–23; King to Bathurst, 10 July 1824 in Leverton, *Records*, 1:40–42. Regarding King's connection to 3rd Earl Bathurst via Lord James Townshend, Maclean notes that Townshend, a naval officer and MP, was "King's friend and patron": "Loss," February 1853, 79. Isaacs records the survivors of the *Mary* naming a spot at Port Natal for Townshend: *Travels*. 1:53. See also King, "Chart of Port Natal," KCC, Stuart Papers, 20, where a "Townsend Dock Yard" appears.

56. *Salisbury*'s navigation: notarial deed, 4 December 1823, in Leverton, *Records*, 1:19–21. For Salisbury Island, see map by W. T. Haddon (1835), TNA, CO 700/NATAL1. Quoted descriptions: see map by N. Isaacs (ca. 1828), TNA, CO 700/South Africa 3, and map by Haddon, TNA, CO 700/NATAL1. Further quotations: J. King to H. Bathurst, 10 July 1824, GHA 1/39 1824; King, "Chart of Port Natal," KCC, Stuart Papers, 20.

57. Formation of company: notarial deed, 17 April 1824, in Leverton, *Records*, 1:28–29. Wylie identifies Peterssen as Farewell's father-in-law in *Myth of Iron*, 299. Warning against claiming territory: P. G. Brink (governor's secretary) to F. Farewell, 5 May 1824, in Leverton, *Records*, 1:36. Settlement party: F. Farewell to Governor Somerset, 1 May 1824, in Leverton, *Records*, 1:35–36.

58. Khoekhoe as Jewish: Parfitt, *Hybrid Hate*, 164. Quote from Gray, *The Natal Papers of "John Ross,"* 168–69.

59. Disgruntled settlers: J. Hoffman and J. Peterssen to Commander Moorsom, 9 March 1825, in Leverton, *Records*, 1:48. Mackeurtan claims the *Julia* burned at sea in December in *Cradle Days of Natal*, 108, though a contemporary report places the event in September: J. Hoffman and J. Peterssen to Commander Moorsom, 9 March 1825, in Leverton, *Records*, 1:48. Another source also mentions December but does not mention a fire: E. Hawes to Commander Moorsom, 16 May 1825, in Leverton, *Records*, 1:49–50. Lost cargo: notarial deed, 7 June 1825, in Leverton, *Records*, 1:52–54. Remains of the *Julia*: *Cape Town Gazette and African Advertiser*, 4 June 1825.

60. Details of letter and quotations: J. King to H. Bathurst, 10 July 1824, GHA, 1/39 1824. King's efforts to access the Admiralty: Owen, *Narrative of Voyages*, 1:254. See also Maclean, "Loss," February 1853, 79. McLean takes the side of King.

61. King's supposed dismissal: Owen, *Narrative of Voyages*, 1:254. Defense of King: Maclean, "Loss," February 1853, 79. Fear over Farewell's fate: King quoted in Maclean, "Loss," January 1853, 33.

62. King's intention: King to Governor Somerset, 9 August 1825, in Leverton, *Records*, 1:67–68. Ammunition request: notarial deed, 22 August 1825, in Leverton, *Records*, 1:73–74. Timeline of arrival: Isaacs, *Travels*, 1:4. Quotation: Isaacs, *Travels*, 1:5.

63. Isaacs, *Travels*, 1: 5–7.

64. Descriptions from Isaacs, *Travels*, 1:8–9; King quoted in Isaacs, *Travels*, 1:14–16.

65. King quoted in Isaacs, *Travels*, 1:12–13. King's account of the wreck is corroborated in the *SA Commercial Advertiser*, 4 January 1826. Quotations: Isaacs, *Travels*, 1:10.

66. Isaacs, *Travels*, 1: 20–23.

67. Dibdin, *The Jew and the Doctor*, 27. Performance of play announced in *Cape Town Gazette and African Advertiser*, 28 April 1826.

68. Fynn's father: Mackeurtan, *Cradle Days of Natal*, 101. On Fynn as criminal: Wylie, *Myth of Iron*, 261.

69. Isaacs' description of animals: *Travels*. 1:24. For another version see Maclean, "Loss," January 1853, 35. Rachel's behavior: Maclean, "Loss," February 1853, 75. Quotations on food and shelter: Isaacs, *Travels*, 1:25.

CHAPTER 2. STRANGE SURPRISING ADVENTURES

1. Descriptions: Isaacs, *Travels*, 1:25–26, 29, 33. Quotation: Cowper, "Verses, Supposed to Be Written by Alexander Selkirk, during His Solitary Abode in the Island of Juan Fernandez" (1782). Literary quotes and allusions scattered throughout Isaacs' work are likely the additions of an editor. Isaacs was certainly literate and capable of lively written expression, but he was no aesthete or scholar.

2. Maclean, "Loss," January 1853, 29.

3. Maclean's "Loss" was published serially from January 1853 to March 1855 in the *Nautical Magazine and Naval Chronicle*. The first entry makes it clear that Maclean began drafting his recollections no later than 1840, since he alludes to the "present Chief Dingan" (Dingane), who was overthrown in 1840: "Loss," January 1853, 29. The existence of Isaacs' diary is attested to by Isaacs' cousin, Saul Isaac, and the sister of Isaacs' apprentice, Ben Moss: letters in *Jewish Chronicle*, 3 May and 24 May 1895, cited in Jolles, *Isaacs*, 66–67. Perhaps a copy of Isaacs' diary survives, as Saul Isaac claimed to have had a copy in his possession.

4. Hamilton, "'The Character and Objects of Chaka,'" 39, 41. Golan's *Inventing Shaka* discusses competing nationalist interests. The secondary literature on Shaka and his reputation is voluminous: see Eldredge's *Creation of the Zulu Kingdom*, Hamilton's *Terrific Majesty*, Wylie's *Myth of Iron*, and J. Wright's contributions to the collected volumes of the *James Stuart Archive of Recorded Oral Evidence Relating to the History of the Zulu and Neighbouring Peoples* [hereafter JSA]. Stuart's informants had competing political and personal interests that shaped their presentation of events. Overview of key issues in Shakan history: Wright, "Reconstituting Shaka Zulu for the Twenty-First Century."

5. Hamilton and Wright, "Moving beyond Ethnic Framing," 670.

6. Eldredge's *Creation of the Zulu Kingdom* treats Isaacs' account as generally accurate. Hamilton's *Terrific Majesty* is prefaced with a quote from Isaacs' *Travels and Adventures* but goes on to censure Isaacs for his vilification of Shaka. Nonetheless Hamilton asserts that *Travels and Adventures* remains one of the most valuable sources for Shaka history (177–78). In *Myth of Iron*, Wylie judges Isaacs to have been a bigot, though he often relies on Isaacs, if only to criticize him. Wylie also claims—falsely and without evidence—that Isaacs was functionally illiterate. Wylie illogically concludes that Isaacs did not write much or most of *Travels and Adventures* and then condemns him for his written account of Shaka. Either Isaacs wrote and revised the vast majority of *Travels and Adventures* and is therefore guilty of certain biases, or he did not, in which case he cannot be guilty of the biases Wylie finds.

7. Isaacs' desolation: *Travels*, 1:32. Recovered materials: Isaacs, *Travels*, 1:28; Maclean, "Loss," March 1853, 140. On disgust: Maclean, "Loss," February 1853, 77–78.

8. Isaacs on nudity: *Travels*, 1:29. Farewell on nudity: Extract, from *Cape Town Gazette and African Advertiser*, 6 January 1826, in Leverton, *Records*, 1:79–80.

9. Drunkenness and desertion: Maclean, "Loss," March 1853, 141; Isaacs, *Travels*, 1:81. Shipbuilding: Maclean, "Loss," March 1853, 141. Tools: Isaacs, *Travels*, 1:32; King, quoted in Isaacs, *Travels*, 1:57. Maclean claims the carpenter's tools were rescued: "Loss," March 1853, 141.

10. Isaacs, *Travels*, 1:35, 1:51–52.

11. Zulu scrutiny: Isaacs, *Travels*, 1:30. On "simianization," see McClintock, *Imperial Leather*, 216; Parfitt, *Hybrid Hate*, 27–28, 160; Gilman, *Jewish Self-Hatred*. 8. Hogarth image in the collection of the Metropolitan Museum, New York, www.metmuseum.org/art/collection/search/401162.

12. Pritchard, *Researches*, 185–86. See also Gilman, *The Jew's Body*, 100–101, 173–74; Wahrman, *The Making of the Modern Self*, 117.

13. Isaacs transliterates the word as "silguaner." Maclean reports that the Zulu referred to the sailors as *isilwane* ("animals" or "beasts"), which may have sounded similar to Isaacs: "Loss," July 1853, 79. Zulu opprobrium: Isaacs, *Travels*, 1:40. Zulu views of Europeans: Maclean, "Loss," July 1853, 349. On hair color: interview with Makewu, 8 October 1899, KCC, Stuart Papers, file 6, MS 1059, For epithets: Mbovu ka Mtshumayeli, 7 February 1904, JSA 3:26; Tununu ka Nonjiya, 1 June 1903, JSA 6:281. Maclean records that whites were called "wild beasts: "Loss," June 1853, 303.

14. Isaacs, *Travels*, 1:41.

15. Enslopee: Isaacs, *Travels*, 1:41. King on the Zulu: "Sketch of Lieut. Farewell's Settlement at Port Natal—Concluded." Maclean on the Zulu: "Loss," July 1853, 350.

16. Significance and symbology of homestead organization: Kuper, "The 'House' and Zulu Political Structure." Descriptions of journey: Isaacs, *Travels*, 1:57, 68.

17. Isaacs, *Travels*, 1:63–64.

18. For timeline: Isaacs, *Travels*, 1:65. Farewell's ivory: "New Colony at Natal." Ivory imports: Walker, *Ivory's Ghosts*, 83–106. Political liberalism and free trade were inextricably linked in the period after the Napoleonic Wars: Bourne, *Foreign Policy*, 4.

19. Isaacs' task: *Travels*, 1:65. Social status: Hamilton and Wright, "Moving beyond Ethnic Framing," 666. On *amalala*: Wright, "Making Identities," 197. White control: Maclean, "Loss," April 1853, 197. Isaacs' views on Zulu women: *Travels*, 1:66–67, 69, 77.

20. Isaacs, *Travels*, 1:70.

21. Farewell's report to Governor Charles Somerset, 6 September 1824, in Bird, *Annals of Natal*, 191. Quotation of Zulu elder: Mnguni, 28 March 1903, KCC, Stuart Papers, file 30, item 2, 7. Governor's warning: P. G. Brink to F. Farewell, 5 May 1824, in Leverton, *Records*, 1:36. Farewell's flag raising: F. Farewell to Governor Somerset, 6 September 1824, in Leverton, *Records*, 1:37–38.

22. Text of grant from 8 August 1824 in Bird, *Annals of Natal*, 193–94. Grant reproduced in Leverton, *Records*, 1:38–40. On preliterate marking in the Native American context, see Lyons, *X-Marks*, 1–3.

23. Grant as a swindle: Wylie, *Myth of Iron*, 314–21. Shaka's expansionism: Okoye, "Tshaka and the British Traders, 1824–1828," 21–22. Emerging nature of Zulu identity and rule: Hamilton, "'The Character and Objects of Chaka,'" 59; Hamilton and Wright, "Moving beyond Ethnic Framing," 666; Hamilton and Leibhammer, "Tribing and Untribing the Archive," 25. Importance of trade to Shaka: Hamilton, "'The Character and Objects of Chaka,'"45, and Mahoney, *The Other Zulus*, 31.

24. Shaka as empire builder: Mbovu ka Mtshumayeli, 25 September 1904, JSA 3:44. Zulu amalgamation of conquered groups: Eldredge, *The Creation of the Zulu Kingdom*, 6–7; Mahoney, *The Other Zulus*, 31–32. Eyewitness account of this policy: Maclean, "Loss," April 1853, 201. Zulu perception of the grant: Ndukwana, quoted in Wylie, *Myth of Iron*, 232–33. On the differences between

Zulu and European understanding, see also Wylie, *Myth of Iron*, 319–20; Guy, "The Tribal History Project," 231. On Shaka's possible understanding of agreements: Okoye, "Tshaka and the British Traders, 1824–1828," 18, 25.

25. Isaacs, *Travels*, 1:72.

26. Isaacs' account of Shaka: Isaacs [?], "Letter from Graham's Town," emphasis in original. The language used in this piece suggests that Isaacs either wrote it or served as the main informant. King's description: Isaacs, *Travels*, 1:61. Farewell thought Shaka was no older than thirty-eight: see Owen, *Narrative*, 1:389.

27. Illustration in Isaacs, *Travels*, vol. 1, facing p. 57. Quotation: Isaacs, *Travels*, 1:61. Bagg was well known as an anatomical draftsman. His sketches had been praised by Charles Darwin: C. Darwin to J. Murray, 21 October 1861, in Burkhardt, *Correspondence of Charles Darwin*, 9:310.

28. Zulu elders on Shaka's appearance: Maquza ka Gawushane, 2 February 1905, JSA 2:232; Ngidi ka Mcikaziswa, 14 August 1904, JSA 5:40; Madikane ka Mlomowetole, 27 June 1905, JSA 2:60; Melapi ka Magaye, 27 April 1905, JSA 3:72; Mayinga ka Mbekuzana, 8 July 1905, JSA 2:248.

29. Recollections of those who knew Isaacs: "In the Days of Chaka," *Natal Mercury*, 7 July 1927, quoted in Jolles, *Isaacs*, 280; A. E. Kennedy to secretary of state, 24 August 1854, Governor's Dispatches to Secretary of State, 24 July 1854–17 September 1855, letter 144, SLNA.

30. Shaka beholding his reflection: Maclean, "Loss," June 1853, 302. Shaka as ugly: Singcofela ka Mtshungu, 29 March 1910, JSA 5:339. Shaka's fascination with Europeans: Melapi ka Magaye, 27 April 1905, JSA 3:73. Maclean's hair: "Loss," July 1853, 73.

31. Isaacs, *Travels*, 1:73; Maclean, "Loss," July 1853, 354 Fynn, *Diary*, 290–91.

32. Isaacs, *Travels*, 1:75, 1:250.

33. Executions in England: Gatrell, *The Hanging Tree*. Punishments in the Royal Navy: Malcomson, *Order and Disorder*. Quotation: MacLean, "Loss," March 1855, 130.

34. Maclean, "Loss," July 1853, 350; Maclean, "Loss," April 1853, 197.

35. Isaacs' motivations and timeline: *Travels*, 1:85, 1:108. Gifts: King, "Lieut. Farewell's Settlement at Port Natal"; Maclean, "Loss," July 1853, 350.

36. Biographical details: Wright, "Henry Francis Fynn," 15. Fynn as thief: Wylie, *Myth of Iron*, 261. Isaacs' description of Fynn: *Travels*, 1:39. Image of Fynn in KCC, album D61/001-125, Personalities and Groups, BRN 310917. Fynn's near nudity: Isaacs, *Travels*, 1:39–40. Adoption of African dress: "New Colony at Natal." Isaacs' regard for Fynn: *Travels*, 1:48, 63.

37. Isaacs, *Travels*, 1:100–101.

38. Ambivalence and strategies of emotional management: Reddy, "Against Constructionism."

39. Isaacs, *Travels*, 1:105, 1:107–9; Kipling, "The White Man's Burden," 78.

40. Shaka's disdain: Isaacs, *Travels*, 1:109–12. On Brown Bess: Ferguson, "'Trusty Bess,'" 50–54. Surprise: Isaacs, *Travels*, 1:113. Hlambamanzi's name: Maclean, "Loss," June 1853, 299–300; Wylie, *Myth of Iron*, 304–5.

41. Isaacs, *Travels*, 1:114, 1:110–11.

42. Wylie is the primary critic of Isaacs' depiction: see *Myth of Iron*, esp. 362–69.

43. Isaacs, *Travels*, 2:218, 2:200, 2:214, 1:146, 1:148.

44. Isaacs, *Travels*, 1:119.

45. Isaacs, *Travels*, 1:161–62, 108, 128.

46. Isaacs, *Travels*, 1:152–53.

47. Isaacs, *Travels*, 1:153, 164, 142, 163.

48. Origin of Farewell and King's enmity and quotation: notarial deed [signed by Hutton and Isaacs], 24 May 1828, document 33, annexure 1 in Leverton, *Records*, 2:37–38. Disagreement over gifts: Isaacs, *Travels*, 1:154, 165. See also notarial deed [signed by Hutton and Isaacs], 24 May 1828. Isaacs' journey: *Travels*, 1:165. Shaka's disappointment with feathers: Isaacs, *Travels*, 1:155. Details on headdress: Melapi ka Magaye, 27 April 1905, JSA 3:72.

49. Quotation: F. Farewell to W. B. Dundas, 10 September 1828, GHA 1/39, 438. Isaacs, *Travels*, 1:165, 177, 172.

50. Isaacs, *Travels*, 1:176, 245, 184.

51. See Cesarani, *Port Jews*; Dubin, *The Port Jews of Habsburg Trieste*.

52. See Cohen, *Global Diasporas*, 177–91; Kopytoff, "The Internal African Frontier," 22.

53. Isaacs, *Travels*, 1:184.

54. Wright, "Making Identities," 197.

55. Isaacs, *Travels*, 1:188; Deposition of John Cane, 13 November 1828, GHA 1/39, 376.

56. Isaacs, *Travels*, 1:188; Maclean, "Loss," February 1855, 66–67.

57. Quotations: Isaacs, *Travels*, 1:189–90. Nasopongo's intercession: King, "Private Correspondence," 2 May 1827, 5.

58. Isaacs, *Travels*, 1:192.

59. Shaka ordering whites to fight: Isaacs, *Travels*, 1:193. Quotations: King, "Private Correspondencl," 2 May 1827, 5.

60. Isaacs, *Travels*, 1:193–94.

61. Isaacs, *Travels*, 1:195. Possibly Isaacs thought he saw the Zulu chewing crocodile dung when in fact they were chewing the scales of crocodiles. See Fynn's recollection, in *Diary*, 309.

62. Isaacs, *Travels*, 1:199–201.

63. Isaacs, *Travels*, 1:201–3.

64. Isaacs, *Travels*, 1:203–6.

65. Isaacs, *Travels*, 1:206–7.

66. Quotation: Isaacs, *Travels*, 2:287. Quotation from King: "Sketch of Lieut. Farewell's Settlement at Port Natal—Concluded." Quotation from Maclean: "Loss," August 1853, 430. Women fleeing to the Europeans: King, "Sketch of Lieut. Farewell's Settlement at Port Natal—Concluded"; Isaacs, *Travels*, 2:290–91. Treatment of women: Isaacs, *Travels*, 2:288–89.

67. Social structure: Kuper, "The 'House' and Zulu Political Structure," 482–83. Zulu informants: Mnguni, 28 March 1903, KCC, Stuart Papers, file 30, item 2, 7; Maziyana ka Mahlabeni, 20 April 1905, JSA 2:267.

68. Kay, *Travels and Researches in Caffraria*, 401.

69. Missionary regulation: Darwin, *Unfinished Empire*, 285. Livingstone quoted in Groves, *The Planting of Christianity*, 176.

70. Remarks on theology: see Hilton. *A Mad, Bad, and Dangerous People?*, 183–85. Quotations: Cooke, "Ancient Faith and Modern Skepticism," March 1866, 145–46.

71. Quotation from Maclean: "Loss," January 1854, 23. Quotations from Isaacs: *Travels*, 1:214–15. Cf. Holden, *History of the Colony of Natal*, 44.

72. Kirby, "Unpublished Documents," 67–68.

73. Isaacs, "Port Natal (from the journal of Mr N Isaacs)," 6 June 1832, n.p., recorded in KCC, Stuart Papers, file 3, extract no. 2, 125.

74. Robert Garden quoted in Jolles, *Isaacs*, "Postscript 11."

75. Maclean, Loss," August 1853, 431–32.

76. Isaacs, *Travels*, 1:210. Isaacs' praise name: Maziyana ka Mahlabeni, 20 April 1905, JSA 2:267, 303n17.

77. Isaacs, *Travels*, 1:211. In this volume Isaacs several times represents Shaka laughing or smiling from a sense of superiority (71, 111, 114, 120), in the midst of atrocity (161, 336), and in dissimulation (115–16, 211, 248).

78. Isaacs' hunting activity: *Travels*, 1:220, 222, 234–35. Shaka quoted in Fynn, *Diary*, 141. Fynn's explanation to Shaka: *Diary*, 142n1; see also KCC, Fynn Family folder 3, diary 1. Fynn on Shaka's obsession: *Diary*, 143.

79. Isaacs on Shaka's invitation: *Travels*, 1:231. Shaka quoted: Isaacs, *Travels*, 1:233. Shaka's determination to send a Zulu emissary to the British appears as extract from *Cape Town Gazette and African Advertiser*, 6 January 1826, in Leverton, *Records*, 1:79–80. Shaka's views on King George: Isaacs, *Travels*, 1:256. Quotations and Shaka's demand for secrecy: Isaacs, *Travels*, 1:233.

80. Quotations: Isaacs, *Travels*, 1:295–96. Fynn on Shaka's motives and gift of ivory: *Diary*, 143.

81. Isaacs, *Travels*, 1:296; Rowland, *An Essay on the Cultivation and Improvement of Human Hair*, 29–30; Glenk, "Macassar Oil," 89.

CHAPTER 3. BLACK NAPOLEON

1. Descriptions drawn from Rose, *Four Years in Southern Africa*. 216–25; Gordon-Cumming, *A Hunter's Life in South Africa*, 319.

2. Shaka's trade goals: Fynn, *Diary*, 131n3. Lists of items desired by Shaka: "Statement of Monagali," 8 October 1828, document 16, annexure 2, in Leverton, *Records*, 2:20. Visit to kwaBulawayo and Nandi: Fynn, *Diary*, 132.

3. Shaka's birth: Isaacs, *Travels*, 1:321–22. Fynn records a different version: *Diary*, 139–40. On Nandi's ailment, Isaacs records this as "Chekery, or dysentery": *Travels*, 1:322. Fynn records this similarly: *Diary*, 12. Eldredge on Shaka's name: *Creation of the Zulu Kingdom*, 49. A Zulu source claims that Shaka derives from a praise name, "He Who Beats but Is Not Beaten": Ngidi kaMcikaziswa, 21 October 1905, JSA, 5:78. Nandi and Shaka's exile: Isaacs, *Travels*, 1:353. Cf. Maclean's version: "Loss," January 1854, 25. For competing histories, see Eldredge, *Creation of the Zulu Kingdom*, 49–50, and Wylie, *Myth of Iron*, 85–96.

4. Shaka and Ngomane: Wylie, *Myth of Iron*, 129–30. Eldredge and Fynn date Shaka's ascension to 1816: Eldredge, *Creation of the Zulu Kingdom*, 76; Fynn, *Diary*, 13–14. Hamilton places it sometime in the 1810s: *Terrific Majesty*, xi. Isaacs suggests that Shaka became king no earlier than 1818: "Immense Field for Emigration," memorandum from Isaacs to G. L. Cole. 1832[?], KCC. Based on interviews with Zulu informants, Stuart dates Shaka's ascension to 1814: "Notes for Tshaka's Life," July 1905, Stuart Archives, file 53, 10. Wylie believes Shaka took control in 1812: *Myth of Iron*, 149. Quotation: Isaacs, *Travels*, 1:159–60.

5. Quotations: Fynn, *Diary*, 133n4. Some informants claim Shaka killed his mother: see Eldredge, *Creation of the Zulu Kingdom*, 190–97. Wylie insists these reports are fantasies: *Myth of Iron*, 410–17. None of the European eye-witnesses who acknowledge Shaka's brutality—Isaacs, Fynn, and Maclean—claim that he killed Nandi. Nandi's name: Fynn, *Diary*, 136n4. Journey to condole Shaka: Isaacs, *Travels*, 1:237–40.

6. Darwin, writing in 1832, quoted in Kuper, *The Reinvention of Primitive Society*, 33. On civilization versus savagery, see Kuper, *The Reinvention of Primitive Society*, 29–31. English depictions of Africans: Brantlinger, "Victorians and Africans," 170.

7. Conrad, *Heart of Darkness*, 61–62.

8. Census figures: "Returns of Population from 1825 to 1829," 2 February 1831, Governor's Correspondence, TNA, CO 714/36. Quotations: *SA Commercial Advertiser*, 27 December 1828, 10 December 1828. Christian humanitarianism in South Africa: Frye, "*The South African Commercial Advertiser* and the Eastern Frontier," 2.

9. Fairbairn and Philips: Bank, "Losing Faith in the Civilizing Mission," 367. Buxton, Philips, and contemporary humanitarianism: Lester, *Imperial Networks*, 43, 24.

10. Quotation: Isaacs, *Travels*, 2:285. Representations of Jews: Gilman, *The Jew's Body*, 64. Valman, "Muscular Jews," traces feminization and passivity in English literature.

11. Quotation: Smith, *An Inquiry into the Nature and Causes of the Wealth of Nations*, 298. Quotation: Unsigned editorial notice, *SA Commercial Advertiser*, 31 December 1828. Quotation: Philips, *Researches in South Africa*, 317.

12. Fynn's message and report on violence: Isaacs, *Travels*, 1:242, 1:244. Mourning displays: Fynn, *Diary*, 134. Zulu reports of excesses: Eldredge, *Creation of the Zulu Kingdom*, 192–95.

13. Fynn, *Diary*, 136.

14. Isaacs, *Travels*, 1:245, 246–47.

15. Isaacs, *Travels*, 1:248.

16. Shaka's interest in Europeans and King George: Maclean, "Loss," June 1853, 300. Shaka's proposed visit to King George and diplomatic mission: Isaacs, *Travels*, 1:248, 256. According to a sworn deposition, King initiated the idea of a Zulu mission to the Cape: "Re-examination of John Cane," 11 November 1828, document 30 in Leverton, *Records*, 2:32. Shaka's desire for British recognition: Maclean, "Loss," June 1853, 300–301.

17. Maclean, "Loss," June 1853, 300.

18. Quotations: Isaacs, *Travels*, 1:248, 250. King and ivory: F. Farewell to W. B. Dundas, 10 September 1828, document 12, annexure 1 in Leverton, *Records*, 2:10. Shaka's preference for King: Isaacs, *Travels*, 1:256.

19. J. Bell to Governor Somerset, 5 February 1828, document 94 in Leverton, *Records*, 2:97.

20. Quotation: Smith. *An Inquiry into the Nature and Causes of the Wealth of Nations*, 172. Shaka's character: Isaacs, *Travels*, 1:191, 246. Shaka as Napoleon: Maclean, "Loss," June 1853, 298.

21. "Tshaka: His Life and Reign," 18 August 1905, KCC, James Stuart Papers, 52/4, 30.

22. Quotation: Haggard, *Nada the Lily*, ix. Haggard notes the influence of Fynney's "Zululand and the Zulus" and Bird's *Annals of Natal* in his preface to the novel (x–xi). Fynney had been a government interpreter among the Zulu. Bird had been a magistrate and government treasurer in Natal.

23. Haggard, *King Solomon's Mines*, iii.

24. Quotation: Haggard, *Nada the Lily*, ix–x. On inversions: Chrisman, *Rereading the Imperial Romance*, 119.

25. Cane testified that Zulu warriors were "not allowed to throw their assegaays on pain of death": "Statement of John Cane," 13 November 1828, document 31 in Leverton, *Records*, 2:35. Wylie on Shaka's innovation or lack thereof: *Myth of Iron*, 214–19.

26. Isaacs on use of spies: *Travels*, 1:249. Quotation: F. Farewell to W. B. Dundas, 10 September 1828, document 12, annexure 1 in Leverton, *Records*, 2:12. Shaka's tactics: Isaacs, *Travels*, 1:249. Zulu soldiers: Isaacs, *Travels*, 2:95, 278–79, 285.

27. "Zoolacritical" government: Isaacs, *Travels*, 2:295. Shaka's troops: Isaacs, *Travels*, 2:278. Celibacy: John Cane stated under oath that the three thousand warriors with Shaka at kwaDukuza were all unmarried: "Statement of John Cane," 13 November 1828, document 31 in Leverton, *Records*, 2:34. Shaka's women: Isaacs, *Travels*, 1:120. Women as tribute: Hamilton, "Restructuring within the Zulu Royal House," 99.

28. Shaka's militarization: Eldredge, *Creation of the Zulu Kingdom*, 76–87. Quotation: Isaacs, *Travels*, 1:325. Shaka's shaping of features of the Zulu Kingdom: Eldredge, *Creation of the Zulu Kingdom*, esp. 231–51; Mahoney, *The Other Zulus*, 21–46.

29. On Bellow's question: Foster, *Transnational Tolstoy*, 142–46; Bellow, "Papuans and Zulus."

30. Msomi, "The Historic Similarity," 73–74.

31. Quotations: Kinene, *Emperor Shaka the Great*, 289, 392, 407.

32. Timeline and dysentery: Isaacs, *Travels*, 1:252–53. One possible reason why Isaacs does not mention this February meeting is that it never occurred, and the resulting document from February 1828 is a forgery put forward by James King, as Wylie believes: *Myth of Iron*, 467–69. I see no reason to assume the grant to be forged, since by this time Shaka considered Farewell, the previous grantee, to be an "old woman," and King was now in his favor. Even Cane, who was hostile to King and in Farewell's employ, admitted under oath that at least part of the February 1828 document "might have been explained to

Chaka": "Statement of John Cane," 13 November 1828, document 31 in Leverton, *Records*, 2:34. Quotation: "At Chaka's Principal Residence . . . February 1828," 473, GHA, 1/39. I have used the generally accepted spelling "Sotobe" for consistency. The original reads "Sotoby," while Bird's version reads "Sotobi." The version of this document in Bird's *Annals of Natal*, 94, differs slightly from the copy of the original in GHA.

33. Shaka's dispatch of diplomats: Isaacs, *Travels*, 1:258. Zulu homestead: Kuper, "The 'House' and Zulu Political Structure."

34. This refutes Wylie's suggestion that in May 1828 James King invented news of a potential attack by Shaka to serve his own purposes: *Myth of Iron*, 455. Notice of Shaka's westward expansion: "Governor's Correspondence," 15 October 1827, TNA, CO 714/36.

35. Quotations: "Private Correspondence at King Chaka's Kraal," 3 January 1828.

36. Buthelezi, "We Need New Names Too," 587; Hamilton and Wright, "Moving beyond Ethnic Framing," 667; Hamilton and Leibhammer, "Tribing and Untribing the Archive," 40.

37. Shaka's grant to King: "At Chaka's Principal Residence," 473, GHA, 1/39. See also Bird. *Annals of Natal*. 94; Cane's testimony, 13 November 1828, reprinted as document 31 in Leverton, *Records*, 2:34.

38. Schooner's names: "Statement of John Cane," 10 November 1828, document 28 in Leverton, *Records*, 2:29. The ship is also referred to as *Chaka*: see D. Campbell to J. Bell, 19 December 1828, document 53 in Leverton, *Records*, 2:52; Holden, *History of the Colony of Natal*. 50; Ingram, *The Story of an African Seaport*. 16. Ingram's account of the vessel's being named *Chaka* is based on information relayed by Fynn. Mackeurtan suggests the origin of the name *Elizabeth and Susan* in *Cradle Days of Natal*, 140. According to the author, genealogist, and presumed descendant Dr. Susan Grundy, King's mother was named Susannah O'Grady. Schooner's voyage: "Memorial of James Saunders King," J. King to R. Bourke, 6 June 1828, GHA, 1/39. Crew and passenger list: R. Bourke to Comm. Skipsey, 13 July 1828, in Leverton, *Records*, 1:218–19.

39. By 1849, the total population of Port Elizabeth was estimated at four thousand: Lucas, *A Historical Geography*, 4:243–44. On clothes: Isaacs, *Travels*, 1:261.

40. Quotations: D.P. Francis to Bell, 9 May 1828, in Leverton, *Records*, 1:154; Isaacs, *Travels*, 1:263.

41. Placards and response: Samuel Hudson, quoted in McKenzie, *Imperial Underworld*, 231. For defamation: Somerset, "Proclamation by His Excellency," 203.

42. Reports on Shaka: "Cape Town (Items)," *Cape Town Gazette and African Advertiser*, 4 June 1825, 28 April 1826, 15 June 1826; King, "Lieut. Farewell's Settlement at Port Natal," 11 July 1826, 2; King, "Sketch of Lieut. Farewell's Settlement at Port Natal—Concluded," 18 July 1826, 2; King, "Private Correspondence at King Chaka's Kraal," 3 January 1828. Bourke's instructions quoted in in J. Bell [secretary to governor] to J.W. van der Riet, 15 May 1828, in Leverton, *Records*, 1:158–59. On the government's supposed largesse: Isaacs, *Travels*, 1:262. Governor's warning to King: Bell to J.W. van der Riet, 15 May

1828; Isaacs, *Travels*, 1:262. King's insistence on escorting diplomats and waiting: J. King to J. W. van der Riet, 24 May 1828, in Leverton, *Records*, 1:163–64.

43. Cloete's appointment: J. Bell to A. J. Cloete, 14 June 1828, in Leverton, *Records*, 1:172–73. Cloete's duel: Du Preez and Dronfield, *Dr. James Barry*, 154–59. Cloete's father, Pieter Laurens Cloete, held at least forty-four slaves at the time slavery was outlawed in the Cape Colony. He is listed in the University College of London database of slave compensation claims (no. 5067, www.ucl. ac.uk/lbs/claim/view/2120016516) as having filed for nearly £1,700. A "P. L. Cloete" at the Cape submitted a claim (no. 3879, www.ucl.ac.uk/lbs/claim /view/2120015330) a month later in excess of £1,300 for the loss of forty-three slaves. Freed slaves themselves were not compensated for their years of toil and suffering—only their owners. King George quoted in Cole, *Reminiscences of My Life*, 32.

44. Contempt towards traders: Robinson and Gallagher, *Africa and the Victorians*, 20. Quotation from Philips: Lester, *Imperial Networks*, 39. Interrogation: A. J. Cloete to Lt. Col. Bell, 27 June 1828, in Leverton, *Records*, 1:185–87.

45. Description of Sotobe: Melapi ka Magaye. 29 April 1905, JSA 3:81; Mayinga ka Mbekuzana. 8 July 1905, JSA 2:247. Feathers: Isaacs, *Travels*, 1:264. Description of ambassadors: Isaacs[?], "Letter from Graham's Town," 19 August 1828. Cloete on Zulu: A. J. Cloete to Lt. Col. Bell, 4 July 1828 and 29 July 1828, in Leverton, *Records*, 1:191–93, 1:244–45.

46. Interrogation: Isaacs, *Travels*, 1:263–65. Cloete's efforts to divide King from the Zulu: A. J. Cloete to Lt. Col. Bell, 27 June 1828, in Leverton, *Records*, 1:185–87; Isaacs, *Travels*, 1:267. Maclean judged the colonial government's treatment of the mission "a shameful failure": "Loss," June 1853, 300.

47. Farewell's letter: F. Farewell to Tonkin, 25 April 1828, TNA, CO 49/22. Given the timing, the letter was likely sent with someone aboard the *Elizabeth and Susan* to be sent onward to London. Competing territorial grants: Isaacs, *Travels*, 1:188. Quotation: Farewell to Tonkin, 25 April 1828.

48. The *Elizabeth and Susan* arrived 4 May 1828: D. P. Francis to Lt. Col. Bell, 9 May 1828, in Leverton, *Records*, 1:154. King's grant and protest: J. King to A. J. Cloete, 29 July 1828, in Leverton, *Records*, 1:246–48. Cloete's rejection of King's documents: A. J. Cloete to Lt. Col. Bell, 29 July 1828, in Leverton, *Records*, 1:244–45; Lt. Col. Bell to Aitchison, 28 November 1828, KCC, Stuart Papers, file 5, extracts 7, 312.

49. Border concerns: Maclean, "Loss," June 1853, 300. Fear of refugees: Hamilton, "The Character and Objects of Shaka," 55. Estimations of troop strength: Shrewsbury to Lt. Col. Somerset, 12 June 1828, in Leverton, *Records*, 1:173–74. Fynn claimed Shaka's forces numbered between thirty and forty thousand "armed men": H. Fynn to Lt. Col. Somerset, 9 September 1828, document 78, annexure 1, enclosure 1 in Leverton, *Records*, 2:75. Quotations: W. B. Dundas to R. Bourke, 20 June 1828, GHA, 1/39, 88. Shaka's military offensive: Fynn, *Diary*, 143; Eldredge, *Creation of the Zulu Kingdom*, 259–60.

50. H. F. Fynn to Lt. Col. Somerset, 9 September 1828, document 78, annexure 1, enclosure 1, in Leverton, *Records*, 2:75. See also Fynn, *Diary*, 144–45.

51. Quotations: Shrewsbury, "Extract of a Letter from Mr. Shrewsbury, dated June 30, 1828," 203, emphasis in original; W. J. Shrewsbury to Lt. Col.

Somerset, 2 July 1828, in Leverton, *Records*, 1:203–4. Wylie highlights Fynn's culpability: *Myth of Iron*, 460–61. See references to Fynn by these names: Sijewana ka Mjanyelwa, 15 November 1899, JSA 5:332; Tununu ka Nonjiya. 1 June 1903, JSA 6:281. On the iziNkumbi: Ndongeni kaXoki, August 1905, JSA 4:243, which makes clear that Fynn's followers were known as the iziNkumbi; *Report on Proceedings*. 424; Wylie, *Myth of Iron*, 370; Fynn, *Diary*, 148.

52. Quotation: W.B. Dundas to Governor Bourke, 15 August 1828, reproduced in Msebenzi, *A History of Matiwane and the Amangwane Tribe*, 243. Bourke's sense of Fynn: Governor Bourke to W. Huskisson. 26 August 1828, in Leverton, *Records*, 1:268–69. Later accounts of authorities convinced of Fynn's involvement: J. Bell to Lt. Col. Somerset, 5 February 1829, document 94 in Leverton, *Records*, 2:94; Endorsement 2 by J. Bell to letter from F. Farewell to J. Bell, 28 February 1829. document 104 in Leverton, *Records*, 2:102. See also endorsement noting that Fynn "used his fire-arms": F. Farewell to W.B. Dundas, 10 September 1828, document 12, addendum 1 in Leverton, *Records*, 2:14. Governor Cole belatedly acknowledging Fynn's subterfuge: J. Bell to H. Somerset, 5 February 1829, in Leverton, *Records*, 2:97. Later damning reports of Fynn's "fire-arms" appear in J. Bell's endorsements to F. Farewell to J. Bell, 19 February 1829, document 98, in Leverton, *Records*, 2:107.

53. Quotations from Sotobe: *Travels*, 1:266–68; A.J. Cloete to Lt. Col. Bell, 11 July 1828, in Leverton, *Records*, 1:209–11; J. King to A.J. Cloete, n.d. [18 July 1828?], Leverton, *Records*, 1:229–30; Isaacs, *Travels*, 1:269.

54. Cloete and Sotobe quoted in Isaacs, *Travels*, 1:269. Isaacs' account corroborates orders Cloete received that instructed him "to acquaint them [the Zulu] that Mr King had no authority to bring them here": J. Bell to A.J. Cloete, 14 June 1828, in Leverton, *Records*, 1:172–73.

55. Bribe and Sotobe's response: Isaacs, *Travels*, 1:269–70. Cloete's fears: A.J. Cloete to Lt. Col. Bell, 27 June 1828, in Leverton, *Records*, 1:185–87. Bourke's orders: Lt. Col. Bell to A.J. Cloete, 4 July 1828, in Leverton, *Records*, 1:194.

56. Diplomats quoted in A.J. Cloete to Lt. Col. Bell, 27 June 1828, in Leverton, *Records*, 1:185–87. Complaints of diplomats: Isaacs[?], "Letter from Graham's Town," 19 August 1828. Diplomats' resentment: A.J. Cloete to Lt. Col. Bell, 27 June 1828, in Leverton, *Records*, 1:185–87. On Sotobe and threats: A.J. Cloete to Lt. Col. Bell, 4 July 1828, GHA, 1/30, 120. Isaacs' sympathy for diplomats: *Travels*, 1:271.

57. Isaacs on Zulu: *Travels*, 2:292. Fynn on Shaka: *Diary*, 152.

58. Assessment of diplomats' intentions: Gov. Bourke to W. Huskisson. 1 August 1828, in Leverton, *Records*, 1:251. Escape attempt: A.J. Cloete to Lt. Col. Bell, 18 July 1828, in Leverton, *Records*, 1:227–29. Cloete on King: A.J. Cloete to Lt. Col. Bell, 27 June 1828, in Leverton, *Records*, 1:185–87. Cloete's dismissal of King: A.J. Cloete to J. King, 30 July 1828, in Leverton, *Records*, 1:248.

59. Ivory for medicine: A.J. Cloete to Lt. Col. Bell, 25 July 1828, in Leverton, *Records*, 1:237–39. Bourke's beliefs: Lt. Col. Bell to A.J. Cloete, 18 July 1828, in Leverton, *Records*, 1:232–33. Assessment of Farewell and King: document 92 in Leverton, *Records*, 2:87.

60. Quotations: Bourke to Secretary Huskisson, 29 June 1828, KCC, Stuart Papers. file 5, extracts 7, 284–85.

61. Quotation: Governor's Correspondence, 29 June 1828, TNA, CO 714/36. Injuries: Hart, *New Annual Army List for 1849*, 259, 277. For more on web of connections, see Lester, *Imperial Networks*. 13.

62. Hamilton notes that published reports on Shaka prior to 1828 did not emphasize his brutality. She indicates that reports highlighting Shaka's power may have filtered into colonial consciousness from the Zulu (or other peoples): "The Character and Objects of Shaka," 40–41, 46. King's published report ("Private Correspondence at King Chaka's Kraal") was instrumental in depicting Shaka as a threat. Maclean seems to have been the first to make a published comparison between Shaka and Napoleon. Quote from Haggard: "An Incident of African History," 588. See also Mackeurtan, *Cradle Days of Natal*, 122. Others who picked up on this phrase or variants (e.g., "African Napoleon") include James Stuart: "Tshaka: His Life and Reign," lecture 18, August 1905, KCC, Stuart Papers, file 52, 30.

63. Gifts: J. King to A.J. Cloete. 31 July 1828, GHA, 1/39, n.p. Hlambamanzi's clothes: J. King to J.W. van der Riet and Account of Expenditure, 1 July 1828, in Leverton, *Records*, 1:200–201. Quotation: military dispatch entries for 1 August, 5 August, and 10 August 1828, TNA, CO 51/13.

64. Timeline: Lt. Col. Somerset to R. Bourke, 1 August 1828; W.B. Dundas to Lt. Col. Somerset, 1 August 1828, both in TNA, CO 51/13. Dundas' report on battle: Dundas to Somerset, 1 August 1828. This battle is also described by a missionary: Shaw, "Extract of a Letter," 130. See also the appendix by N.J. Van Warmelo in Msebenzi, *A History of Matiwane and the Amangwane Tribe*, 236. Contemporaneous account of Kei River: Rose, *Four Years in South Africa*, 171. Dundas on Zulu troops: W.B. Dundas to Lt. Col. Bell, 5 August 1828, TNA, CO 51/13; Dundas reports on attack in Minutes of Council, 11 August 1828, TNA, CO 51/12. Bourke celebrates attack: Governor's Correspondence, 26 August 1828, TNA, CO 714/36. Bourke's orders: Governor's Correspondence, 3 August 1828, TNA, CO 714/36. See also Isaacs, *Travels*, 1:281.

65. Bourke on Farewell and King: Governor's Correspondence, 26 August 1828, TNA, CO 714/36. Interrogation quoted in in A.J. Cloete to Lt. Col. Bell, 25 July 1828, in Leverton, *Records*, 1:237–39. Bourke on King and Natal settlers: R. Bourke to Huskisson, 26 August 1828, KCC, Stuart Papers, file 5, extracts 7, 289. Dundas' supposed failure: W.B. Dundas to Lt. Col. Bell, 5 August 1828, TNA, CO 51/13.

66. Shaka's orders quoted in Isaacs to G.L. Cole [governor], 19 December 1828, document 58 in Leverton, *Records*, 2:57. Isaacs uses the same language in *Travels*, 1:277–78. Corroboration of Fynn's actions: W.B. Dundas to R. Bourke, 15 August 1828, in Msebenzi, *A History of Matiwane and the Amangwane Tribe*, 243, which indicates Fynn had been in the area as late as 8 July 1828; see also Fynn, *Diary*, 144. Missionary witnesses include Shrewsbury, "Extract of a Letter from Mr. Shrewsbury, dated June 30, 1828," 203; Shaw, "Extract of a Letter," 130; Shrewsbury, "Extract of a Letter from Mr. Shrewsbury, dated Butterworth, Sept. 30, 1828," 269. For identity of invaders: Eldredge, *Creation of the Zulu Kingdom*, 280–81; Wylie, *Myth of Iron*, 406–

97, 442–43. Isaacs notes they belonged to "Maduane" (Matiwane), the amaN-gwane chief: *Travels*, 1:278.

67. Field report quoted in J. S. van Wyck to Lt. Col. Somerset, 21 August 1828, in Leverton, *Records*, 1:261–62. Junior officer report: W. D. Warden to H. Somerset, 7 September 1828, TNA, CO 51/13. Quotation from Somerset: Lt. Col. Somerset to Gov. Bourke, 11 September 1828, TNA, CO 51/13. Somerset later claimed again that he had battled the Zulu in this skirmish: D. Campbell and Lt. Col. Somerset to Acting Secretary, 25 July 1834, document 205 in Leverton, *Records*, 2:270. Quotations: Bannister, *Humane Policy*, 155–56, 158, 160. See also Elbourne, "The Bannisters," 61.

68. Shaka's military movements: Isaacs, *Travels*, 1:278–79. Wylie on Fynn: *Myth of Iron*, 466.

69. Description of gifts: "Statement of John Cane," 10 November 1828, document 28 in Leverton, *Records*, 2:29. Isaacs' report on drowning: *Travels*, 1:276. Episode confirmed in "Statement of John Cane," 10 November 1828, document 28 in Leverton, *Records*, 2:29.

70. Isaacs, *Travels*, 1:281.

71. Quotations and descriptions: Isaacs, *Travels*, 1:286. Cat episode: Maquza kaGawushane, 2 February 1905, JSA 2:235. The informant does not identify Isaacs as the speaker, but it seems plausible given Isaacs' several mentions of mice nibbling at his feet.

72. Shaka's criticism of Fynn and suspicions of poisoning: Isaacs, *Travels*, 1:288. Hlambamanzi on King: "From a Correspondent at Port Elizabeth," 31 December 1828, 1.

73. Quotations: Isaacs, *Travels*, 1:288–89. On mirrors: Carpenter, "The Tribal Terror," 482–83.

74. Isaacs, *Travels*, 1:290.

75. Medicine chest episode: Isaacs, *Travels*, 1:290–92. Fynn on Shaka's rage: *Diary*, 155.

76. Quotation: GHA, 1/39, n.d., 366. This scrap is reprinted as "Message from Chaka," document 16, annexure 3, in Leverton, *Records*, 2:20. A version of this document dated 10 September 1828, showing Shaka's signature, is reproduced in Msebenzi, *A History of Matiwane and the Amangwane Tribe*, plate 2. Bullock horn: "Statement of Managarda." 10 November 1828, in Leverton, *Records*, 2:27. Quotation from Monagali: "Statement of Monagali," 8 October 1828, document 16, annexure 2 in Leverton, *Records*, 2:20. On Shaka's desire for Macassar oil, see also D. Campbell [probably] to J. Bell, 10 October 1828, document 15 in Leverton, *Records*, 2:17. Cane quotation: "Statement of John Cane," 10 November 1828, document 28 in Leverton, *Records*, 2:30.

77. Shaka's threats: H. Fynn to Somerset, 9 September 1828, document 78, annexure 1, enclosure 1 in Leverton, *Records*, 2:75; Fynn, "Graham's Town—January 16th"; Isaacs, *Travels*, 1:293. Hlambamanzi's claims: Isaacs, *Travels*, 1:296; Fynn to Somerset, 9 September 1828. Treatment and diagnosis of King: Isaacs, *Travels*, 1:280, 298. Medicines administered: Fynn, "Captain King's Illness."

78. King quoted in Isaacs, *Travels*, 1:299–300. King's death throes: Fynn, "Captain King's Illness." King's burial: Isaacs, *Travels*, 1:305.

79. Shaka's suspicion: Isaacs, *Travels*, 1:309–10. Shaka's grief: Isaacs to G. L. Cole [governor], 19 December 1828, document 58 in Leverton, *Records*, 2:58. Quotation from Shaka: Isaacs, *Travels*, 1:310–11.

80. On "kingship": Lemarchand, "Introduction," 5–7. Extent of grant: Isaacs, *Travels*, 1:312. Grant reproduced in "The Colonization of South Africa," 9. There is reason to suspect this grant was forged, since Isaacs does not recall it in his letter to Governor Cole: Isaacs to G. L. Cole, 19 December 1828, document 58 in Leverton, *Records*, 2:58. Shaka's annoyance at Hlambamanzi's signature: Isaacs, *Travels*, 1:312.

81. Isaacs, *Travels*, 1:313, 316.

82. Assassination: Isaacs, *Travels*, 1:314, emphasis in original. Mbopha's appearance: Magidigidi kaNobebe, 9 May 1905, JSA 2:93. Mbopha as trusted: Maclean, "Loss," January 1854, 26. Shaka's stabbing: H. Fynn to Somerset, 28 November 1828, document 78, annexure 1 in Leverton, *Records*, 2:74. Shaka's death speech: Dinya kaZokozwayo, 7 February 1905, JSA 1:96.

83. Quotation: Isaacs, *Travels*, 1:315–16. Cause of assassination: Magidigidi kaNobebe, 9 May 1905, JSA 2:96; Sijawane, n.d., KCC, Stuart Papers, file 6, KCM 23464, MS 1058, 44, emphasis in original. Zulu informant on Shaka's gray hair: Makewu, 8 October 1899, KCC, Stuart Papers, file 6, KCM 23464, MS 1059, 4.

CHAPTER 4. APPETITE FOR CONSUMPTION

1. Fynn, *Diary*, 157–58.

2. Isaacs, *Travels*, 1:353.

3. Cane's interrogation: "Statement of John Cane," 11 November 1828, document 28 [dated 10 November 1828], in Leverton, *Records*, 2:28. Officials' quotation: Despatch, G. L. Cole [governor] to G. Murray [secretary of war and colonies]. 31 January 1829, document 92 in Leverton, *Records*, 2:87. The precise location of the territory noted by Cane is unknown. Cane calls the western boundary the Kiangarooboo River, which may be the Kandandhlovu. If this is correct, Shaka's grant would have provided a much greater expanse of coastline than previously granted to Farewell, King, or Isaacs: report from John Cane, 8 October 1828, GHA, 1/39[?], 360. Cane later reported that Dingane sought to grant land for white settlement from St. John's River (the Mzimvubu) to the mouth of the Tugela River, about 250 miles: Bell to Cole, 11 January 1831, *Papers Relative to the Condition and Treatment*, 58. Shaka's concerns: D. Campbell to J. Bell [from reporting by Cloete], 10 October 1828, document 15 in Leverton, *Records*, 2:16.

4. On seal and ring: deposition of John Cane, 11 November 1828, GHA, 1/39, 429. The *SA Commercial Advertiser*, 29 November 1828, reported that "an official seal" had been sent to Shaka. Cane on Macassar oil: D. Campbell to J. Bell [from reporting by Cloete], 10 October 1828, document 15 in Leverton, *Records*, 2:17. Shaka's motives: A. J. Cloete to Lt. Col. Bell, 25 July 1828, in Leverton, *Records*, 1:237–39.

5. Amount of ivory: D. Campbell to J. Bell [from reporting by Cloete], 10 October 1828. document 15 in Leverton, *Records*, 2:16–17. Sale of ivory by

King: Bell to Aitchison, 24 November 1828, KCC, Stuart Papers, file 5, extracts 7, 310. See also memorandum, J. Bell, 26 November 1828, document 41 in Leverton, *Records*, 2:43. King quoted in Isaacs, *Travels*, 1:299. Shaka quoted in "Statement of John Cane," 11 November 1828, document 28 [dated 10 November 1828] in Leverton, *Records*, 2:30.

6. Murray quoted: Hay [undersecretary to Murray] to Barrow, 29 November 1828, document 45, annexure 1 in Leverton, *Records*, 2:47. Intelligence provided by Sotobe and Mbozamboza: A.J. Cloete to Lt. Col. Bell, 25 July 1828, in Leverton, *Records*, 1:237–39.

7. Quotations: Hay [undersecretary to Murray] to Barrow, 29 November 1828, document 45, annexure 1 in Leverton, *Records*, 2:47.

8. Funeral details: Isaacs, *Travels*, 1:307. Several documents make Farewell's aim clear: F. Farewell to W.B. Dundas, 10 September 1828, document 12, annexure 1 in Leverton, *Records*, 2:10–13; F. Farewell to Lt. Col. Somerset, 15 December 1828, document 49 in Leverton, *Records*, 2:49–50; W. Beddy to F. Farewell, 6 March 1829, document 112 in Leverton, *Records*, 2:112.

9. Isaacs. *Travels*, 2:278; Fynn, *Diary*, 164.

10. Hutton's illness and gift of ivory: Isaacs, *Travels*, 1:354–55. Dingane overturning unpopular decrees: Isaacs, *Travels*, 2:278; Fynn, *Diary*, 164. Dingane's killings: N. Isaacs to G.L. Cole [governor], 19 December 1828, document 58 in Leverton, *Records*, 2:58.

11. Hutton's death: KCC, Fynn Family folder 3, diary 1. 86. Dommana and quote: Maclean, "Loss," August 1853, 433.

12. Hutton's effects and Isaacs notifying King's friends: *SA Commercial Advertiser*, 13 June 1832, in KCC, Stuart Papers, file 3, extract 2, 135. Timeline: Isaacs, *Travels*, 2:2. Another document records their arrival date as 16 December: T. Williamson to Cain[?], 16 December 1828, document 50 in Leverton, *Records*, 2:50. Isaacs drafts report: *Travels*, 2:3. Impounding of schooner: D.P. Francis to J. Bell, 19 December 1828, document 54 in Leverton, *Records*, 2:54; Isaacs, *Travels*, 2:3.

13. Dispute over schooner's registration: Lt. Col. Bell to D.P. Francis. 30 May 1818, in Leverton, *Records*, 1:165–66; W. Wilberforce to Custom House, 29 May 1828, in Leverton, *Records*, 1:165; Memorial, J.S. King to R. Bourke, 6 June 1828, in Leverton, *Records*, 1:167–68; Lt. Col. Bell to J.S. King, 12 June 1828, in Leverton, *Records*, 1:171. King's fears: Memorial, J.S. King to R. Bourke, 6 June 1828, in Leverton, *Records*, 1:167–68. Reasons for delaying departure: A.J. Cloete to Lt. Col. Bell, 4 July 1828, in Leverton, *Records*, 1:191–93.

14. Isaacs' protest: *Travels*, 2:3; *SA Commercial Advertiser*, 13 June 1832. Governor's demand: Lt. Col. Bell to A.J. Cloete. 18 July 1828, in Leverton, *Records*, 1:232–33. Francis's assurances and Isaacs' view: *Travels*, 2:3–4. Francis also claimed as much in D.P. Francis to J. Bell, 19 December 1828, document 54 in Leverton, *Records*, 2:54. Francis's threat of dismissal: D.P. Francis to J. Bell, 2 January 1829, document 76 in Leverton, *Records*, 2:72; J. Bell to D.P. Francis. 9 January 1829, document 81 in Leverton, *Records*, 2:77. Maclean's antislavery: "Loss," February 1855, 65; Maclean. "The Liberty of British Subjects," 68–69. Possibly Maclean's *Susan King* was named for James Saunders King's mother.

15. Isaacs, *Travels*, 2:4.

16. Bannister's efforts: S. Bannister to Governor Cole, 12 May 1829, in Bannister, *Humane Policy*, appendix 1. xcvi–cv. Collaboration between Farewell and Bannister: F. Farewell to J. Bell, 19 February 1829, document 98, annexure 1, in Leverton, *Records*, 2:100–102. Quotation from Bannister's plans: "Contents of Notes upon the Policy of Great Britain towards the Natives of the Colonies," in S. Bannister to J. Bell, 19 May 1829, document 122, annexure 1, enclosure 1 in Leverton, *Records*, 2:122. Bannister on Indigenous peoples and opposition to existing policies: S. Bannister to G. Murray, 12 May 1829, document 122, annexure 2 in Leverton, *Records* 2:127–28. Imperialist logic: Kennedy, *Mungo Park's Ghost*, 90.

17. Bannister. *Humane Policy*, appendix 1, cv.

18. Quotation: R. Aitchison to J. Bell, 20 December 1828, document 65 in Leverton, *Records*, 2:64. Cole on Farewell: Cole to G. Murray, 31 January 1829, KCC, Stuart Papers, file 5, extracts 7, 302. Murray quoted: Hay [undersecretary to Murray] to Tonkin, 10 October 1828, TNA, CO 49/22. Cole's assessment of Natal: Governor's Correspondence, 30 January 1829 TNA, CO 714/36. Informal empire: see Bourne, *The Foreign Policy of Victorian England.* 5.

19. Cole resisting Farewell and Fynn's schemes: J. Bell to F. Farewell, 6 March 1829, document 113 in Leverton, *Records*, 2:113. Gunpowder: F. Farewell to J. Bell, 11 March 1829, document 116 in Leverton, *Records*, 2:117; J. Bell to Colonial Office, 13 March 1829, document 117 in Leverton, in *Records*, 2:118.

20. Isaacs' return to St. Helena: *Travels*, 2:5. 1:351–52. Fast schooners: *SA Commercial Advertiser*, 4 January 1826. Venereal disease on St. Helena: Heilmann, *Neo-/Victorian Biographilia and James Miranda Barry.* 32.

21. Solomon employs Isaacs: *SA Commercial Advertiser*, 13 June 1832, in KCC, Stuart Papers, file 3, extract 2, 136. Healthiness of St. Helena: Isaacs, *Travels*, 2:5; SHA, *The St. Helena Calendar and Directory for 1830*, "Bill of Mortality, for 1829," 75. Statistics for death by disease (1830): Wade, *History of the Middle and Working Classes*, table 12, 555–56.

22. Solomon considers South Africa: Isaacs, *Travels*, 2:9. American captain's visit: *SA Commercial Advertiser*, 13 June 1832, in KCC, Stuart Papers, file 3, extract 2, 136. In the article, Isaacs calls the captain Wilkins and the ship *Frances*. In his book he calls the captain Williams and the vessel *Francis*. Quotations: Isaacs, *Travels*, 2:6.

23. Isaacs on America: *Travels*, 2:6. Isaacs misrepresents conversation: "Immense Field for Emigration," Memorandum, Isaacs to Cole, 29 September 1832, KCC, 27131, Nathaniel Isaacs Uncat. Mss. Isaacs confesses: *Travels*, 2:7.

24. Nqeto: F. Farewell to D. Campbell, 17 September 1829, document 129 in Leverton, *Records*, 2:139. Cargo: Fynn. *Diary*, 168; Kay quoted in Isaacs, *Travels*, 2:19. Dingane's requests: H. Fynn to Lt. Col. Somerset, "List of Articles," 14 August 1829, document 137, annexure 1, enclosure 1 in Leverton, *Records*, 2:148.

25. Details of attack: J. Cane to W. Shepston. 12 October 1829, in Leverton, *Records*, document 132, annexure 1, 2:142; Fynn. *Diary*. 169–70; Isaacs, *Travels*, 2:15–16.

26. Report of Farewell's demise: "Deaths," *United Service Journal and Naval and Military Magazine.* April 1830, 519. Brig *St. Michael*: *New York Evening Post*, 10 November 1829. Isaacs on Captain Page: "Immense Field for Emigration," Memorandum, Isaacs to Cole, 29 September 1832, KCC, 27131. Nathaniel Isaacs Uncat. Mss.; Isaacs, *Travels*, 2:10.

27. Isaacs, *Travels*, 2:11–12.

28. Isaacs, *Travels*, 2:92.

29. Dingane and *African Adventurer*: Fynn. *Diary.* 299; Isaacs, *Travels*, 2:13.

30. Isaacs, *Travels*, 2:26–27.

31. Isaacs, *Travels*, 2:27–28.

32. Stuart's drawing of uMgungundlovu, with explanations (before index): Fynn, *Diary*, n.p.

33. Isaacs, *Travels*, 2:281–82.

34. Descriptions of Dingane: Isaacs, Travels, 2:37, 280–81; Gardiner. *Narrative of a Journey to the Zoolu Country*, frontispiece, 42, 57. Chinese visitor: Isaacs, *Travels*, 2:36; Fynn, *Diary*, 179.

35. Isaacs, *Travels*, 2:32 (transliteration altered). 33, 37, 39.

36. Dingane presents tusks: Isaacs, *Travels*, 2:40. Fynn offered kingship: *SA Commercial Advertiser*, 13 June 1832, in KCC, Stuart Papers, file 3, extract 2, 136; Isaacs, *Travels*, 2:37, 41. Isaacs and Fynn formalize partnership: *Travels*, 2:43.

37. Cargo loaded on the *St. Michael*: Isaacs, *Travels*, 2:42–43. Zulu passenger: "Immense Field for Emigration," Memorandum, Isaacs to Cole, 29 September 1832, KCC, 27131, Nathaniel Isaacs Uncat. Mss. Vessels sailing from Philadelphia: Disturnell and Clayton. *New York as It Is, in 1835*, 146. Farnham & Fry appears to have operated in New York, Philadelphia, and Salem, Massachusetts.

38. Trade with Zulu and quotations: "Immense Field for Emigration," Memorandum, Isaacs to Cole, 29 September 1832, KCC, 27131, Nathaniel Isaacs Uncat. Mss.; Isaacs, *Travels*, 2:150. Quotation from early traveler: Thompson, *Travels and Adventures in Southern Africa*, 201.

39. Wakefield, "Notes on Chap. II Book 1," 59–60n64.

40. Theological discussion: Isaacs, *Travels*, 2:134 (transliteration altered). Fynn on belief: *Diary*, 263–65.

41. Endelman, *The Jews of Georgian England*, esp. chapter 8.

42. Isaacs, *Travels*, 2:45–48.

43. Episode and quotations: Isaacs, *Travels*, 2:49; Fynn, *Diary*, 173. Dingane's arms request: Isaacs, *Travels*, 2:58.

44. Isaacs, *Travels*, 2:51.

45. Isaacs, *Travels*, 2:64.

46. Isaacs' homestead: *Travels*, 2:70, 77. Hlambamanzi's move: Fynn, *Diary*, 188n; Isaacs, *Travels*, 2:258.

47. Isaacs, *Travels*, 2:77–78, 81–83.

48. Isaacs, *Travels*, 2:65.

49. Isaacs, *Travels*, 2:69–70.

50. Isaacs, *Travels*, 2:93. Isaacs never identified his illegitimate progeny in his published work. It is possible that Porter was not his son, but it is unlikely that a child who was not Isaacs' own son would have been paid a ceremonial visit and been mentioned so often.

51. Isaacs, *Travels*, 2:141, 120.

52. Isaacs, *Travels*, 2:132, 229.

53. Isaacs, *Travels*, 2:133–34, 281, 220, 233. This may be the same episode of Dingane asking for European guns and assistance recorded in H. Fynn to W. Thompson, 21 July 1831, document 177, annexure 1, enclosure 1 in Leverton, *Records*, 2:202.

54. On British concern: "It is pretty generally known that the American government has for several years past been on the look out for a favourable port between their own shores and India, and it is also known that De la Goa Bay has been recommended to Congress with that view." John Philip to P. G. Brink, 13 April 1824, in Leverton, *Records*, 1:30–33. On claims of weapons: *SA Commercial Advertiser*, 11 December 1830, in KCC, Stuart Files, file 18, extracts 4, 9. On rumors reaching governor: *Papers Relative to the Condition and Treatment*, part 2, enclosure 2, D. Campbell to G. Murray. 30 November 1830, 59; also D. Cambpell to J. Bell, 26 November 1830, document 164, annexure 1 in Leverton, *Records*, 2:172. Isaacs' refutation: N. Isaacs to editor, *SA Commercial Advertiser*, 2 June 1832, quoted in Rochlin, "Nathaniel Isaacs and Natal," appendix 3, 265.

55. Isaacs, *Travels*, 2:211–13.

56. Isaacs, *Travels*, 2:256–57.

57. Hlambamanzi's plot: Isaacs, *Travels*, 2:220–21; H. Fynn to W. Thompson, 21 July 1831, document 177, annexure 1, enclosure 1 in Leverton, *Records*, 2:202. Cane: Isaacs, *Travels*, 2:215; see also W. McDowell Fynn's account in statement to D. Campbell, 2 July 1831, document 176, annexure 1 in Leverton, *Records*, 2:198–200; H. Fynn to W. Thompson, 21 July 1831, document 177, annexure 1, enclosure 1 in Leverton, *Records*, 2:202–8.

58. Isaacs, *Travels*, 2:223–29.

59. Isaacs, *Travels*, 2:227–31.

60. Destruction of Cane's homestead: Isaacs, *Travels*, 2:237. Dingane's instructions: W. McDowell Fynn, statement to D. Campbell, 2 July 1831, document 176, annexure 1 in Leverton, *Records*, 2:200. Flight of Europeans: Isaacs, *Travels*, 2:240–41. Fynn quotation: *Diary*, 191–94, probably from 21 April 1831.

61. Quotation: Isaacs, *Travels*, 2:242. Arrival of *St. Michael*: Isaacs, *Travels*, 2:243; H. Fynn to W. Thompson, 21 July 1831, document 177, annexure 1, enclosure 1 in Leverton, *Records*, 2:203.

62. Fynn hiding items: Isaacs, *Travels*, 2:244–45. Use of cannons: Isaacs, *Travels*, 2:246; H. Fynn to W. Thompson, 21 July 1831, document 177, annexure 1, enclosure 1 in Leverton, *Records*, 2:202–3.

63. Dingane's claims, Isaac and Fynn's wariness: Isaacs, *Travels*, 2:249–51. Isaacs and Fynn's concession to Dingane: Isaacs, *Travels*, 2:250, 259; H. Fynn to W. Thompson, 21 July 1831, document 177, annexure 1, enclosure 1 in Leverton, *Records*, 2:203–4.

64. Isaacs, *Travels*, 2:261–64.

65. Dingane's reproach, Hlambamanzi's speech, and Dingane's alarm: Isaacs, *Travels*, 2:264–65 (transliteration altered). Hlambamanzi is said to have reported much the same thing: see H. Fynn to W. Thompson, 21 July 1831, document 177, annexure 1, enclosure 1 in Leverton, *Records*, 2:204.

66. Fynn's fears: Isaacs, *Travels*, 2:266; H. Fynn to W. Thompson, 21 July 1831, document 177, annexure 1, enclosure 1 in Leverton, *Records*, 2:204. Lukilimba's confession: *SA Commercial Advertiser*, 10 October 1832, in KCC, Stuart Papers, file 17, extracts 3, 129 (transliteration altered). Fynn calls this individual Kebimba or Kebinba in H. Fynn to W. Thompson, 21 July 1831, document 177, annexure 1, enclosure 1 in Leverton, *Records*, 2:202–8. Fynn's shooting of Lukilimba: *Diary*, 205. Isaacs falsely claims "Kelimba" [Lukilimba] intended to kill Fynn: *Travels*, 2:27. Fynn indicates he was never in danger: H. Fynn to W. Thompson, 21 July 1831, document 177, annexure 1, enclosure 1 in Leverton, *Records*, 2:205–7.

67. Fynn's message: *Diary*, 204, probably 29 May 1831. See also Isaacs, *Travels*, 2:263; *SA Commercial Advertiser*, 12 September 1832. in KCC, Stuart Papers, file 17, extracts 3, 123. Flight of Fynn and Cane: Isaacs, *Travels*, 2:264. Isaacs sets sail: *Travels*, 2:270.

68. Cane's account of Dingane's sanction: letter to editor from J. Cane, *Graham's Town Journal*, 4 July 1832, in KCC, Stuart Papers, file 3, extract 1, 42. Murder of Hlambamanzi: Isaacs, *Travels*, 2:274; Another settler, Ogle, lured Hlambamanzi into a trap and actually killed him. Dingane after the purge: Isaacs, *Travels*, 2:276–78.

69. Isaacs, *Travels*, 2:341, 353, 360–62, 370.

70. Ramanataka: Isaacs, *Travels*, 2:367. Quotation: Memorandum, Isaacs to G.L. Cole. 29 September 1832, document 189, annexure 1 in Leverton, *Records*, 2:229.

71. Ramanataka's plot: Isaacs, *Travels*, 2:374–75. Quotations: Memorandum N. Isaacs to G.L. Cole. 29 September 1832, document 189, annexure 1 in Leverton, *Records*, 2:229–30.

72. Journey up East African coast: Isaacs, *Travels*, 2:381, 392–93; "East Coast of Africa—Brig *St. Michael's*."

73. Isaacs, letter to editor of *SA Commercial Advertiser*, April 1832, in Rochlin, "Nathaniel Isaacs and Natal," appendix 3. 265.

74. Isaacs' address: Memorandum, N. Isaacs to G.L. Cole, 29 September 1832, document 189, annexure 1 in Leverton, *Records*, 2:230. Isaacs with Cole: N. Isaacs to E.G. Stanley [secretary of state], 5 September 1833, document 190 in Leverton, *Records*, 2:231. Quotations on America and slavery: "Immense Field for Emigration," Memorandum, Isaacs to Cole, 29 September 1832, KCC, 27131, Nathaniel Isaacs Uncat. Mss. A photograph of this letter was found in the second volume of Fynn's signed copy of Isaacs' *Travels and Adventures*. KCC, SR 968.4, ISA.

75. Quotation: G.L. Cole to Hay, 11 January 1831, in *Papers Relative to the Condition and Treatment*, 57. On diplomatic games: Hoppen. *The Mid-Victorian Generation*. 156.

76. Quotations: N. Isaacs to G.L. Cole, December 1832, document 183 in Leverton, *Records*, 2:213; J. Bell [secretary to governor] to N. Isaacs, 28 December 1832, document 183 in Leverton, *Records*, 2:213.

77. Quotations: N. Isaacs to H. Fynn, 10 December 1832, in Kirby, "Unpublished Documents," 66–67.

CHAPTER 5. FEVERISH TRADE

1. "In the Days of Chaka," 7 July 1927, quoted in Jolles, *Isaacs*, 280 (recollections from either 1833 or 1836).

2. Wellington: Stanton, *Sketches of Reforms and Reformers*, 174. On electorate: Tombs, *The English and their History*, 443.

3. Right to hold title, retail, admission to bar: Henriques, *The Jews and the English Law*, 191–94, 200–201, 203. Jews could operate wholesale businesses within London earlier. See also Endelman, *The Jews of Georgian England*, 281.

4. Grey and systematic colonization: Heartfield, *The Aborigines' Protection Society*, 6. Colonial Office: Thomas, *The Philosophic Radicals*, 372. Quotation from Wakefield, *Plan of a Company*, 4.

5. Wakefield and his influence: Mills, *The Colonization of Australia*, 76–89; Stern, *Empire, Incorporated*, 219–29, 230–41; Burroughs, *Colonial Reformers and Canada*, xii–xvi. On Mill's support: Semmel, *The Rise of Free Trade Imperialism*, 108. Bentham's contributions and George Grey: Bentham. *Writings on Australia*, 4n15.

Grey, colonization, Native peoples: Heartfield, *The Aborigines Protection Society*, 15–17. Undersecretary Howick presumably had the approval of his father: Semmel, *The Rise of Free Trade Imperialism*. 108; Temple, *A Sort of Conscience*, 146. Members of the society: "National Colonization Society," *Times*, 17 June 1830, 3. Wakefield's theories later served as a foil for Karl Marx: see Piterberg and Veracini, "Wakefield, Marx, and the World Turned Inside Out."

6. Reformers and colonialism: Lydon, "'Mr. Wakefield's Speaking Trumpets,'" 86–87; Heartfield, *The Aborigines' Protection Society*, 66–68; Darwin, *Unfinished Empire*, 106; Burroughs, *Colonial Reformers and Canada*, 43–45.

7. Pressure by Isaacs: N. Isaacs to E. G. Stanley, 5 September 1833, document 190 in Leverton, *Records*, 2:231. Details and quotations: "Prospectus," KCC, file 15, James Stuart's Misc. Papers on the Early History of Natal, vol. 1. The precise date of this document is unclear: it was presumably drawn up between autumn 1833, when Cole returned to the United Kingdom, and October 1835, when the solicitors listed in the document, Gregory and Price, dissolved their partnership. I suggest the document was composed in the aftermath of an 1834 memorial submitted to the governor by Cape Town merchants in favor of annexing Natal. Governor Benjamin D'Urban endorsed this position: B. D'Urban to E. G. Stanley, 17 June 1834, document 199 in Leverton, *Records*, 2:250. Description of Wakefield's proposal: *Plan of a Company*, 3.

8. Quotations: "Prospectus," KCC, Stuart Papers. file 15, James Stuart's Misc. Papers on the Early History of Natal, vol. 1.

9. Quotations: "Prospectus," KCC, Stuart Papers. file 15, James Stuart's Misc. Papers on the Early History of Natal, vol. 1. On Fynn: KCC, Fynn Family folder 3, diary 1. Fynn's document gives the name as Manard or Menard. On Maynard's initial trade in Port Natal: C. Maynard and H. Maynard to D.P. Francis, 9 June 1831, document 172 in Leverton, *Records*, 2:191. More on Maynard brothers: Lloyd, "Religion, Same-Sex Desire," 56. Maynard's withdrawal of support: KCC, Fynn Family folder 3, diary 1.

10. The map was perhaps used as supporting evidence for Isaacs' 1828 grant from Shaka. The anonymous author of "The Colonization of South Africa" indicates that the map appears on the reverse side of the page bearing the text of this grant, 26 July 1895, 9. For map: "Africa: South-Eastern Coast," TNA, CO 700/South Africa 3. The map includes mention of a Zulu skirmish in 1831, and so the date of 1828 appearing beneath Isaacs' name cannot be accurate to the year of the map's production. Jolles (re)discovered this map: see *Isaacs*, 241–42.

11. Quotation: "Africa: South-Eastern Coast," TNA, CO 700/South Africa 3. Land as wealth: Tombs, *The English and Their History*, 487.

12. Quotations from legend: "Africa: South-Eastern Coast," TNA, CO 700/ South Africa 3. Isaacs notes "uninhabited land" to the southwest and land occupied by "fragments of broken tribes"—likely referring to the remnants of groups conquered by other, more powerful entities. These notations impinge on the Mfecane ("devastation") debate in South African historiography. A complicated and explosive issue, the Mfecane debate pits scholars who view the social, political, and military consolidations and revolutions of the late eighteenth and early nineteenth centuries in eastern South Africa as causing widespread dislocation, famine, and death against those who contend that the Mfecane was a later invention to cover the depredations that followed the incursion of white colonialists. Partisans have conventionally blamed either the expansion of the Zulu Kingdom under Shaka or European rapaciousness. See Hamilton, *The Mfecane Aftermath*. "Zoolo country": "Prospectus, KCC, Stuart Papers, file 15, James Stuart's Misc. Papers on the Early History of Natal, vol. 1.

13. "Africa: South-Eastern Coast," TNA, CO 700/South Africa 3.

14. Disraeli, *The Wondrous Tale of Alroy*, 15. Owen, *Narrative of Voyages*, 255.

15. N. Isaacs to H. Fynn, 10 December 1832, in Kirby, "Unpublished Documents," 67. Isaacs writes "Dingarn" in the original.

16. Quotation: N. Isaacs to H. Fynn, 10 December 1832, in Kirby, "Unpublished Documents," 68. On Fynn's diary, see *Diary*, xii–xiii. The diary was not edited and published until 1950.

17. Isaacs to H. Fynn, 10 December 1832, in Kirby, "Unpublished Documents," 67.

18. Isaacs on Fynn's lack of response: N. Isaacs to H. Fynn, 10 December 1832, and N. Isaacs to H. Fynn, 7 September 1840, in Kirby, "Unpublished Documents," 67, 76. Commercial goals: Isaacs, *Travels*, 1:xiii. Contract: Contract, R. Bentley and N. Isaacs, 27 January 1834 BL, Add MS 46612, f. 58.

19. Petition of Merchants to "His Majesty the King in Council," document 199, annexure 1 in Leverton, *Records*, 2:251–52.

20. "East India Dock Road, North Side: Nos 1–301 (and Nos 2–50)," in Hobhouse, *Survey of London*, 127–47.

21. Details borrowed from or inspired by Henry Mayhew's "Letters III–IV," *Morning Chronicle*, 26 October and 30 October 1849. Mayhew details the London Docks, slightly to the west of the East India Company docks. See also Leigh, *Leigh's New Picture of London*, 86–87, 374–75.

22. Ben Moss notes that he was at sea on 6 January 1835, so Isaacs probably set sail in December 1834. Moss records his deceased wife's name as Alice in S.

Moss to brother, 18 January 1836, MFA-J. Saul's brother, Joseph Solomon, who also resided for a time in Jamestown, married Hannah Moss. Samuel Moss was one of Hannah's brothers and Ben his son.

23. Description drawn from Leigh, *Leigh's New Picture of London*, 66–67. For Mayhew's image of the "urban savage," see Qureshi, *Peoples on Parade*, 23–28.

24. Jews as criminal: Leigh, *Leigh's New Picture of London*, 67–68. Leigh reanimates a 1796 charge and statistics provided by Patrick Colquhoun in *A Treatise on the Police of the Metropolis*, 172–74. Mayhew's stereotypes and comparisons: *London Labour*, 1:1–3, 87.

25. Kirsch, *Benjamin Disraeli*, xv.

26. S. Moss to B. Moss, 19 November 1836, MFA-J. The letters run from January 1835 until July 1837.

27. B. Moss to S. Moss, 6 January 1835; S. Moss to B. Moss, 7 January 1835; B. Moss to S. Moss, 23 February 1835; N. Isaacs postscript to B. Moss to S. Moss, 23 February 1835, all in MFA-J.

28. St. Mary's Island is now Banjul Island, site of the Gambia's capital.

29. Sephardic settlements: Mark and da Silva Horta, *The Forgotten Diaspora*. Items and descriptions of structures: Gray and Dochard, *Travels in Western Africa*, appendix, article 1, 365–66, 368.

30. French attack: "Evidence: Obstruction of the 'Eliza,'" TNA, FO 403/1, division 3, no. 2, 2–3. Death of Baboukas: Lieutenant Governor Rendall to Lord Glenelg, 9 July 1836, TNA, CO 87/24.

31. Claims of Traza revenge: Lieutenant Governor Rendall to Lord Glenelg, 9 July 1836, TNA, CO 87/24. Account of *Medusa*: Miles, *The Wreck of the Medusa*. French refusal to pay compensation: G. C. Redman, Harrison, Forster & Smith to Viscount Palmerston, 6 April 1836, TNA, FO 403/1, division 3, 59–60.

32. Timeline: G. C. Redman, Harrison, Forster & Smith to Viscount Palmerston, 6 April 1836, TNA, FO 403/1, division 3, 60; N. Isaacs to Lieutenant Governor Rendall, 27 August 1835, TNA, FO 403/1, division 3, 70. Items: "Goods and Vessels Ordered for Portendic in Account with G. C. Redman," 5 September 1835, TNA, FO 403/1, division 3, 31. Clan names: Isaacs to Rendall, 27 August 1835, 70. Isaacs' agreement: "Literal Translation of the Agreements between N. Isaacs and the Moors at Portendic," 23 August 1834, TNA, FO 403/1, division 3, 18.

33. Isaacs' request: N. Isaacs to Lieutenant Governor Rendall, 19 February 1835 TNA, FO 97/185; N. Isaacs to Lieutenant Governor Rendall, n.d., MFA-J. Redman's assessment: "Statement of Losses Incurred by the Brig 'Eliza.'" TNA, FO 403/1, division 3, 13. Isaacs quoted: Isaacs to Rendall, 19 February 1835.

34. Quotations: B. Moss to S. Moss, 23 February 1835; B. Moss to S. Moss, 9 August 1835; B. Moss to S. Moss, 14 September 1835; B. Moss to J. Druitt [friend], n.d., all in MFA-J.

35. Quotations: B. Moss to S. Moss, 22 May 1835; B. Moss to cousin, 12 March 1835, both in MFA-J. Ben's father thanks God in eight of fourteen letters; Isaacs never appeals to God or employs Jewish phraseology anywhere in his writings.

36. Quotation and information: S. Moss to B. Moss, 30 May 1835, MFA-J; chronology by Amelia Marks, appended to letters in MFA-J.

37. Timeline, journey, and collection of parrots: B. Moss to S. Moss, 10 June 1835, 30 June 1835, and 14 September 1835, all in MFA-J. Isaacs on his travels: N. Isaacs to Lord Palmerston, 29 March 1854 TNA, CO 87/58. British expeditions had been launched up the Niger River decades earlier: see Kennedy, *Mungo Park's Ghost*; Mouser, "Forgotten Expedition into Guinea."

38. B. Moss to S. Moss, 29 July 1835, 14 September 1835, 9 August 1835, 13 September 1835, all in MFA-J.

39. Isaacs' efforts to dispose of cargoes: Claims of G. C. Redman. n.d., TNA, FO 97/185; Deposition of N. Isaacs, 7 September 1835, TNA, FO 97/185; "Copy of Account Sales," 6 June 1835, TNA, FO 403/1, division 3, 23.

40. B. Moss to S. Moss, 9 August 1835, 17 August 1835; B. Moss to uncle [G. Bagshaw?], 13 September 1835; B. Moss to S. Moss, 27 December 1835, 14 September 1835, 29 October 1835. Shoes: N. Isaacs to S. Moss, 18 February 1836, all in MFA-J.

41. N. Isaacs to S. Moss, 28 December 1835, MFA-J.

42. N. Isaacs to B. Moss, 26 January 1836, emphasis in original; N. Isaacs to S. Moss, 18 February 1836; B. Moss to S. Moss, 10 April 1836, all in MFA-J.

43. *Travels and Adventures* probably appeared in June 1836. Notice and price: *Bent's Literary Advertiser* 376, 10 June 1836, 64. Contract cancellation: Contract, R. Bentley and N. Isaacs, 27 January 1834, Canceled 22 February 1834, BL, Add MS 46612, f. 58.

44. Frye, "*The South African Commercial Advertiser* and the Eastern Frontier," 55–60.

45. "Review," *Athenaeum*, 451, 18 June 1836, 425; "Travels in Eastern Africa [Review]," *Metropolitan*, April 1837, 101–2; "Art. I. [Review]," *Quarterly Review*, February 1837, 7; "Travels in Eastern Africa [Review]," *Gentleman's Magazine*, December 1836, 630. See also "Travels and Adventures in Eastern Africa [Review]," *Court Magazine*, July 1836, 41; "The Zoolus of Eastern Africa [Review]," *Mirror of Literature, Amusement, and Instruction* 799, 8 October 1836, 1; "Review," *Athenaeum*, 451, 18 June 1836, 426.

46. "Art. X. [Review]," *Monthly Review*, July 1836, 416.

47. Gardiner, *Narrative of a Journey to the Zoolu Country*, 384, 411. Gardiner likens Zulu customs to Jewish practices: 95–96; 280.

48. Quotation: "Review [of *Narrative of a Journey to the Zoolu Country*]," *Athenaeum*, 445, 7 May 1836, 328. The meeting convened by Cane is mentioned as occurring on 20 June 1836: Mackeurtan, *Cradle Days of Natal*, 194–95. The beginnings of present-day Durban had been taking shape since 1835: "Regulations of the Town of 'D'Urban' Port Natal," in Gardiner, *Narrative of a Journey to the Zoolu Country*, 399–402. Cane was present at this meeting on 23 June 1835, and H. Fynn is listed as a "subscriber" to the efforts. An illustration titled "A Plan of the Town of D'Urban," including "Farewell Square," appears at the end of Gardiner's book. Erroneous claims that N. Isaacs attended Cane's meeting: Mackeurtan, *Cradle Days of Natal*, 194–95. Isaacs' whereabouts: journal of B. Moss, 20 June 1836, MFA-J.

49. Ben's itinerary: journal of B. Moss, 17 May 1836. Ben on Bah Fall: journal of B. Moss, 3 May 1836. Seasickness: journal of B. Moss, 3–10 May 1836. Ben on Isaacs: journal of B. Moss, 3 May 1836. Quotations from Ben: B. Moss to Flowers, n.d. [likely June 1836]; journal of B. Moss, 17, 27, 29 May 1836. Rationing: journal of B. Moss, 27 May 1836. All in MFA-J.

50. Dismal view: B. Moss to S. Moss, 3 August 1836, MFA-J. Use of gum arabic in textiles: Webb, "The Trade in Gum Arabic," 152–55. Traza control: *British and Foreign State Papers, 1845–1846*, 34:1073n.

51. Quotations: Isaacs, "Memorandum Book, 1836," extract, MFA-J. The extract is unsigned, but the language and handwriting appear to be Isaacs'. Profitability of gum arabic: W. Forster, "An Account Showing the Loss," 7 July 1835, TNA, FO 403/1, division 3, 43. Other brokers testified that a hundredweight (112 pounds) of gum sold for about £4 10s. in London in 1834–35: Johnson & Renny to Mssrs. Forster and Smith, 24 October 1836, TNA, FO 403/1, division 3, 32. Isaacs traded six hundred pieces of bafts—blue cotton cloth in pieces fourteen to eighteen yards long—for 134,400 pounds of gum at Portendick in 1835: "Certificate as to Barter of Gum for Goods," TNA, FO 403/1, division 3, 27. Isaacs' itinerary: journal of B. Moss, 16 June 1836, MFA-J. There was no governor of the Gambia at the time; Rendall was lieutenant governor.

52. Quotations: B. Moss to S. Moss, 27 June 1836. Isaacs quoted: memo 2 from journal of N. Isaacs, n.d. [June 1836?]. Ben's activities recorded: journal of B. Moss, 13, 17, 21,and 25 June 1836. High prices paid: journal of B. Moss, 20 June 1836. Items bartered: memorandum book, 1836, probably in Isaacs' hand. All in MFA-J.

53. Ben Moss quoted: journal of B. Moss, 3 July 1836, MFA-J. Agreement: Convention between Sultan Mahomed Lahabebe and Governor Rendall, 4 July 1836, TNA, CO 87/24. Sultan's request and dispatch of princes: Lieutenant Governor Rendall to Lord Glenelg, 9 July 1836 TNA, CO 87/24. Ben Moss meets retinue: journal of B. Moss, 10 July 1836, MFA-J. Cargo on *Matchless*: journal of B. Moss, 18 July 1836, MFA-J. Moss presumably uses the imperial ton (2,240 pounds, 240 pounds more than a US ton).

54. Journal of B. Moss, 16 July 1836, MFA-J.

55. Quotations: Memo 1 from journal of N. Isaacs, n.d. [June 1836?] MFA-J. Slight alterations of spelling have been made.

56. Quotation: Memo 2 from journal of N. Isaacs, n.d. [June 1836?], MFA-J. Description of weather: Brooks, *Yankee Traders*, 80–81. Sandstorm: journal of B. Moss, 24 July 1836, MFA-J.

57. Port and Isaacs' survey: Lieutenant Governor Rendall to Lord Glenelg, 9 July 1836, TNA, CO 87/24. Abdullah: Journal of B. Moss, 8 August 1836, MFA-J.

58. Isaacs' escape: Journal of B. Moss, 9 August 1836. Isaacs' purchases: Journal of B. Moss, 10–12 August 1836. Challenging conditions: Journal of B. Moss, 15 August 1836. All in MFA-J.

59. Isaacs' journey: Journal of B. Moss, 16 August 1836, MFA-J. Isaacs confirms French subterfuge: "Supplemental Evidence to General Case: Declaration of Mr. Nathaniel Isaacs," 8 December 1836, TNA, FO 403/1, division 3,

72. Isaacs' claim and description of landscape: Journal of B. Moss, 2 September 1836, MFA-J. Tallying accounts: journal of B. Moss, 3 September 1836, MFA-J.

60. Information on Isaacs' relations with Nanette Guey are courtesy of the Senegalese genealogical researcher Xavier Ricou, email, 1 May 2022. Hannah Isaacs' mother had earlier and likely erroneously been identified as Madeleine Diole: see Jones, *The Métis of Senegal*, 220n4. Photo of Hannah in Jones, *The Métis of Senegal*, 67.

61. Journal of B. Moss, 3–4 October 1836, 12–15 October 1836, MFA-J.

62. Journal of B. Moss, 25–28 October 1836, MFA-J. Goldsmith quotation: *History of the Earth*. 309, 311. Fumigation techniques: Brooks, *Yankee Traders*, 83.

63. Journal of B. Moss, 31 October, 1 November, 7 November, 14 November 1836, MFA-J.

64. B. Moss to Uncle [G. Bagshaw], 20 November 1836, MFA-J.

65. Hannah Isaacs and marriage: Jolles, *Isaacs*, 254–55. Redman's accounting: "Interest: Account on George C. Redman's claims," 9 November 1836, TNA, FO 403/1, division 3.

66. See Benjamin Robert Hadon, *The Anti-slavery Society Convention, 1840*, National Portrait Gallery, NP 599, www.npg.org.uk/collections/search/portrait /mw00028/The-Anti-Slavery-Society-Convention-1840. Bannister: Belmessous. *Assimilation and Empire*. 70–71: Laidlaw, "'Aunt Anna's Report,'" 20.

67. Bannister, Hodgkin, and Aborigines Protection Society: Belmessous, *Assimilation and Empire*, 73. The society's founding date is murky. According to Fox Bourne, the society arose as an "outside committee" to the parliamentary select committee: *Aborigines Protection Society*, 3. Stern ties the society to Wakefield's influence: *Empire, Incorporated*, 240–41.

68. Laidlaw, "'Aunt Anna's Report,'" 4, 14.

69. Quotations: *Report of the Parliamentary Select Committee on Aboriginal Tribes*. 81, 103, 139. Laidlaw describes in "'Aunt Anna's Report,'" 1–28, how Buxton's daughter and an unmarried female cousin composed much of the select committee report.

70. Fox Bourne, *Aborigines Protection Society*, 24; Heartfield, *Aborigines' Protection Society*, 66–68.

71. Parfitt, *Hybrid Hate*, 1–31.

72. Prichard, *Researches into the History of Mankind* [1813], 67n(a), 236; Prichard, *Researches into the History of Mankind* [1837], 2:286.

73. Buxton, *The African Slave Trade*, 483. Buxton attributes the phrase to an earlier missionary. On divergent positions regarding colonization, see Hall, *Civilising Subjects*, 48–49.

74. Darwin, *Unfinished Empire*, 276.

75. Heartfield, *The Aborigines' Protection Society*, 23; *Report of the Parliamentary Select Committee on Aboriginal Tribes*, n.p. [iii].

76. Christian missionizing of Jews: Endelman, *The Jews of Georgian England*, 68–78. Quotations: Gilbert, "The London Jews," 868. In this 1864 article, Gilbert considers "improvements" made by Jews during previous decades.

77. Lushington's speech in the Commons: "Seizure by the French Government," 8 May 1838, *Hansard Parliamentary Debates*, 3rd series, vol. 42, https://

hansard.parliament.uk/Commons/1838-05-08/debates/d47139b5-a70e-45fb-beoc-c85b84722325/SeizureByTheFrenchGovernment. On speech impediment: "Lushington, Stephen (1782–1873), of Mery-hill, nr. Watford, Herts. and 2 Great George Street, Mdx," in *The History of Parliament: The House of Commons 1820–1832*, ed. D. R. Fisher, available at www.historyofparliamentonline.org/research.

78. French aggression and Ben's pastimes: B. Moss to S. Moss, 25 April 1837, MFA-J. Jewish trade in birds: Mayhew, *London Labour*, 2: 70–71, 118. Ben's collection and report on McCormack: B. Moss to S. Moss, 6 May 1837, MFA-J.

79. On epidemic: B. Moss to S. Moss, 26 June 1837, MFA-J; see also Boyce, "The History of Yellow Fever in West Africa," 182; Clarke, *Sierra Leone*, 86. Quotations: B. Moss to S. Moss, 26 June 1837; B. Moss to S. Moss, 29 July 1837, emphasis in original. Both in MFA-J.

80. B. Moss to S. Moss, 15 May 1837, 29 July 1837, MFA-J. Isaacs' earnings: "Goods and Vessels Ordered for Portendic in Account with G. C. Redman," 5 September 1835, TNA, FO 403/1, division 3, 31.

81. "Records of the Family of Mrs. Henry Marks," n.d., MFA-J.

82. Bannister quotation: *Memoir Respecting the Colonization of Natal*, 6. The *Memoir* was presented by the Cape of Good Hope Trade Society to the secretary of state for the colonies. The Trade Society and the South African Land and Emigration Association were perhaps overlapping entities. Isaacs mentioned: *Memoir Respecting the Colonization of Natal*, 7–8. Supporters listed: J. G. Thomson and T. Bagshaw to R. V. Smith, 24 March 1840, in Kirby, *Andrew Smith and Natal*, 206–7. Association's plans rejected: Kirby, *Andrew Smith and Natal*, 182.

83. Offer: Fynn, *Diary*, 313. Hlambamanzi is quoted in H. Fynn to G. L. Cole [governor], 21 July 1831, document 177, annexure 1, enclosure 1 in Leverton, *Records*, 2: 204.

84. Holden, *History of the Colony of Natal*, 72–73, 65.

85. Quotations: N. Isaacs to H. Fynn, 7 September 1840, in Kirby, "Unpublished Documents," 76. Documents related to association's ongoing efforts: Kirby, *Andrew Smith and Natal*, 186–205.

86. Quotations: N. Isaacs to H. Fynn, 7 September 1840, in Kirby, "Unpublished Documents," 77.

87. N. Isaacs to H. Fynn, 7 September 1840, in Kirby, "Unpublished Documents," 76. Benjamin Isaacs may have been present at the Cape meeting presided over by Cane. Volumes (presumably sent by Isaacs to Fynn) in KCC, SR 968.4 ISA.

CHAPTER 6. RAILROAD CHRISTIANIZATION

1. Newton, *Letters, Sermons, and a Review of Ecclesiastical History*, letter 6, 18 January 1763. 47–48, 50, emphasis in original.

2. Newton first publicly spoke against the slave trade in 1788: Brown, *Moral Capital*, 337; Hochschild, *Bury the Chains*, 75–77, 130–31.

3. See Anderson, *Abolition in Sierra Leone*, xiii.

4. Quotation: Clarkson, *The History of the Rise*, 2:228. Amelioration as a waystation to abolition: Brown, *Moral Capital*, 55–59, 70–75.

5. Brown, *Moral Capital*, 184–86; Hilton, *A Mad, Bad, and Dangerous People?*, 187–88.

6. Quotation: Gidney, *The History of the London Society*, 44. Newton and Wilberforce: Clarkson, *The History of the Rise*, 1:241; see also Schama, *Rough Crossings*, 174–75.

7. Peckard quoted: *Am I Not a Man?*, 1. Peckard's defense of Jews: *The Popular Clamour*. On emancipation: Feldman, "Conceiving Difference," 171, 185n55.

8. Subscribers: Schwarz, "Commerce, Civilization and Christianity," 252–76. Multiracial aspects of settlement: Schama, *Rough Crossings*, 195–98.

9. On Maroons: Scanlon, *Freedom's Debtors*. 30. Ethnic, national, and other distinctions vary in reference to the peoples resettled in and around Freetown: see Anderson, *Abolition in Sierra Leone*, 127–33.

10. Quotation and context: Clarkson, *The History of the Rise*, 2:344, 1:8. Philips quotation: *Researches in South Africa*. 355–56.

11. Knox: *The Races of Men*. 181, 131, 134. Quotations from Beddoe: *The Races of Britain*, 218–19; "On the Physical Characteristics of the Jews," 236.

12. Jews and "savages": see Penslar, *Shylock's Children*, 41–42.

13. Endelman, *The Jews of Georgian England*, 70–72. Wilberforce's activities: Gidney, *The History of the London Society*, 37, 147.

14. Freetown as a mission city: Gough, "Rethinking the Colonial City," 169. On administration: Everill, "Bridgeheads of Empire?," 791.

15. Smith quotation: *An Inquiry into the Nature and Causes of the Wealth of Nations*. 37. Congress of Vienna aims: "Declaration of the Eight Powers Relative to the Universal Abolition of the Slave Trade" (1815), reprinted in Herstlet, *Map of Europe by Treaty*, 60–61; see also Vick, *Congress of Vienna*, 161–93.

16. Clarkson's meetings with the tsar: Taylor, *A Biographical Sketch of Thomas Clarkson*, 117–21. Way's meetings with the tsar: Rovner, *In the Shadow of Zion*, 20–21. Moral prestige: Brown, *Moral Capital*, 457–58.

17. Clarke, *Sierra Leone*, 84–85.

18. Quotations: Clarke, *Sierra Leone*, 40. Accounts of Sierra Leone's abolitionist origins and practical shortcomings: Gilbert, *Black Patriots and Loyalists;* Scanlan, *Freedom's Debtors;* Schama, *Rough Crossings*.

19. Quotation from Bridge, *Journal of an African Cruiser*, 183. Cholera: Tombs, *The English and Their History*, 437.

20. Details from Alexander, *Narrative of a Voyage of Observation*, 96; Burton and Cameron, "The Market at Freetown," 62–63; Clarke, "Sketches," 321, 328; Melville, *A Residence at Sierra Leone*, 111–12.

21. Description of chapels: Alexander, *Narrative of a Voyage of Observation*, 90. Sunday observance: Simpson. *A Private Journal Kept during the Niger Expedition*, 18. Contemporaries quoted: Thomas, *Adventures and Observations*, 75; F. Teal to general secretaries, n.d. [ca. 1853–54], WMMS, West Africa, Correspondence. box 282. On religious schools: N. Macdonald to Sir John Packington, Bart., "Report on the Annual Blue Book for the Colony of

Sierra Leone for the year 1851," 26 June 1852, SLNA, Governor's Despatches to Secretary of State, 1 January 1852–20 January 1853, no. 83.

22. Missionary quoted: T. Raston to General Secretaries, 3 June 1843, WMMS, West Africa, Correspondence, box 281. "Degraded" native peoples: F. Teal to general secretaries, n.d. [ca. 1853–54], WMMS, West Africa, Correspondence, box 282. Police ordinance: Crooks, *History of Sierra Leone*, 184; Clarke, "Sketches of Sierra Leone and Its Inhabitants," 334. Display details: *Routledge's Guide*, 118.

23. Census figures: N. Macdonald to Sir John Packington, Bart., "Report on the Annual Blue Book for the Colony of Sierra Leone for the year 1851," 26 June 1852, SLNA, Governor's Despatches to Secretary of State, 1 January 1852–20 January 1853, no. 83. Lifespan of Europeans in West Africa: Temperley, *White Dreams, Black Africa*, 45. Epithets for Sierra Leone: Rankin, *The White Man's Grave*; Alexander, *Narrative of a Voyage of Observation*, 90. Doomed churchmen: Mayhew, "The Church in Danger," 24 July 1847, 23.

24. Disease and treatment: E. Ware, "Health Hazards," 88–94; Clarke, *Sierra Leone*, 90–103. Quotation: A. E. Kennedy to secretary of state, 24 August 1854, SLNA, Governor's Dispatches to Secretary of State, 24 July 1854–17 September 1855, letter 144. Fyfe, *History of Sierra Leone*, 275 notes that Isaacs owned property on Gloucester Street. Another source indicates Isaacs that lived on Cross Street: Agreement between William Gabbidon, John Lawson, and George Alexander Kidd, 11 March 1844, Memorial, appendix A2, TNA, CO 267/245. Isaacs likely owned more than one property.

25. Solomon as consular agent: Hearl, "Saul Solomon of St. Helena," 170. Solomon's accounting from "The Government of St. Helena to S. Solomon," 20 October 1840, signed document that hangs on the wall of Longwood House in St. Helena.

26. Redman's accounting: "Statement of Loss sustained by Mr. George C. Redman," 18 July 1840, TNA, FO 403/1, division 5, enclosure no. 16, 17. Redman, Conneau (Canot), and slaving ship: Conneau, *A Slaver's Log Book*, 292, 309, 55.

27. Amphibious Jews: Conneau, *A Slaver's Log Book*, 28. The phrase is used to describe Cuban traders, but Jewish merchants in Sierra Leone are also noted: 77, 97. Deed reproduced: Conneau, *A Slaver's Log Book*, 309; see also Holsoe, "Theodore Canot at Cape Mount," appendix 3, 176. Namina Lahay quoted by J. Jeremie in "Report of Select Committee on West Coast of Africa," appendix 19, 287.

28. Canot's career: Holsoe, "Theodore Canot at Cape Mount," 164–65, 172. British assessment of Canot and hose system: N. Macdonald to Grey, 12 May 1847, SLNA, Governor's Despatches to Secretary of State 2, October 1846–12 September 1847.

29. Expedition vessels: Reid, "Dr. Reid on the Ventilation of the Niger Steam Ships," parts 1 and 2. Percentage of steam vessels: Groves, *The Planting of Christianity*, 7n3. Expedition's stop in Sierra Leone: Allen and Thomson, *A Narrative of the Expedition*, 77; Temperley, *White Dreams, Black Africa*, 76–77; Groves, *The Planting of Christianity*, 9. Captain dining with expedition leaders: Simpson, *A Private Journal*, 23.

30. Quotations: Buxton. *The African Slave Trade*, 10, 306. Commerce and Christianity among Krios: Porter, *Religion versus Empire?*, 99.

31. Livingstone expedition and quotation: Groves, *The Planting of Christianity*, 8, 13. Casualty figures: Groves, *The Planting of Christianity*, 10; Temperley, *White Dreams, Black Africa*, 142.

32. Hebrew Bibles: Temperley, *White Dreams, Black Africa*, 69. Letters and quotation [my translation]: Simpson, *A Private Journal*, appendix, 129–31n(b). Possibly the rabbis were motivated by eighteenth-century reports on West African "Loango Jews": see Parfitt, *Hybrid Hate*, 64–75.

33. Parfitt, *Hybrid Hate*, 170.

34. Fynn quoted: KCC, Fynn Family folder 3, diary 1, 17. Fynn informing Bishop: Gray, *A Journal of the Bishop's Visitation Tour*, 90. Myth of Zulu origins among Jews: Fynney, *Zululand and the Zulus*; J.W. Matthews, *Incwadi Yami* (1887); various documents compiled by the colonial service official and ethnographer James Stuart; "Information on Shaka" and "Notes for Tshaka's Life," July 1905, KCC, Stuart Papers, file 53; "Annotations to *Annals of Natal*" and "Customs of Jews Similar to Zulu," KCC, Uncat. Mss., 24; Samuelson, *Long Long Ago*; Mokoena, "'The Black House,'" 401–11. On Haggard: Kaufman, "*King Solomon's Mines?*," 517–39.

35. Quotations: Winterbottom, *An Account of the Native Africans*, 129–30, 187. For more on this dynamic, see Parfitt, "The Use of the Jew," 51–53.

36. Quotations: Clarke, *Sierra Leone*, 22. On the *Amistad*: Rediker, *The Amistad Rebellion*, 217–22.

37. Customs: A.E. Kennedy to H. Merivale, 23 January 1855, TNA, CO 267/252; partial Custom House (London) notice, 21 May 1845, TNA, CO 267/249. Isaacs' fine: S. Hill to J. Carr, 20 November 1855 TNA, CO 267/249. Redman's indemnity: *Hansard's Parliamentary Debates*, 3rd series, vol. 113, 22 July 1850, 93.

38. Isaacs' relocation: A.E. Kennedy to secretary of state, 18 September 1854, SLNA, Governor's Despatches to Secretary of State, 24 July 1854–17 September 1855, letter 161. Kennedy dates Isaacs' "fraud on Revenue" to 5 January 1844 and attributes his move to Matakong to the desire to avoid further charges and prosecution. Isaacs purchased Matakong from the holder of the mortgage, John Dawson, on 4 August 1844: Memorial, appendix A3, TNA, CO 267/245. On Gabbidon, see Walker, *The Black Loyalists*, 311. Tangled legal saga of title to Matakong: CRL, MI (1st and 2nd acquisition); Memorial, appendix A1, 11 March 1842; A2, 11 March 1844; A3, 4 August 1844; A4, 15 January 1850; A5, 28 September 1852, TNA, CO 267/245.

39. Descriptions from "Matacong," n.d. [ca. 1867], CRL, MI, no. 178.

40. Thanks to Dr. J. Michael Daniels for information on rock composition.

41. Thoughts on islands inspired by Dening, "The Theatricalities of Derring-do," 159–76.

42. Quotations from Isaacs and as chieftain: N. Isaacs [unsigned] to H.H. Molyneux (secretary of state for the colonies), 30 November 1867, CRL, MI, no. 135. Pier and wharf: "Matacong, West Coast of Africa," 2 December 1854, 551. Tramway: "Matacong," n.d. [ca. 1867], CRL, MI, no. 178. On landlord-and-stranger compacts, see Dorjahn and Fyfe, "Landlord and Stranger," 391–97.

43. Memorial, appendix A7, 14 December 1850, TNA, CO 267/245.

44. Large trees: Gough, "Rethinking the Colonial City," 191–92. Map notation: "Forikaria, Mellakori, and Tanna Rivers," map by T. Boteler, 1829, corrected 1873, CRL, MI, no. 238. Walls and compound: "Matacong, West Coast of Africa," 2 December 1854, 552. Report on establishment of school: "Sierra Leone District: Report of the Religious State of the Societies in the Freetown Circuit for the Year Ending Dec. 31st 1853," WMMS, West Africa, Synod Minutes, box 289. Battlements and natural spring: "Matacong," n.d. [ca. 1867], CRL, MI, no. 178.

45. Reports by superintendent: "Extract of a Letter from Rev. Gilbert, December 24, 1852," 47–48. Two visits: *Report of the Wesleyan-Methodist Missionary Society for the Year Ending April, 1853*, and "Sierra Leone District: Report of the Religious State of the Societies in the Freetown Circuit for the Year ending Dec. 31st 1853," both in WMMS, West Africa, Synod Minutes, box 289. For conclusion: L. Reay to general secretaries, 21 April 1853, WMMS, West Africa, Correspondence, box 282.

46. Hearl, "Saul Solomon of St. Helena," 170–73.

47. Thanks to Solomon & Company director Mandy Peters for granting access.

48. Importance of peanuts, and quotation: Brooks, *Yankee Traders*, 182. Isaacs as French consular agent: S. Hill to J. Carr, 20 November 1855, TNA, CO 267/249; *Journal de Jurisprudence Commerciale et Maritime* 30 (1851): 122.

49. Fyfe, *History of Sierra Leone*, 258; Lewis, *Slaves for Peanuts*, 17.

50. Dickens (unsigned), "The Literary Examiner: Review of *Narrative of the Expedition*," 533; Hoppen, *The Mid-Victorian Generation*, 390.

51. Dynamic of collaboration: Bangura, *The Temne of Sierra Leone.* 7, 16, 20. Praise of Isaacs: Governor L. Smyth-O'Connor to Duke of Newcastle, 13 March 1854, TNA, CO 87/57. Isaacs' knowledge of Mende: Deposition of Bangah. 5 July 1854, TNA, CO 267/243. The testimony indicates Isaacs spoke "Cassoo," the language now called Mende. For more on this designation, see Anderson, *Abolition in Sierra Leone*, 133–35. Isaacs peppers both volumes of *Travels and Adventures* with isiZulu phrases and records his growing ability to speak the language. Ben Marks recorded that he learned to speak basic Wolof for the purpose of trading, and it is reasonable to assume that Isaacs did as well: B. Moss to S. Moss, 27 December 1835, MFA-J.

52. Mouser, "Trade and Politics"; Coifman, "The Western West African and European Frontier," 273, 276; Mouser. *American Colony*, 50.

53. Isaacs does not appear to have had a relationship with Mary Ann Skelton prior to her father's death in 1843. See Ware, "Enoch Richmond Ware's Voyage," 10 February 1843, 308–9. Mouser, "The Nunez Affair," 740n78, indicates, based on a genealogy compiled by C. Fyfe, that Mary Ann Skelton married S. R. Lightbourn in 1846. Skelton and Fraser families: Schafer, "Family Ties That Bind," 1–2, 13. Elizabeth Skelton's parents were John Fraser, a Scottish American, and Phenda, a free African. William Skelton Jr.'s parents were William Skelton Sr. and a free African woman whose name was not recorded. On William Skelton Jr: Mouser, *American Colony*, 26. Schafer states that Isaacs married Emma Skelton, which does not accord with other sources: "Family Ties That Bind," 15.

54. Quotation and timeline: Ware, "Enoch Richmond Ware's Voyage to West Africa," 15 January 1843, 304, 307. Children: Nathaniel Isaacs' will, in Jolles, *Isaacs*, 293–94.

55. Mary Ann's second marriage: Mouser, "The Nunez Affair," 721–22. Family relationships: Mouser, "The Lightbourn Family," 25–27. Neria Bely: Niane, "Africa's Understanding of the Slave Trade," 85–86; Barrow, *Fifty Years in Western Africa*, 59–60; Lewis, *Slaves for Peanuts*, 33.

56. William Skelton Sr.'s wife, Elizabeth Fraser Skelton, was the daughter of John Fraser, a white Scottish-American slaver who operated plantations in South Carolina. He later moved to Spanish Florida to avoid the interdiction of his human cargo once the United States outlawed the slave trade. Fraser's vast holdings were supplied with slaves from a West African depot managed by his free African wife, Phenda, who was Elizabeth Fraser Skelton's mother and hence Mary Ann's grandmother. Later, as the only heir to her slave-trading father's estate, Elizabeth Fraser inherited a fortune, which was the subject of a fascinating court case. See Schafer, "Family Ties That Bind." On legitimacy and illegitimacy: Law, "Introduction," 26n1.

57. Quotation: Ware, "West African Trading Voyage, Part II," 27 November 1844, 55. Examples of how the trade was conducted: Journal of B. Moss, 16 July 1836, MFA-J; Gray and Dochard, *Travels in Western Africa*, appendix, article 1, 368.

58. Slave labor: Brooks, "Peanuts and Colonialism," 49–50. Quotation: Maclean, "The Liberty of British Subjects," 1 May 1846, 68. Threat of force: Maclean, "Loss," February 1855, 65.

59. Quotations: N. Isaacs and J. Waddell to Governor Macdonald, 21 October 1847, Memorial, appendix C1, TNA, CO 267/245; N. Isaacs and J. Waddell to Governor Macdonald, 4 December 1847, Memorial, appendix C2, TNA, CO 267/245. Isaacs employing King Demba's son: "Matacong, West Coast of Africa," 2 December 1854, 552. Dubreka was about fifty miles by sea northeast of Matakong.

60. Report and quotation: N. Isaacs and J. Waddell to Governor Macdonald, 4 December 1847, Memorial, appendix C2, TNA, CO 267/245.

61. Drawing made January and February 1866 by artillery lieutenant D. Hirtz, supplied by genealogical researcher Xavier Ricou, 2 May 2022.

62. Memorial, appendix D3, TNA, CO 267/245.

63. See Wolf, "Aspects of Group Relations."

64. Domingo, "The Caulker Manuscript," 38, notes that George Stephen Caulker worked for Isaacs, as did Stephen Jasper Caulker, Joseph Caulker, and David Caulker: S. J. Caulker to Governor A. E. Kennedy, n.d., Memorial, appendix M1, TNA, CO 267/245. Another source notes that Joseph Caulker was a shipwright for Isaacs on Matakong: TNA, CO 267/252, 23 August 1854; Memorial, Appendix J, 29 March 1853, TNA, CO 267/245. H. Christian's report on Matakong notes that David Caulker, the son of Joseph Caulker, worked for Isaacs on Matakong. See also Browne-Davies, "Jewish Merchants in Sierra Leone, 1831–1934." On desirability of Ba Tha's territory: Domingo, "The Caulker Manuscript," 44. On civil war: Acting Governor Price to H. G.

Grey, 15 September 1848, SNLA, Governor's Despatches to Secretary of State, 20 September 1847–27 September 1848.

65. Murder of informant: Domingo, "The Caulker Manuscript," 33–35. Image: Clarke, "Sketches" 357, plate 5, figure 22. The accompanying text incorrectly identifies the chief as "Candibar"; details indicate this is Canray Bah.

66. Domingo, "The Caulker Manuscript," 39–40.

67. Domingo, "The Caulker Manuscript," 41.

68. Domingo, "The Caulker Manuscript," 41–42.

69. Acting Governor Price to H. G. Grey, 15 September 1848, SNLA, Governor's Despatches to Secretary of State, 20 September 1847–27 September 1848.

70. The nine treaties are: (1) 2 November 1847, Treaty with Dembah, King of Kaloom Country, Dubreka & Tombo Island; (2) 15 November 1847, Treaty with Alimami Ali King of the Fouricaria Country; (3) 16 November 1847, Treaty with Alimami Sarleah, Chief of the Bareira Country; (4a) 2 August 1851, Treaty with Quia Foday, King of the Fouricaria Country & (4b) 7 December 1851, Treaty with Quia Foday, King of the Fouricaria Country; (5) 23 December 1851, Treaty with Bey Ingar, King of the Small Scarcies River; (6) 26 December 1851, Treaty with Bey Farima, King of Macbatee; (7) 26 December 1851, Treaty with Sattan Lahai, King of Kambia; (8) 17 January 1852, Treaty with Bala Bango, King of the Rio Pongas; (9) 29 January 1852, Treaty with Stephen, King of Wonkafong. Treaties 3, 4b, and 5–9 are extant in TNA, CO 267/249. Treaties 1 and 2 are extant in an uncatalogued folder marked "Treaties 1847," SNLA; Treaty 4a is extant in an uncatalogued folder marked "Treaties 1851," SNLA. Gifts exchanged: Governor N. Macdonald to King Tongoh, 12 March 1851, SLNA, Governor's Letterbook 1848–1851. Lenie Isaacs' tombstone is barely legible today. For a map showing the graves of Lenie and Nathaniel, see Jolles, *Isaacs*, 286.

71. Quotations: Extract from minutes of governor's report, 13 February 1853, Memorial, appendix G3, TNA, CO 267/245; E. Dillet, N. Isaacs, and F. E. Dennis to Governor N. Macdonald, 2 February 1852, Memorial, appendix G2, TNA, CO 267/245; Governor N. Macdonald to Colonial Secretary H. G. Grey, 17 February 1852, SLNA, Governor's Despatches to Secretary of State, 1 January 1852–20 January 1853. Difficulties of negotiation efforts and praise of trade: Dillet, Isaacs, and Dennis to Macdonald, 2 February 1852.

72. Gratitude conveyed: H. R. Searle (clerk) to N. Isaacs, 12 February 1852, Memorial, appendix G3, TNA, CO 267/245. Governor's commendation: Governor N. Macdonald to Colonial Secretary H. G. Grey, 17 February 1852, SLNA, Governor's Despatches to Secretary of State, 1 January 1852–20 January 1853. Governor's private thanks: N. Macdonald to F. E. Dennis, E. Dillet, and N. Isaacs, 13 February 1852, Memorial, appendix G4, TNA, CO 267/245. Fyfe identifies Dillet as "Afro-West Indian" in *History of Sierra Leone*, 277. Electric fish: Governor N. Macdonald to Colonial Secretary H. G. Grey, 16 April 1847, SLNA, Governor's Despatches to Secretary of State, 2 October 1846–12 September 1847, unnumbered.

73. N. Macdonald to Colonial Secretary H.G. Grey, 17 February 1852, SLNA, Governor's Despatches to Secretary of State, 1 January 1852–20 January 1853, no. 32.

74. 1852 statistics: Memorial, appendix B1, TNA, CO 267/245. 1853 statistics: Memorial, appendix B2, TNA, CO 267/245. Marseille: Daumalin, "Commercial Presence," 212–14. On value, see Brooks, "Peanuts and Colonialism," 44n49. Further research is required into the economics of the peanut trade to determine Isaacs' possible revenue and the reliability of the figures he supplied. Governor's estimate: Governor L. Smyth-O'Connor to Duke of Newcastle, 13 March 1854, TNA, CO 87/57.

75. Parley, *A Grammar of Modern Geography*, 318–21.

76. Cruikshank's illustration appeared as the frontispiece to Henry Mayhew's satiric novel *1851; or, The Adventures of Mr and Mrs Sandboys* (1851). See British Museum, Department of Prints and Drawings, www.britishmuseum.org/collection/object/P_1978-U-3003. Items displayed at Exhibition: *Alphabetical and Classified Index*.

77. See Scanlan, *Freedom's Debtors*, 19–25.

78. Philippa, Hannah's mother, married the English trader William Hayes, active at Cape Mesurado. This strategic headland was the site of the American Colonization Society's early efforts at African "repatriation" and is now part of Monrovia. See Christopher, *Freedom in White and Black*, 64; Fyfe, *History of Sierra Leone*, 146. "Housekeepers" and their offspring: Walker, *The Black Loyalists*, 310–11.

79. See Coventry Patmore, "Angel in the House," 1854.

80. Lewis served as chief commissary judge of the Mixed Commission Court of Sierra Leone. See "Death Notices," *Spectator*, 26 March 1842, 308. On governor's gratitude to Hayes: N. Macdonald to H. Hayes, 17 February 1852, SLNA, Local Letter Book, 30 June 1851–30 December 1852.

81. Missionary report: "Extract from a Letter from the Rev. Richard Fletcher, March 17, 1853," February 1854, 29. Sunday school enrollment: "Sierra Leone District: Report of the Sunday Schools in the Freetown Circuit for the Year ending Dec. 31st 1854," WMMS, West Africa, Synod Minutes, box 289. For those Isaacs employed, dry dock, population figures, and governor's gratitude: Governor L. Smyth-O'Connor to Duke of Newcastle, 13 March 1854, TNA, CO 87/57. Descriptions of hazards: Brooks, *Yankee Traders*, 81.

82. Quotations: "Matacong, West Coast of Africa," December 1854, 551–52. Cannon: Governor L. Smyth-O'Connor to Duke of Newcastle, 13 March 1854, TNA, CO 87/57.

83. Quotation: "Matacong, West Coast of Africa," 2 December 1854, 551. Dues: Jolles, *Isaacs*, 270.

84. Descriptions drawn from "Matacong, West Coast of Africa," 551–52. On coins: "London Numismatic Club Meeting (5 June 2007)," 26–27.

85. For quotations: Governor N. Macdonald to W. Gabbidon, 29 September 1852, Memorial, appendix G4, TNA, CO 267/245. See also Governor N. Macdonald to King Tombo Mahmodoo, 29 September 1852, Memorial, appendix H2, TNA, CO 267/245.

86. Gabbidon's response: W. Gabbidon to Governor N. Macdonald, 12 October 1852, Memorial, appendix H3,TNA, CO 267/245. King Tombo's response and quote: King Tombo Mahmodoo to Governor N. Macdonald, 25 October 1842, Memorial, appendix H4,TNA, CO 267/245. On free trade and freedom: Tombs, *The English and Their History*, 475.

87. Quotation: King Tombo Mahmodoo to Governor N. Macdonald, 25 October 1852, Memorial, appendix H4,TNA, CO 267/245. Protest of colonists: "Memorial of the Undersigned Citizens of Freetown," TNA, CO 267/244, 2, emphasis in original. This document complains of Governor Kennedy's overreach.

88. Kennedy's antislavery efforts: Governor A. E. Kennedy to secretary of state, 24 August 1854, SLNA, Governor's Despatches to Secretary of State, 24 July 1854–17 September 1855, letter 144. Here he describes rumors of Isaacs' slave trading. Characterization of Kennedy: Gilliland, "The Early Life and Governorships of Sir Arthur Edward Kennedy," 290, 307. By February 1853, Kennedy actively attempted reform: see Gilliland, 288–89. Isaacs' misdeeds: Governor A. E. Kennedy to colonial secretary, 13 February 1854, SLNA, Governor's Notebook 1853–1868; TNA, CO 267/234, 13 August 1853. (Special thanks to Dr. Milo Gough for providing this citation.) Kennedy on Isaacs: Governor A. E. Kennedy to H. Merivale, 23 January 1855, TNA, CO 267/252. Gabbidon informs governor: Governor A. E. Kennedy to secretary of state, 16 September 1854, SLNA, Governor's Despatches to Secretary of State, 24 July 1854–17 September 1855, letter 160.

89. On Momodu Yeli: Fyfe, *History of Sierra Leone*, 270. Image: Clarke, "Sketches," plate 7, figure 34. In the accompanying text this man is identified as "Mahomadoo Yelly." Details indicate that this is Momodu Yeli, whom Clarke compares to ancient Israelites, apparently referencing Psalms 92:10–11 (357).

90. Dispatch of steamer: Governor A. E. Kennedy to secretary of state, 16 September 1854, SLNA, Governor's Despatches to Secretary of State, 24 July 1854–17 September 1855, letter 160. Kennedy slights Isaacs' African ways: Governor A. E. Kennedy to H. Merivale, 23 January 1855, TNA, CO 267/252, 41.

91. Isaacs as deceitful: Governor A. E. Kennedy to H. Merivale, 23 January 1855, TNA, CO 267/252, 23 January 1855, 22. Kennedy on Isaacs and another Jewish resident: Governor A. E. Kennedy to secretary of state, 24 August 1854, SLNA, Governor's Despatches to Secretary of State, 24 July 1854–17 September 1855, letter 144. The individual referred to was either Abraham Lemon or Edward Lemon. See Browne-Davies, "Jewish Merchants in Sierra Leone," n.p. [18–19]. Kennedy on Aku quoted in Anderson, *Abolition in Sierra Leone*, 193.

92. Ikey Solomon: Pelham, *The Chronicle of Crimes*, 2:235–41. Dickens, *The Adventures of Oliver Twist*, 39–40, 43.

93. Tillotson, *Novels of the Eighteen-Forties*, 75–76. Sikes's "savage" manner is referred to three times in *Oliver Twist*. Quotations: Dickens, *The Adventures of Oliver Twist*, 285, 287–88.

94. N. Isaacs to Colonial Secretary J. Russell, 21 July 1855, TNA, CO 267/251.

95. Isaacs on battle: N. Isaacs to Governor A. E. Kennedy, 21 March 1853 Memorial, appendix H12, TNA, CO 267/245. Chieftains' statement: Statement

of Mamadoo Turee, minister of Bamba Mamy Lahi, king of Malageah, 15 April 1853, Memorial, appendix A6, TNA, CO 267/245. Isaacs on Gabbidon's machinations: N. Isaacs to Governor A.E. Kennedy, 29 March 1853, Memorial, appendix H12, TNA, CO 267/245. United States coverage: "From Sierra Leone," *Newark Daily Advertiser*, 6 June 1853; "From Sierra Leone," *Maine Farmer*, 15 June 1853, both reprinted from *Journal of Commerce* (New York).

96. Christian quoted in Report of H. Christian, 29 March 1853, Memorial, appendix J, TNA, CO 267/245.

CHAPTER 7. THE QUEEN VERSUS THE KING

1. Population statistics: Governor Macdonald to Sir John Packington, Bart., 4 July 1852, SLNA, Governor's Despatches to Secretary of State, 1 January 1852–20 January 1853, letter 91. Other sources indicate a lower population of about 16,500: see Crooks, *A History*, 184. Cannabis: Clarke, "Short Notice of the African Plant *Diambe,*" 10–11.

2. Slave ship masts: Crooks, *A History*, 180; Groves, *The Planting of Christianity*, 17. Details from this and preceding paragraph adapted from Clarke, "Sketches," 323, 325–26, 330, 338, 342, 346.

3. Colony's progress: Governor Macdonald to Sir John Packington, Bart., "Report on the Annual Blue Book for the Colony of Sierra Leone for the Year 1851," 26 June 1852, SLNA, Governor's Despatches to Secretary of State, 1 January 1852–20 January 1853, no. 83. On languages: Hair, "Colonial Freetown," 561. Biblical Hebrew: Groves, *The Planting of Christianity*, 18. Poor homes: F. Teal to general secretaries, n.d. [ca. 1853–1854], WMMS, West Africa, Correspondence, box 282. Weather and cockroaches: Clarke, *Sierra Leone*, 13–14, 21, 122–23. Captivity for tobacco or rice: J.G. Smyth to R. Armstrong, 1 November 1851, SLNA, Colonial Secretary's Letterbook. 1 October 1851–8 October 1852, no. 709; Governor Macdonald to Sir John Packington, Bart., 4 July 1852, SLNA, Governor's Despatches to Secretary of State, 1 January 1852–20 January 1853, letter 91. Missionary quote: H. Badger to general secretaries, 9 November 1843, WMMS, West Africa, Correspondence, box 281.

4. Crooks, *A History*, 189.

5. Domestic servitude: Colonial Secretary Charles Grant, Baron Glenelg, to Lieutenant Governor Campbell, n.d., SLNA, Secretary of State's Dispatches, 27 January 1836–31 December 1836, Coastal societies: Lewis, *Slaves for Peanuts*, 32. Treatment of domestic slaves: Scanlan, *Freedom's Debtors*, 211; Clarke, "Sketches," 359.

6. Quotation: W. Molesworth to Governor. Hill, 24 September 1855, SLNA, Secretary of State's Dispatches, 3 April 1855–31 December 1855, letter 24. Proclamation: Governor Hill, Proclamation, 25 September 1855, TNA, 267/248. West Africa Squadron instructions: Scanlan, *Freedom's Debtors*, 23. Praise of Kennedy: Secretary Newcastle to Governor Kennedy, 27 April 1853, SLNA, Secretary of State's Dispatches, 2 January 1852–20 December 1853.

7. Isaacs' letter: N. Isaacs to A.E. Kennedy, 15 May 1854, Memorial, appendix K, TNA, CO 267/245. Claims about Gabbidon: Smyth to B. Campbell, 23 October 1850, SLNA, Colonial Secretary's Letterbook, 7 September 1850–4

December 1850, letter 152. Campbell's impartiality is questionable; he married the widowed Elizabeth Skelton, mother of Isaacs' former paramour, Mary Ann: see Brooks. *Yankee Traders*, 203. Twelve chiefs allied with Isaacs submitted an affidavit claiming Gabbidon was a "notorious Slave Dealer": Memorial, appendix O, 26 August 1855, TNA, CO 267/245.

8. A. E. Kennedy to G. Grey, 16 September 1854, SLNA, Governor's Despatches to Secretary of State, 24 July 1854–17 September 1855, letter 160.

9. Quotations: Deposition of Seeray Moodoo, 15 July 1854, TNA, CO 267/241. Another witness on Nah Watah: Deposition of Bah Dougalah, 24 August 1854, TNA, CO 267/241.

10. Kennedy's belief: A. E. Kennedy to G. Grey, 16 September 1854, SLNA, Governor's Despatches to Secretary of State, 24 July 1854–17 September 1855, letter 160. Kennedy reports that Lieutenant Commander Christian "was certainly overreached and deceived by Mr. Isaacs." Copy of report and Isaacs as "sole monarch": notation to Governor L. Smyth-O'Connor to Duke of Newcastle, 13 March 1854, TNA, CO 87/57.

11. Deposition of Bangah, 5 July 1854; deposition of Barracah, 19 October 1854; deposition of Langoh, 8 December 1854, all in TNA, CO 267/243.

12. Malageah as "royal residence": "Forikaria, Mellakori, and Tanna Rivers," map by T. Boteler, 1829, corrected 1873, CRL, MI, no. 238. Lahai's defense of Isaacs: "News of Gabbidon's War," 15 April 1853, Memorial of N. Isaacs, appendix A6, TNA, CO 267/24. Kennedy's reported threat: Statement by Mahamadoo Touray, n.d., memorial of N. Isaacs, appendix L, TNA, CO 267/245. Placard incident: A. E. Kennedy to H. Merivale, 23 January 1855, TNA, CO 267/252, 33.

13. Quotation: Warrant, 9 August 1854, TNA, CO 267/243. Weather report: A. E. Kennedy to Secretary of State, 24 August 1854, SLNA, Governor's Despatches to Secretary of State, 24 July 1854–17 September 1855, letter 144. Storms: Clarke, "Sketches," 322.

14. Lyons and quotations: A. E. Kennedy to Secretary of State, 24 August 1854, SLNA, Governor's Despatches to Secretary of State, 24 July 1854–17 September 1855, letter 144. Kennedy indicates that Lyons came from Canterbury. Browne-Davies notes Lyons as Isaacs' nephew: "Jewish Merchants in Sierra Leone," 30n113. See also Jolles, *Isaacs*, 275–76. Demand that Isaacs turn himself in: Superintendent of police to N. Isaacs, 18 August 1854, TNA, CO 267/252. On Thomas Lewis: A. E. Kennedy to H. Merivale, 23 January 1855, TNA, CO 267/252. Lewis's complaint: T. Lewis to Capt. A. Heseltine, 18 August 1854, memorial of N. Isaacs, appendix N, TNA, 267/245.

15. Nineteen rescued: A. E. Kennedy to Secretary of State, 24 August 1854, SLNA, Governor's Despatches to Secretary of State, 24 July 1854–17 September 1855, letter 144. Testimony: Deposition of Ballah, 7 August 1855; deposition of Beahmah, 8 December 1854; deposition of Jeremiah (Isaacs), 23 October 1854; deposition of Daniel (Isaacs), 23 October 1854; deposition of Leah (Isaacs), 26 October 1854; deposition of Eliza, 26 October 1854; deposition of Ephraim (Isaacs, alias Anthony Rookes), 9 November 1854; deposition of Nathan (Isaacs, alias Cecil Rookes), 9 November 1854, all in TNA, CO 267/243. Leah (Isaacs) had previously been interviewed by Lieutenant Commander Henry Christian:

"Examination of Persons Called before Me and Questioned without Previous Notice Being Given," 29 March 1853, memorial of N. Isaacs, appendix J, TNA, CO 267/245. Quotations from Spence: deposition of William Spence, 29 December 1854, TNA, CO 267/243. On dispatches: documents marked "The Queen vs. Isaacs for Slave Dealing," TNA, CO 267/247.

16. Deposition of Dembah Bambayah, 18 July 1855, TNA, CO 267/249. Shebar: Arrowsmith, "Sierra Leone and the Adjoining Coasts."

17. Deposition of Dembah Bambayah, 18 July 1855, TNA, CO 267/249.

18. Report in G. Lawson to Governor Kennedy, 4 August 1854, TNA, CO 267/252. Don Luis and Pedro Blanco: B. Campbell to Earl of Clarendon, 3 February 1855, Sessional Papers, vol. 19, class B, "Correspondence with British Ministers and Agents in Foreign Countries Relating to the Slave Trade, April 1 1854–March 31 1855," no. 45, p. 46. Testimony of *Amistad* captives: Barber, *History of the Amistad Captives*, 11–14. A missionary to the region later identified "a Jew by name of ISAACS [as] the person often referred to in Captain Canot's book, under the title of 'my Israelitish friend,' who gave Canot . . . assistance in the slave trade": Thompson, *The Palm Land*, 90. This report is likely incorrect, as the dates for Canot's recollections and Isaacs' presence in the colony do not align.

19. Don Pedro and quotation: Hall, "Abolition of the Slave Trade of Gallinas," 33, 35; also Hall, "Voyage to Liberia," 338. IOU: Bill of exchange, P. Alvaraz Limidel to N. Isaacs, 22 April 1846, SLNA, Governor's Despatches to Secretary of State, 24 July 1854–17 September 1855.

20. Increased profitability: Leveen, "A Quantitative Analysis," 64–70 (esp. table 4, 65), 81. Increased demand: Hochschild, *Bury the Chains*, 311.

21. Testimony: Deposition of Ballah, 7 August 1855; deposition of Daniel, 23 October 1854; deposition of Jeremiah, 23 October 1854; deposition of Ephraim, 9 November 1854, all in TNA, CO 267/243. Matakong chapel: A. E. Kennedy to H. Merivale, 23 January 1855, TNA, CO 267/252. Thomas Reader: Thompson, *The Palm Land*, 90.

22. A. E. Kennedy to G. Grey, 24 August 1854, SLNA, Governor's Despatches to Secretary of State, 24 July 1854–17 September 1855, letter 144.

23. British population, urban growth, and free trade: Tombs, *The English and Their History*, 436, 483, 558–59. Commerce as remedy for slave trade: Robinson and Gallagher, *Africa and the Victorians*, 34. Empire and enterprise: Darwin, *Unfinished Empire*, 389, 394. Evangelical revival and Church of England: Tombs, *The English and Their History*, 465. "Jew Question": Campbell, "Substance of a Speech on the Jewish Question."

24. The merchant in Shakespeare's play is Antonio, not Shylock. I claim poetic license, but good on you for checking.

25. See Lester, *Imperial Networks*, 143.

26. D'Israeli's mockery: *The Genius of Judaism*, 178. Disraeli on conversion efforts and appeal of Judaism: *Tancred*, 68–70, 79. Shift from humanitarianism to imperialism: Bank, "Losing Faith in the Civilizing Mission," 380.

27. A. E. Kennedy to G. Grey, 16 September 1854, SLNA, Governor's Despatches to Secretary of State, 24 July 1854–17 September 1855, letter 160.

28. Macdonald quotation: Governor Macdonald to Sir John Packington, Bart., "Report on the Annual Blue Book for the Colony of Sierra Leone for the

Year 1851," 26 June 1852, SLNA, Governor's Despatches to Secretary of State, 1 January 1852–20 January 1853, no. 83. The phrase *capitalisme sauvage* appears to have been coined by the historian Marianne Debouzy in the late 1960s, in reference to American industrial rapaciousness beginning in the 1860s.

29. A. E. Kennedy to H. Merivale, 23 January 1855, TNA, CO 267/252.

30. Quotations: Governor Macdonald to Sir John. Packington, Bart., 5 July 1852, SLNA, Governor's Despatches to Secretary of State, 1 January 1852–20 January 1853.

31. My analysis of Isaacs is based on Scanlon's point that British antislavery policies in Sierra Leone were gradualist, acquisitive, and militarized: see *Freedom's Debtors*, 20. Isaacs' landscape: Statement by P. Manning, n.d., CRL, "Matacong Island Papers."

32. Parker, "Justice and Conscience," 84–85.

33. Scanlon, *Freedom's Debtors*, 112–13.

34. Kennedy, "Loss of the *Forerunner*," 15 and 21 November 1854; Bedingfeld, "Wreck of the 'Forerunner' African Mail Steamer," 522.

35. Quotation from sailor: Evans, "The Loss of the *Forerunner* Mail Boat." List of those lost: "Report of an Investigation into the Circumstances Attending the Loss of the Steam Ship Forerunner," 320–21. Official investigation into the disaster: "The Wreck of the African Mail Steamer *Forerunner*." Forerunner's seaworthiness: "Statement from African Steam Ship Company," 28 November 1854, TNA, CO 267/244. Captain's negligence: "Report of an Investigation," esp. 316–17.

36. Quotation: H. Merivale to C. E. Trevelyan, 18 November 1854, TNA, TS 25/827. Lack of original documents noted: J. Greenwood to C. E. Trevelyan, 22 November 1854, TNA, CO 267/243.

37. Dougan's assurances and addendum: R. Dougan to Sir G. Grey, 27 December 1854, TNA, CO 267/243.

38. Isaacs' opposition to Kennedy and remarks on local chiefs: N. Isaacs to G. Grey, 20 December 1854, TNA, CO 267/245.

39. N. Isaacs to G. Grey, 20 December 1854, TNA, CO 267/245.

40. Memorial of N. Isaacs, TNA, CO 267/245, 1, 4.

41. Memorial of N. Isaacs, 6–7, appendix O, TNA, CO 267/245.

42. Memorial of N. Isaacs, TNA, CO 267/245, 8, 10–12, 14–15, 20–22, 25.

43. A. E. Kennedy to H. Merivale, 23 January 1855, TNA, CO 267/252, 3–6, 24.

44. A. E. Kennedy to H. Merivale, 23 January 1855, TNA, CO 267/252, 14–17, 20–21, 25–27, 38, 44, 51.

45. N. Isaacs to G. Grey, 7 February 1855, TNA, CO 267/251.

46. N. Isaacs to G. Grey, 7 February 1855, TNA, CO 267/251.

47. N. Isaacs to S. Herbert, n.d., [February 1855], TNA, CO 267/251.

48. Internal memorandum to S. Herbert, 19 February 1855, TNA, CO 267/251, Benson Isaacs' date of death is from the gravestone; photograph in author's collection.

49. Quotations from Merivale: H. Merivale to W. Ball, 21 May 1855, TNA, CO 267/250, emphasis in original. Details on secrecy: unsigned to R. Dougan, 25 May 1855, TNA, CO 267/246.

50. W. Molesworth to Governor Hill, 22 September 1855, SLNA, Secretary of State's Dispatches, 3 April 1855–31 December 1855.

51. Description of mosque based on one destroyed in Sabbajee: Governor L. Smyth-O'Connor to Duke of Newcastle, 8 June 1853, TNA, CO 87/55. HMS *Teazer* also participated in the Sabbajee attack. Details of Malageah attack: W. Molesworth to Governor Hill, 22 September 1855, SLNA, Secretary of State's Dispatches, 3 April 1855–31 December 1855; Letter[?] to H. Merivale, 20 July 1855, TNA, CO 267/247. Effort to destroy town: "Fatal Encounter," 19 July 1855.

52. Descriptions of battle: W. Molesworth to Governor Hill, 22 September 1855, SLNA, Secretary of State's Dispatches, 3 April 1855–31 December 1855; Letter[?] to H. Merivale, 20 July 1855, TNA, CO 267/247; "Fatal Encounter," 12.

53. Quotation from official: Letter[?] to H. Merivale, 20 July 1855, TNA, CO 267/247. Merivale quotation: Memorandum, H. Merivale to J. Ball, 23 July 1855, TNA, CO 267/251.

54. N. Isaacs to J. Russell, 21 July 1855, TNA, CO 267/251.

55. Isaacs' announcement: N. Isaacs to J. Russell, 21 July 1855, TNA, CO 267/251. Merivale quotation: memorandum, H. Merivale to J. Ball, 23 July 1855, TNA, CO 267/251. Dougan informed: J. Ball to R. Dougan, 23 July 1855, TNA, CO 267/251.

56. Hayes as Isaacs' mistress: R. Dougan to secretary of state, 2 August 1855, TNA, CO 267/247. Gabbidon's demise: letter[?] to general secretaries, 13 November 1855, WMMS, West Africa, Correspondence, box 282. Last mention of Gabbidon: H. Labouchere to S. Hill, 23 November 1855, SLNA, Secretary of State's Dispatches, 3 April 1855–31 December 1855.

57. Quotations: W. Molesworth to S. Hill, 22 September 1855, SLNA, Secretary of State's Dispatches, 3 April 1855–31 December 1855. Molesworth's relationship with Bentham and Wakefield: Thomas, *The Philosophic Radicals*, 3, 379–80; Darwin, *Unfinished Empire*, 106.

58. Hill quotation: S. Hill to H. Labouchere, 20 December 1855, TNA, CO 267/249. Palmerston's intent to prosecute Isaacs: S. Hill[?] to H. Labouchere, 13 February 1856, TNA, CO 267/249. Witnesses discharged: Letter[?] to S. Hill, 20 February 1856, TNA, CO 267/249.

59. Epitaph from photograph of gravestone in author's collection. Isaacs' contributions: Jolles, *Isaacs*, 270.

60. Quotation: Milon, "L'Ile de Matakong." Watercolor: CRL, "Matacong Island Papers," undated watercolor titled "Matakong."

61. Uncatalogued folder marked "Treaties 1851," SLNA; Treaty with Quia Foday, king of the Fouricaria Country, 7 December 1851, TNA, CO 267/249.

62. Milon, "L'Ile de Matakong."

CHAPTER 8. SAVAGE INVASIONS

1. Reach, "Chaka—King of the Zulus," 1 January 1853, 299.

2. Maclean, "Loss," January 1853, 29–30, February 1855, 65.

3. Caldecott, *Descriptive History*, frontispiece.

4. Altick. *The Shows of London*, 279.

5. Caldecott, *Descriptive History*, 5–6.

6. Lindfors, *Early African Entertainments*, 91; "Last Three Weeks of the Zulu Kafirs," poster, Wellcome Collection, https://wellcomecollection.org/works/arpcmtev.

7. Caldecott, *Descriptive History*, title page, 3.

8. Caldecott, *Descriptive History*, 7; "Visit of the Zulu Kaffirs."

9. Quotations: "St. George's Gallery," 18 May 1853; "The Caffres at Hyde-Park-Corner," 18 May 1853. For placement: "University Intelligence," 18 May 1853; "United States Exploring and Surveying Expedition," 18 May 1853. On fascination with Zulus: "What They Are Doing in the Great Metropolis," May 1853.

10. Quotations: "Zulu Kaffirs," illustration caption; image patterned on illustration in Caldecott, *Descriptive History*, 29; "Caffre Exhibition," 21 May 1853; "Our Weekly Gossip," 28 May 1853; "St. George's Hall—The Zulu Kafirs," 25 July 1853.

11. Caldecott, *Descriptive History*, 9–11, 13, 18–19, 21.

12. Caldecott, *Descriptive History*, 30.

13. Dickens, "The Noble Savage," 337; Lindfors, "A Zulu View," 5.

14. Dickens, "The Noble Savage," 338–339.

15. Quotations: Dickens, "The Noble Savage," 338; Dickens, *The Adventures of Oliver Twist*, 39. African and Jewish odors: Parfitt, *Hybrid Hate*, 161–62.

16. Depictions of subalterns: McClintock, *Imperial Leather*, 52–53. Quotations: Dickens, "The Noble Savage," 339.

17. Most of the report reads convincingly, though occasionally the text suggests embroidery by the author.

18. *Natal Journal*, July 1858, quoted in Lindfors, "A Zulu View," 9–11.

19. *Natal Journal*, July 1858, quoted in Lindfors, "A Zulu View," 14–15.

20. *Natal Journal*, July 1858, quoted in Lindfors, "A Zulu View," 12, 16–18.

21. Quotation: N. Isaacs, unsigned, to [?], n.d., "Matacong Island Papers (first acquisition)," CRL, no. 179. The account of the ownership transfer is not entirely clear. Thomas Reader as partner and trading volume: N. Isaacs, "Matacong," unsigned report, "Matacong Island Papers (first acquisition)," ca. 1867, CRL, no. 178

22. Schwersensky's speech: "Foundation Stone of Hope Place Synagogue," 1 July 1856, 2, reprinted in Jolles, *Isaacs*, 258. Plaque: P. Ettinger, quoted in in Jolles, *Isaacs*, 261. Isaacs at consecration ceremony: *Jewish Chronicle*, 18 September 1857, reprinted in Jolles, *Isaacs*, 261.

23. Identification of Jews with crime: Mayhew, *London Labour*, 2:117; Disraeli, *Tancred*, 389.

24. Isaacs' contributions: Ledger and minute book, Canterbury Hebrew Congregation, quoted in Jolles, *Isaacs*, 270. Extension of credit: "Matacong Island Papers (first acquisition)," CRL, nos. 189, 190. Removal of Reader and Isaacs' return: N. Isaacs, unsigned, to [?], n.d. "Matacong Island Papers (first acquisition)," CRL, no. 179. Probable murder of Reader: W. Lawson [1887], quoted in Ijagbemi, "The Yoni Expedition of 1887," 248; "Legislative Council Chamber," 22 February 1890, 4. On Lemberg: "Fifty Years Residence," 11 February 1911; Browne-Davies, "Jewish Merchants," 33–37.

25. Isaacs' complaints: N. Isaacs, unsigned, to [?], n.d., "Matacong Island Papers (first acquisition)," CRL, no. 179.

26. Isaacs, quoted in W. Solomon, *Saul Solomon*, 78. The recipient of Isaacs' letter was Saul Solomon, the son of Joseph and Hannah Solomon and nephew of Saul Solomon of St. Helena.

27. 1871 census, quoted in Jolles, *Isaacs*, 272. A 1927 article records Isaacs as having two children, Phillipa and Ben. Possibly the informant confused Phoebe with Phillipa, in which case this Ben would be Benson. The article states that Phoebe/Phillipa and Ben(son) were educated in Canterbury and returned to West Africa once grown: see Jolles, *Isaacs*, 280.

28. Phoebe Anne Isaacs and Manning: 1871 census, quoted in Jolles, *Isaacs*, 272. List of goods sold: receipt from P. Manning, 24 April 1877, "Matacong Island Papers (first acquisition)," CRL, no. 136. Isaacs' residence and shared address: Jolles, *Isaacs*, 263, 271. Registry of vessel: Mayo, *Mercantile Navy List*, 378. *Uncle Nat*'s registry number was 11569. Reader as original owner: Scottish Built Ships, www.clydeships.co.uk, accessed 27 August 2024.

29. State of Matakong: A. Protin to N. Isaacs, 9 June 1867, "Matacong Island Papers (first acquisition)," CRL, no. 188. Quotations and reference in Isaacs' letter: N. Isaacs (unsigned) to Earl of Carnarvon [secretary for colonies], 30 November 1867, "Matacong Island Papers (first acquisition)," CRL, no. 135.

30. Copy of mortgage and conveyance, P. Manning, 16 April 1883, "Matacong Island Papers (second acquisition)," CRL, no. 283.

31. Isaacs' living arrangements: copy of mortgage and conveyance, P. Manning, 16 April 1883, "Matacong Island Papers (first acquisition)," CRL, no. 283. Quotation from will: will of Nathaniel Isaacs, reprinted in Jolles, *Isaacs*, postscript to 286. Cause of death: death certificate for Nathaniel Isaacs, reprinted in Jolles, *Isaacs*, postscript to 273.

32. Lewis's death: Declaration, P. Manning, 16 April 1883, "Matacong Island Papers (first acquisition)," CRL, no. 208. Sale of Matakong: conveyance Mrs. & Mr. P. Manning to R. Philpott, 23 May 1882, "Matacong Island Papers (first acquisition)," CRL, no. 203. Quotation on Matakong's status: official notification no. 84, 3 November 1891,"Matacong Island Papers (first acquisition)," CRL, no. 229.

33. Photo in author's collection.

POSTSCRIPT

1. Travelogues: Brantlinger, *Rule of Darkness*, 180. "Africanism": Miller, *Blank Darkness*, 14–22.

2. Haggard, *Cetywayo and His White Neighbors*. Quotations: Haggard, "A Zulu War Dance," reprinted as appendix 7 to *Cetywayo and His White Neighbors*, 278–94.

3. Quotation: Herzl, *Old New Land*, 193. Uganda plan: Rovner, *In the Shadow of Zion*, 45–77. Herzl on Zangwill: *Complete Diaries*, 26 November 1895, 1:276.

4. On Zangwill, Haggard, and Johnston, see Rovner, *In the Shadow of Zion*, 47. Quotation: Zangwill, "A Land of Refuge," 12.

Bibliography

ARCHIVES CONSULTED

BL   British Library, London

CRL   Cadbury Research Library, University of Birmingham, Birmingham, UK

EIC   East India Company records, digitized at the British Library, London

GHA   Government House Archives, Cape Town, South Africa

KCC   Killie Campbell Collection, held by University of KwaZulu-Natal, Durban, South Africa

MFA-J   Marks Family Archive

The letters and other documents in this family archive were held by Amelia Marks, the sister of Benjamin Moss. The materials passed into the possession of Peter and Barbara Rossiter of Arkley, Hertfordshire. Barbara Rossiter permitted Dr. Michael Jolles to allow me to photograph the contents.

SHGA   St. Helena Government Archives, Jamestown, St. Helena

SLNA   Sierra Leone National Archives, Fourah Bay College, Freetown, Sierra Leone

TNA   The National Archives, London

WMMS   Wesleyan Methodist Missionary Archives, School of Oriental and African Studies, University College London

WORKS CITED

Alexander, James Edward. *Narrative of a Voyage of Observation among the Colonies of Western Africa.* Vol. 1. London: Henry Colburn, 1837.

Allen, William, and T. R. H. Thomson. *A Narrative of the Expedition Sent by Her Majesty's Government to the River Niger in 1841.* Vol. I. London: Richard Bentley, 1848.

Alphabetical and Classified Index to the Official Catalogue of the Great Exhibition of the Works of Industry of all Nations, 1851. London: Spicer Brothers, 1851.

Altick, Richard D. *The Shows of London.* Cambridge, MA: Harvard University Press, 1978.

Anderson, John. *A Practical Essay on the Good and Bad Effects of Sea-Water and Sea-Bathing.* London: C. Dilly, 1795.

Anderson, Richard Peter. *Abolition in Sierra Leone: Re-building Lives and Identities in Nineteenth-Century West Africa.* Cambridge: Cambridge University Press, 2020.

Arendt, Hannah. *The Origins of Totalitarianism.* New York: Harcourt, 1985.

Arrowsmith, John. "Sierra Leone and the Adjoining Coasts, to Accompany Dr. Madden's Report." London: James & Luke Hansard, 1842. Available at Old Maps Online, www.oldmapsonline.org/map/uu/1874-372687.

"Art. I" [Review of Isaacs' *Travels and Adventures*]. *Quarterly Review*, February 1837, 1–29.

"Art. X" [Review of Isaacs' *Travels and Adventures*]. *Monthly Review*, July 1836, 416–21.

Bangura, Joseph J. *The Temne of Sierra Leone: African Agency in the Making of a British Colony.* Cambridge: Cambridge University Press, 2017.

Bank, Andrew. "Losing Faith in the Civilizing Mission: The Premature Decline of Humanitarian Liberalism at the Cape, 1840–1860." In *Empire and Others: British Encounters with Indigenous Peoples, 1600–1850,* 364–83, ed. Martin Daunton and Rick Halpern. Philadelphia: University of Pennsylvania, 1999.

Bannister, Saxe. *Humane Policy: Or, Justice to the Aborigines of New Settlements.* London: T. & G. Underwood, 1830.

———. *Memoir Respecting the Colonization of Natal, in South-Eastern Africa.* London: John W. Parker, 1839.

Barber. John W. *A History of the Amistad Captives: Being a Circumstantial Account of the Capture of the Spanish Schooner* Amistad, *by the Africans on Board* [1840]. Reprinted in *Massachusetts Review* 10, no. 3 (1969): 493–532.

Barrow, Alfred H. *Fifty Years in Western Africa: Being a Record of the Work of the Indian Church on the Banks of the Rio Pongo.* London: Society for Promoting Christian Knowledge, 1900.

Barrow, Ian. *The East India Company, 1600–1858.* Indianapolis, IN: Hackett, 2017.

Beddoe, John. "On the Physical Characteristics of the Jews." *Transactions of the Ethnological Society of London* 1 (1861): 222–37.

———. *The Races of Britain: A Contribution to the Anthropology of Western Europe.* London: Trübner & Co., 1885.

Bedingfeld, Norman. "Wreck of the 'Forerunner' African Mail Steamer." *Illustrated London News*. 18 November 1854.

Bellow, Saul. "Papuans and Zulus." *New York Times*. 10 March 1994. P. A25.

Belmessous, Saliha. *Assimilation and Empire: Uniformity in French and British Colonies, 1541–1954*. Oxford: Oxford University Press, 2013.

Bentham, Jeremy. *Writings on Australia, VII: Colonization Company Proposal*. Prepublication version. Edited by. T. Causer and P. Schofield, Bentham Project, 2018. discovery.ucl.ac.uk/id/eprint/10055306/1/7.%20Colonization%20Company%20Proposal.pdf.

Bicheno, James. *The Restoration of the Jews, the Crisis of All Nations*. London: J. Barfield, 1807.

Bird, John. *Annals of Natal*. Vol 1. Cape Town: T. C. Struik, 1965.

Bourne, Kenneth. *The Foreign Policy of Victorian England, 1830–1902*. New York: Clarendon Press, 1970.

Boyce, Robert. "The History of Yellow Fever in West Africa." *British Medical Journal*. 28 January 1911: 181–85.

Brantlinger, Patrick. *Rule of Darkness: British Literature and Imperialism, 1830–1914*. Ithaca, NY: Cornell University Press, 1988.

———. "Victorians and Africans: The Genealogy of the Myth of the Dark Continent." *Critical Inquiry* 12, no. 1 (Autumn 1985): 166–203.

Bridge, Horatio. *The Journal of an African Cruiser*. Edited by Nathaniel Hawthorne. Aberdeen: George Clark & Son, 1848.

British and Foreign State Papers, 1845–1846. Vol. 34. London: Harrison & Sons, 1860.

Brooks, George. "Peanuts and Colonialism: Consequences of the Commercialization of Peanuts in West Africa, 1830–70." *Journal of African History* 16, no. 1 (1975): 29–54.

———. *Yankee Traders, Old Coasters and African Middlemen: A History of American Legitimate Trade with West Africa in the Nineteenth Century*. Boston: Boston University Press, 1970.

Brown, Christopher Leslie. *Moral Capital: Foundations of British Abolitionism*. Chapel Hill: University of North Carolina Press, 2006.

Browne-Davies, Nigel. "Jewish Merchants in Sierra Leone, 1831–1934." *Journal of Sierra Leone Studies* 6 (2017): n.p. [3–110].

Burg, B. R. *Boys at Sea: Sodomy, Indecency, and Courts-Martial in Nelson's Navy*. New York: Palgrave McMillan, 2007.

Burkhardt, Frederick, ed. *The Correspondence of Charles Darwin*. Vol. 9. Cambridge: Cambridge University Press, 1994.

Burroughs, Peter. *Colonial Reformers and Canada, 1830–1849*. Montreal: McGill-Queen's University Press, 1969.

Burton, Richard Francis, and Verney Lovett Cameron. "The Market at Freetown" [1883]. In *Africa*, ed., F. D. Herbertson, 62–63. London: Adam and Charles Black, 1902.

Buthelezi, Mbongiseni. "We Need New Names Too." In *Tribing and Untribing the Archive*, vol. 2, ed. C. Hamilton and N. Leibhammer, 584–99. Pietermaritzburg, South Africa: University of KwaZulu-Natal Press, 2016.

Buxton, Thomas Fowell. *The African Slave Trade and Its Remedy*. London: John Murray, 1840.

"Caffre Exhibition." *Spectator*. 21 May 1853.

"The Caffres at Hyde-Park-Corner." *Times*. 18 May 1853.

Caldecott, Charles Henry. *Descriptive History of the Zulu Kafirs: Their Customs and Their Country*. London: John Mitchell, 1853.

Campbell, William Frederick. *Substance of a Speech on the Jewish Question*. London: Francis & John Rivington, 1849.

Carpenter, Edmund. "The Tribal Terror of Self-Awareness." In *Principles of Visual Anthropology*, ed. Paul Hockings, 481–91. New York: Mouton de Gruyter, 2003.

Cesarani, David, ed. *Port Jews: Jewish Communities in Cosmopolitan Maritime Trading Centres, 1550–1950*. Portland, OR: Frank Cass, 2002.

Chaucer, Geoffrey. *The Canterbury Tales*. Edited by Paul Ruggiers. Norman: University of Oklahoma Press, 1979.

Chrisman, Laura. *Rereading the Imperial Romance: British Imperialism and South African Resistance in Haggard, Schreiner, and Plaatje*. Oxford: Clarendon Press, 2000.

Christopher, Emma. *Freedom in White and Black: A Lost Story of the Illegal Slave Trade and Its Global Legacy*. Madison: University of Wisconsin Press, 2018.

"Churton's British and Foreign Public Library." *Bent's Monthly Literary Advertiser*, March 1836, 33.

Clarke, Robert. "Short Notice of the African Plant *Diambe*, commonly called Congo Tobacco." In *Hooker's Journal of Botany and Kew Garden Miscellany*. London: Reeve and Benham, 1851.

———. *Sierra Leone: A Description of the Manners and Customs of the Liberated Africans*. London: James Ridgway, 1843.

———. "Sketches of the Colony of Sierra Leone and its Inhabitants." *Transactions of the Ethnological Society of London* 2 (1863): 320–63.

Clarkson, Thomas. *The History of the Rise, Progress, and Accomplishment of the Abolition of the Slave Trade*. Vols. 1– 2. London: Longman, Hurst, Rees, and Orme, 1808.

Cohen, Robin. *Global Diasporas*. London: Routledge, 2008.

Coifman, Victoria. "The Western West African and European Frontier: Contributions from Former Archbishop of Conakry Raymond-Marie Tchidimbo's *Autobiography* for West African History." In *Paths Towards the Past*, ed. Robert W. Harms, Joseph Miller, and David Newbury, 273–92. Atlanta, GA: African Studies Association Press, 1994.

Cole, Alfred Whaley. *Reminiscences of My Life and of the Cape Bench and Bar*. Cape Town, South Africa: J. C. Juta & Co., 1896.

Colledge, J. J., and Ben Warlow. *Ships of the Royal Navy*. Philadelphia: Casemate, 2010.

"The Colonization of South Africa." *Jewish Chronicle*. 26 July 1895.

Colquhoun, Patrick. *A Treatise on the Police of the Metropolis*. London: H. Fry, 1796.

Conneau, Theophilus [Théodore Canot]. *A Slaver's Log Book, or 20 Years' Residence in Africa: The Original Manuscript.* Englewood Cliffs, NJ: Prentice-Hall, [1853] 1976.

Conrad, Joseph. *Heart of Darkness* and Selections from *The Congo Diary.* New York: Random House, [1899] 1999.

Cooke, William. "Ancient Faith and Modern Skepticism." *Methodist New Connexion Magazine.* March 1866, 141–56.

"Correspondence with British Ministers and Agents in Foreign Countries Relating to the Slave Trade, April 1 1854-March 31 1855." *Sessional Papers.* Vol. 29, class B. London: Harrison & Sons, 1855.

Crooks, J. J. *A History of the Colony of Sierra Leone, Western Africa.* Dublin: Brown & Nolan, 1903.

Darwin, John. *Unfinished Empire: The Global Expansion of Britain.* New York: Bloomsbury, 2012.

Daumalin, Xavier. "Commercial Presence, Colonial Penetration: Marseille Traders in West Africa in the Nineteenth Century." In *From Slave Trade to Empire: European Colonization of Black Africa, 1780s–1880s,* ed. Olivier Pétré-Grenouilleau, 209–30. New York: Routledge, 2004.

Dening, Greg. "The Theatricalities of Derring-Do." In *Readings/Writings,* 159–176. Melbourne: Melbourne University Press, 1998.

Dibdin, Thomas. *The Jew and the Doctor: A Farce in Two Acts.* London: Longman, Hurst, Rees, and Orme, 1805.

Dickens, Charles. *The Adventures of Oliver Twist.* London: Chapman and Hall, [1837] 1855.

———. "The Noble Savage." *Household Words.* 11 June 1853, 337–339.

———. [unsigned]. "The Literary Examiner: Review of *Narrative of the Expedition Sent by Her Majesty's Government to the River Niger in 1841.*" *Examiner.* 19 August 1848, 531–533.

Disraeli, Benjamin. *Tancred, or The New Crusade.* New York: Routledge, [1847] 1877.

———. *The Wondrous Tale of Alroy* [1833]. In *The Early Novels of Benjamin Disraeli,* ed. Geoffrey Harvey. London: Pickering and Chatto, 2004.

D'Israeli, Isaac. *The Genius of Judaism.* London: Edward Moxon, 1833.

———. "On the Late Installation of a Grand Sanhedrim of the Jews in Paris." *Monthly Magazine, or, British Register.* August 1807, 34–38.

Disturnell, John, and Edwin Clayton. *New York as It Is, in 1835.* New York: J. Disturnell, 1835.

Domingo, George Maximillan. "The Caulker Manuscript." Edited by J. De Hart. *Sierra Leone Studies* 4 (1920): 17–48.

Dorjahn, V. R., and Christopher Fyfe. "Landlord and Stranger: Change in Tenancy Relations in Sierra Leone." *Journal of African History* 3, no. 3 (1962): 391–97.

Dubin, Lois. *The Port Jews of Habsburg Trieste: Absolutist Politics and Enlightenment Culture.* Stanford, CA: Stanford University Press, 1999.

Du Preez, Michael, and Jeremy Dronfield. *Dr. James Barry: A Woman Ahead of Her Time.* London: One World, 2016.

"East Coast of Africa—Brig *St. Michael's.*" *Shipping and Commercial List and New-York Price Current.* 25 April 1832.

Eilberg-Schwartz, Howard. *The Savage in Judaism: An Anthropology of Israelite Religion and Ancient Judaism.* Bloomington: Indiana University Press, 1990.

Elbourne, Elizabeth. "The Bannisters and Their Colonial World: Family Networks and Colonialism in the Early Nineteenth Century." In *Within and Without the Nation,* ed. Karen Dubinsky, Adele Perry, and Henry Yu, 49–75. Toronto: University of Toronto Press, 2018.

Eldredge, Elizabeth. *The Creation of the Zulu Kingdom, 1815–1828: War, Shaka, and the Consolidation of Power.* New York: Cambridge University Press, 2014.

Endelman, Todd. *The Jews of Georgian England, 1714–1830.* Philadelphia: Jewish Publication Society, 1979.

Evans, James. "The Loss of the *Forerunner* Mail Boat." *Times.* 20 November 1854.

Everill, Bronwen. "Bridgeheads of Empire? Liberated African Missionaries in West Africa." *Journal of Imperial and Commonwealth History* 40, no. 5 (2012): 789–805.

"Extract of a Letter from Rev. Robert Gilbert, December 24, 1852." In "Remarkable Movement in Sierra Leone." *Wesleyan Missionary Notices,* March 1853, 42–48.

"Extract from a Letter from the Rev Richard Fletcher, March 17, 1853." In *Wesleyan Missionary Notices.* February 1854, 28–29.

"Fatal Encounter on the West Coast of Africa." *Times.* 19 July 1855.

Feldman, David. "Conceiving Difference: Religion, Race and the Jews in Britain, c. 1750–1900." *History Workshop Journal* 176 (2013): 160–86.

Felsenstein, Frank. *Anti-Semitic Stereotypes: A Paradigm of Otherness in English Popular Culture, 1660–1820.* Baltimore, MD: Johns Hopkins University Press, 1995.

Ferguson, Jonathan. "'Trusty Bess': the Definitive Origins and History of the term 'Brown Bess.'" *Arms and Armour.* 14, no. 1 (2017): 49–69.

"Fifty Years Residence in West Africa." *Sierra Leone Weekly News.* 11 February 1911.

Foster, John Burt Jr. *Transnational Tolstoy: Between the West and the World.* New York: Bloomsbury, 2013.

Fox Bourne, Henry Richard. *The Aborigines Protection Society: Chapters in Its History.* London: P. S. King and Son, 1899.

"From a Correspondent at Port Elizabeth." *South African Commercial Advertiser.* 31 December 1828.

"From Sierra Leone." *Maine Farmer.* 15 June 1853.

"From Sierra Leone." *Newark Daily Advertiser.* 6 June 1853.

Frye, J. "The *South African Commercial Advertiser* and the Eastern Frontier, 1834–1847." MS thesis, Rhodes University (Grahamstown, South Africa), 1968.

Fyfe, Christopher. *A History of Sierra Leone.* Oxford: Oxford University Press, 1962.

Fynn, Henry Francis. "Captain King's Illness." *South African Commercial Advertiser.* 17 January 1829.

———. "The Diary of Henry Francis Fynn." Edited by James Stuart and D. McK. Malcolm. Pietermaritzburg, South Africa: Shuter & Shooter, 1986.

———. "Graham's Town—January 16th." *South African Commercial Advertiser.* 24 January 1828.

Fynney, Fred B. *Zululand and the Zulus.* Pretoria: State Library, [ca. 1880] 1967.

Gardiner, Allen F. *Narrative of a Journey to the Zoolu Country, in South Africa.* London: William Crofts, 1836.

Gatrell, V. A. C. *The Hanging Tree: Execution and the English People, 1770–1868.* Oxford: Oxford University Press, 1994.

Gidney, W. T. *The History of the London Society for Promoting Christianity amongst the Jews, 1809–1908.* London: London Society for Promoting Christianity amongst the Jews, 1908.

Gilbert, Alan. *Black Patriots and Loyalists: Fighting for Emancipation in the War for Independence.* Chicago: University of Chicago Press, 2012.

Gilbert, William. "The London Jews." *Good Words for 1864.* 864–870.

Gilliland, Henry C. "The Early Life and Governorships of Sir Arthur Edward Kennedy." Thesis, University of British Columbia, 1951.

Gilman, Sander. *Jewish Self-Hatred: Anti-Semitism and the Hidden Language of the Jews.* Baltimore, MD: Johns Hopkins University Press, 1986.

———. *The Jew's Body.* New York: Routledge, 1991.

Glenk, Robert. "Macassar Oil." *Western Druggist* 41, no. 3 (March 1894): 89.

Golan, Daphna. *Inventing Shaka: Using History in the Construction of Zulu Nationalism.* Boulder, CO: Lynne Rienner, 1994.

Goldsmith, Oliver. *History of the Earth and Animated Nature.* Vol. 1. Edited by Thomas Browne. Manchester: J. Gleave, 1814.

Gordon-Cumming, Roualeyn. *A Hunter's Life in South Africa.* Vol. 1. London: John Murray, 1850.

Gostling, William. *A Walk in and about the City of Canterbury.* Canterbury: William Blackley, 1825.

Gough, Milo. "Rethinking the Colonial City: A Spatial History of Freetown, Sierra Leone." PhD diss., Goldsmiths, University of London, 2022.

Gray, Robert. *A Journal of the Bishop's Visitation Tour through the Cape Colony, in 1850.* Part 2. London: Society for the Propagation of the Bible, 1851.

Gray, Stephen, ed., *The Natal Papers of "John Ross."* Pietermaritzburg, South Africa: University of Natal Press, 1992.

Gray, William, and Staff Surgeon Dochard. *Travels in Western Africa, in the Years 1818, 1819, 1820, and 1821, from the River Gambia.* London: John Murray, 1825.

Green, Geoffrey. *The Royal Navy and Anglo-Jewry, 1740–1820.* London: Naval and Maritime Bookshop, 1989.

Grose, Francis. *Classical Dictionary of the Vulgar Tongue.* London: Sherwood, Neely, and Jones, 1823.

Groves, C. P. *The Planting of Christianity in Africa*, vol. 2, 1840–1878. London: Lutterworth Press, 1954.

Guy, Jeff. "The Tribal History Project, 1862–64." In *Tribing and Untribing the Archives*. Vol. 1. Edited by C. Hamilton and N. Leibhammer, 216–37. Pietermaritzburg, South Africa: University of Kwa-Zulu Natal Press, 2016.

Haggard, H. Rider. "An Incident of African History." *Youth's Companion*, 8 November 1900, 588–589.

———. *Cetywayo and His White Neighbours*. London: Trübner & Co., 1882.

———. *King Solomon's Mines*. New York: F. M. Lupton, [1885] 1889.

———. *Nada the Lily*. London: Macdonald & Co., [1892] 1951.

Hair, P. E. H. "Colonial Freetown and the Study of African Languages." *Africa: Journal of the International African Institute*. 57, no. 4 (1987): 560–65.

Hall, Catherine. *Civilising Subjects: Metropole and Colony in the English Imagination, 1830–1867*. Chicago: University of Chicago Press, 2002.

Hall, James. "Abolition of the Slave Trade of Gallinas." *Annual Report of the American Colonization Society* 33 (1850): 33–36.

———. "Voyage to Liberia." *African Repository* 33 (1857): 336–40.

Hamilton, Carolyn. "'The Character and Objects of Chaka': A Reconsideration of the Making of Shaka as 'Mfecane' Motor." *Journal of African History* 33, no. 1 (1992): 37–63.

———. "Restructuring within the Zulu Royal House." *African Studies* 56, no. 2 (1997): 85–113.

———. *Terrific Majesty: The Powers of Shaka Zulu and the Limits of Historical Invention*. Cambridge, MA: Harvard University Press, 1998.

Hamilton, Carolyn, ed. *The Mfecane Aftermath: Reconstructive Debates in South African History*. Johannesburg: Witwatersrand University Press, 1995.

Hamilton, Carolyn, and John Wright. "Moving Beyond Ethnic Framing: Political Differentiation in the Chiefdoms of the KwaZulu-Natal Region before 1830." *Journal of Southern African Studies*. 43, no. 4 (2017): 663–79.

Hansard's Parliamentary Debates. 3rd series. Vol. 113. 19 July–15 August 1850. London: Hansard, 1850. https://archives.parliament.uk/collections/getrecord/GB61_HAN_3_113.

Hart, H. G., ed. *The New Annual Army List for 1849*. London: John Murray, 1849.

Hasted, Edward. *The Canterbury Guide, or Traveller's Pocket Companion*. Canterbury: Bristow & Cowtan, 1807.

———. *The History of the Ancient and Metropolitical City of Canterbury*. Vol. 1. Canterbury: W. Bristow, 1801.

Hearl, Trevor. "Saul Solomon of St. Helena, 1776–1852." *St. Helena Britannica*, 165–74. London: Society of Friends of St. Helena, 2013.

Heartfield, James. *The Aborigines' Protection Society: Humanitarian Imperialism in Australia, New Zealand, Fiji, Canada, South Africa, and the Congo, 1836–1909*. New York: Columbia University Press, 2011.

Heilmann, Anne. *Neo-/Victorian Biographilia and James Miranda Barry: A Study in Transgender and Transgenre*. New York: Palgrave Macmillan, 2018.

Henriques, Henry Straus Quixano. *The Jews and the English Law*. Oxford: Horace Hart, 1908.

Herstlet, Edward. *Map of Europe by Treaty.* Vol. 1. London: Butterworths, 1875.

Herzl, Theodor. *The Complete Diaries of Theodor Herzl.* Vol. I. Edited by Raphael Patai. Translated by Harry Zohn. New York: Herzl Press and Thomas Yosseloff, 1960.

———. *Old New Land/Altneuland.* Translated by Lotta Levensohn. Minneapolis: Filiquarian Publishing, [1902] 2007.

Hilton, Boyd. *A Mad, Bad, and Dangerous People?: England 1783–1846.* Oxford: Clarendon Press, 2006.

Hobhouse, Hermione, ed. "East India Dock Road, North side: Nos 1–301 (and Nos 2–50)." In *Survey of London,* vols. 43–44, *Poplar, Blackwall and Isle of Dogs.* London: London County Council, 1994. Available at British History Online, www.british-history.ac.uk/survey-london/vols43–4.

Hochschild, Adam. *Bury the Chains: Prophets and Rebels in the Fight to Free an Empire's Slaves.* New York: Mariner Books, 2006.

Holden, William. *History of the Colony of Natal, South Africa.* London: Alexander Heylin, 1855.

Holsoe, Svend. "Theodore Canot at Cape Mount, 1841–1847." *Liberian Studies Journal* 44. no. 2 (1971–72): 163–81.

Hoppen, K. Theodore. *The Mid-Victorian Generation, 1846–1886.* New York: Clarendon Press, 1998.

Ijagbemi, Adeleye. "The Yoni Expedition of 1887: A Study of British Imperial Expansion in Sierra Leone." *Journal of the Historical Society of Nigeria* 7, no. 2 (1974): 241–54.

Ingram, J. F. *The Story of an African Seaport.* Durban, South Africa: G. Coester, 1899.

[Isaacs, Nathaniel?]. "Letter from Graham's Town." *Colonist.* 19 August 1828.

Isaacs, Nathaniel. *Travels and Adventures in Eastern Africa.* 2 vols. London: Edward Churton, 1836.

Jackson, Basil. *Notes and Reminiscences of a Staff Officer.* New York: E. P. Dutton & Co., 1903.

James, William. *Naval History of Britain.* Vol. 5. London: Harding, Lepard, & Co., 1826.

James Stuart Archive of Recorded Oral Evidence Relating to the History of the Zulu and Neighbouring Peoples. Edited by John Wright and Colin de B. Webb. Pietermaritzburg: University of Natal Press, 1976–.

Jolles, Michael. *Samuel Isaac, Saul Isaac and Nathaniel Isaacs.* London: Jolles Publications, 1998.

Jones, Hilary. *The Métis of Senegal: Urban Life and Politics in French West Africa.* Bloomington: Indiana University Press, 2013.

Journal de Jurisprudence Commerciale et Maritime. 30 (1851): 116–24.

Kaufman, Heidi. "*King Solomon's Mines?*: African Jewry, British Imperialism, and H. Rider Haggard's Diamonds." *Victorian Literature and Culture* 33 (2005): 517–39.

Kay, Stephen. *Travels and Researches in Caffraria.* London: John Mason, 1833.

Kennedy, Arthur. "Loss of the *Forerunner.*" *Times.* 15 and 21 November 1854.

Kennedy, Dane. *Mungo Park's Ghost: The Haunted Hubris of British Explorers in Nineteenth-Century Africa*. New York: Cambridge University Press, 2024.

King, James Saunders. "Lieut. Farewell's Settlement at Port Natal." *South African Commercial Advertiser*. 11 July 1826.

———. "Private Correspondence at King Chaka's Kraal." *Colonist*. 3 January 1828.

———. "Sketch of Lieut. Farewell's Settlement at Port Natal—Concluded." *South African Commercial Advertiser*. 18 July 1826.

King, Richard John. *Handbook for Travellers in Kent and Sussex*. London: John Murray, 1868.

Kipling, Rudyard. "The Man Who Would Be King." *The Phantom Rickshaw and Other Eerie Tales*. Allahabad: A. H. Wheeler & Co., 1888.

———. "The White Man's Burden." *The Writings in Prose and Verse of Rudyard Kipling*. Vol. 21. New York: Charles Scribner's Sons, 1903.

Kirby, Percival. *Andrew Smith and Natal: Documents Relating to the Early History of that Province*. Cape Town: Van Riebeck Society, 1955.

———. "Unpublished Documents Relating to the Career of Nathaniel Isaacs, the Natal Pioneer." *Africana Notes and News* 18, no. 2 (June 1968): 63–79.

Kirsch, Adam. *Benjamin Disraeli*. New York: Schocken Books, 2008.

Knox, Robert. *The Races of Men*. Philadelphia: Lea and Lanchard, 1850.

Kopytoff, Igor. "The Internal African Frontier." In *The African Frontier: The Reproduction of Traditional African Societies*, ed. I. Kopytoff, 3–84. Bloomington: Indiana University Press, 1987.

Kunene, Mazisi. *Emperor Shaka the Great: A Zulu Epic*. Trans. Mazisi Kunene. Pietermaritzburg: University of KwaZulu-Natal Press, 2017.

Kuper, Adam. The 'House' and Zulu Political Structure in the Nineteenth-century." *Journal of African History* 34, no. 3 (1993): 469–87.

———. *The Reinvention of Primitive Society: Transformations of a Myth*. London: Routledge, 2005.

Laidlaw, Zoë. "'Aunt Anna's Report': The Buxton Women and the Aborigines Select Committee, 1835–1837." *Journal of Imperial and Commonwealth History* 32, no. 2 (2004): 1–28.

Las Cases, Emmanuel-Auguste-Dieudonné. *Memoirs of the Life, Exile, and Conversations of the Emperor Napoleon*. Vol. 2. London: Henry Colburn, 1835.

Law, Robin. "Introduction." In *From Slave Trade to "Legitimate" Commerce*, ed. Robin Law, 1–31. New York: Cambridge University Press, 1995.

"Legislative Council Chamber." *Sierra Leone Weekly News*. 22 February 1890.

Leigh, Samuel. *Leigh's New Picture of London*. London: Leigh and Son, 1834.

Lemarchand, René. "Introduction." *African Kingships in Perspective*, ed. R. Lemarchand, 1–32. Forest Grove, OR: Frank Cass, 1977

Lester, Alan. *Imperial Networks: Creating Identities in Nineteenth-Century South Africa and Britain*. New York: Routledge, 2001.

"Letters . . . Received by the *Mary*." *South African Commercial Advertiser*. 12 October 1825.

Leveen, E. Phillip. " A Quantitative Analysis of the Impact of British Suppression Policies on the Volume of the Nineteenth-Century Atlantic Slave Trade." In *Race and Slavery in the Western Hemisphere: Quantitative Studies*, ed. Stanley Engerman and Eugene Genovese, 51–81. Princeton: Princeton University Press, 1975.

Leverton, B. J. T. "James Saunders King." *Natalia* 4 (1974): 18.

Leverton, B. J. T., ed. *Records of Natal*, vol. 1, *1823–1828*. Pretoria, South Africa: Government Printer, 1984.

———. *Records of Natal*, vol. 2, *September 1828–July 1835*. Pretoria, South Africa: Government Printer, 1989.

Lewis, Jori. *Slaves for Peanuts: A Story of Conquest, Liberation, and a Crop that Changed History*. New York: New Press, 2022.

Life on Board a Man-of-War: Including a Full Account of the Battle of Navarino. Glasgow: Blackie, Fullarton & Co., 1819.

Lindfors, Bernth. *Early African Entertainments Abroad: From the Hottentot Venus to Africa's First Olympians*. Madison: University of Wisconsin Press, 2014.

———. "A Zulu View of Victorian London." *Munger Africana Library Notes* 48 (1979): 1–19.

Lloyd, Naomi. "Religion, Same-Sex Desire, and the Imagined Geographies of Empire: The Case of Constance Maynard (1849–1935)." *Women's History Review* 25, no. 1 (2016): 53–73.

"London Numismatic Club Meeting (5 June 2007)." *Journal of the London Numismatic Club* 8, no. 11 (2008): 26–27.

Lucas, Charles. *A Historical Geography of the British Colonies*. Vol. 4. London: Oxford University Press, 1913.

Lydon, Jane. "'Mr. Wakefield's Speaking Trumpets': Abolishing Slavery and Colonising Systematically." *Journal of Imperial and Commonwealth History* 50, no. 1 (2022): 81–112.

Lyons, Scott Richard. *X-Marks: Native Signatures of Assent*. Minneapolis: University of Minnesota Press, 2010.

Mackeurtan, Graham. *The Cradle Days of Natal*. London: Longmans, Green & Co., 1930.

Maclean, Charles Rawden. "The Liberty of British Subjects Invaded in the United States." *Anti-slavery Reporter*, 1 May 1846, 68–69.

———. "Loss of the Brig *Mary* at Natal, with Early Recollections of That Settlement." *Nautical Magazine and Naval Chronicle*. January 1853 29–36; February 1853, 74–80; March 1853, 298–303; June 1853, 140–44; July 1853, 74–80; August 1853, 430–34; February 1855, 64–71.

Mahoney, Michael R. *The Other Zulus: The Spread of Zulu Ethnicity in Colonial South Africa*. Durham, NC: Duke University Press, 2012.

Malcomson, Thomas. *Order and Disorder in the British Navy, 1793–1815*. Rochester, NY: Boydell Press, 2016.

Mark, Peter and José da Silva Horta. *The Forgotten Diaspora: Jewish Communities in West Africa and the Making of the Atlantic World*. New York: Cambridge University Press, 2011.

"Marriages." *Cape Town Gazette and African Advertiser.* 24 August 1822.

Marryat, Frederick. *Peter Simple.* New York: Henry Holt, [1834] 1998.

"Matacong, West Coast of Africa." *Illustrated London News.* 2 December 1854.

Matthews, J. W. *Incwadi Yami, or Twenty Years' Personal Experience in South Africa.* New York: Rogers & Sherwood, 1887.

Mayhew, Henry. "Letter III." *Morning Chronicle* (London). 26 October 1849.

———. "Letter IV." *Morning Chronicle* (London). 30 October 1849.

———. *London Labour and the London Poor.* Vol. 1. London: G. Woodfall & Son, 1851.

———. *London Labour and the London Poor.* Vol. 2. London: Griffin, Bohn, and Co., 1861.

Mayhew, Horace. "The Church in Danger." *Punch.* 24 July 1847.

Mayo, John, Ed. *Mercantile Navy List and Maritime Directory for 1866.* London: William Mitchell, 1866.

McClintock, Anne. *Imperial Leather: Race, Gender, and Sexuality in the Colonial Contest.* New York: Routledge, 1995.

McKenzie, Kirsten. *Imperial Underworld: An Escaped Convict and the Transformation of the British Colonial Order.* Cambridge: Cambridge University Press, 2016.

Melville, Elizabeth. *A Residence at Sierra Leone.* Vol. 1. London: John Murray, 1849.

Mendelsohn, Adam. *The Rag Trade: How Jews Sewed Their Way to Success in America and the British Empire.* New York: NYU Press, 2015.

Miles, Jonathan. *The Wreck of the Medusa.* New York: Grove Press, 2007.

Miller, Christopher. *Blank Darkness: Africanist Discourse in French.* Chicago: University of Chicago Press, 1985.

Mills, John. *The British Jews.* London: Houlston & Stoneman, 1853.

Mills, Richard Charles. *The Colonization of Australia (1829–42): The Wakefield Experiment in Empire Building.* London: Sidgwick & Jackson, 1915.

Milon, M. "L'Ile de Matakong." *Le courrier du soir.* 18 April 1879.

Mokoena, Hlonipha. "'The Black House,' or How the Zulus Became Jews." *Journal of Southern African Studies* 44, no. 3 (2018): 401–11.

Morris, Donald. *The Washing of the Spears: The Rise and Fall of the Zulu Nation.* New York: Simon & Schuster, 1965.

Mouser, Bruce. *American Colony on the Rio Pongo: The War of 1812, The Slave Trade, and the Proposed Settlement of African Americans, 1810–1830.* Trenton, NJ: Africa World Press, 2013.

———. "Forgotten Expedition into Guinea, West Africa, 1815–17: An Editor's Comments." *History in Africa* 35 (2008): 481–89.

———. "A History of the Rio Pongo: Time for a New Appraisal?" *History in Africa* 37 (2010): 329–54.

———. "Lightbourn Family of Farenya, Rio Pongo." *Mande Studies* 13 (2011): 21–90.

———. "The Nunez Affair." *Bulletin des séances,*. Académie Royale des Sciences d'Outre-Mer, 4 (1973–74): 697–742.

———. "Trade and Politics in the Nunez and Pongo Rivers, 1790–1865." PhD diss., Indiana University, 1971.

———. "Women Slavers in Guinea Conakry." In *Women and Slavery in Africa*, ed. Claire Robertson, 320–39. Portsmouth, NH: Heinemann, 1997.

Msebenzi. *The History of Matiwane and the Amangwane Tribe as Told by Msebenzi to His Kinsman Albert Hlongwane*, ed. N.J. Van Warmelo. Pretoria: Government Printer, 1938.

Msomi, Welcome. "The Historic Similarity between Shaka Zulu and Mabatha." In *Umabatha*. Pretoria, South Africa: Via Afrika/Skotaville, 1996.

"Napoleon at St. Helena, Part 4." *St. James's Magazine and United Empire Review*. 2 (1876): 534–41.

"National Colonization Society." *Times*. 17 June 1830.

The Navy List for January, 1819. London: John Murray, 1819.

"New Colony at Natal." *Morning Chronicle* (London). 2 February 1826.

Newton, John. *Letters, Sermons, and a Review of Ecclesiastical History*. Vol. 1. Edinburgh: A. Murray & J. Cochran, 1780.

Niane, Djibril Tamsir. "Africa's Understanding of the Slave Trade: Oral Accounts." *Diogenes* 179 (Autumn 1997): 75–90.

Nordhoff, Charles. *Sailor Life on Man of War and Merchant Vessel*. New York: Dodd, Mead, and Company, 1883.

"Notice" [for Isaacs' *Travels and Adventures*]. *Bent's Literary Advertiser*, 10 June 1836, 64.

Okoye, Felix N.C. "Tshaka and the British traders, 1824–1828." *Transafrican Journal of History* 2, no. 1 (1972): 10–32.

O'Meara, Barry. *Napoleon in Exile, or A Voice from St. Helena*. Vol 2. London: W. Simpkin & R. Marshall, 1822.

"Our Weekly Gossip." *Athenaeum*. 28 May 1853, 650.

Owen, William Fitzwilliam. *Narrative of Voyages to Explore the Shores of Africa, Arabia, and Madagascar*. Vol. 1. London: Richard Bentley, 1833.

Papers Relative to the Condition and Treatment of the Native Inhabitants of Southern Africa. Part 2. London: House of Commons, 1835.

Pappalardo, Bruno. *How to Survive in the Georgian Navy: A Sailor's Guide*. London: Osprey Publishing, 2019.

Parascandola, Louis. "Introduction." In Frederick Marryat, *Peter Simple*. New York: Henry Holt, 1998.

Parfitt, Tudor. *Hybrid Hate: Jews, Blacks, and the Question of Race*. New York: Oxford University Press, 2020.

———. "The Use of the Jew in Colonial Discourse." In *Orientalism and the Jews*, ed. Ivan Davidson KalMarch and Derek J. Penslar, 51–67. Waltham, MA: Brandeis University Press, 2005.

Parker, Theodore. "Of Justice and the Conscience." In *Ten Sermons of Religion*, 66–101. Boston: Crosby, Nichols and Company, 1853.

Parley, Peter [Samuel Griswold Goodrich]. *A Grammar of Modern Geography*. London: William Tegg, 1854.

Peckard, Peter. *Am I Not a Man? And a Brother?* Cambridge: J. Archdeacon, 1788.

———. *The Popular Clamour against the Jews Indefensible*. Cambridge: J. Bentham, 1753.

Pelham, Camden. *The Chronicle of Crimes, or the New Newgate Calendar*. Vol. 2. London: Thomas Tegg, 1841.

Penslar, Derek. *Shylock's Children: Economics and Jewish Identity in Modern Europe*. Berkeley: University of California Press, 2001.

Pepys, Samuel. *Diary and Correspondence of Samuel Pepys*. Vol. 2. Philadelphia: David McKay, 1889.

Philips, C.H. *The East India Company, 1784–1834*. Manchester: Manchester University Press, 1961.

Philips, John. *Researches in South Africa; Illustrating the Civil, Moral, and Religious Condition of the Native Tribes*. Vol. 2. London: James Duncan, 1828.

Piterberg, Gabriel, and Lorenzo Veracini. "Wakefield, Marx, and the World Turned Inside Out." *Journal of Global History* 10 (2015): 457–78.

Porter, Andrew. *Religion versus Empire? British Protestant Missionaries and Overseas Expansion, 1700–1914*. New York: Manchester University Press, 2004.

Postans, Mrs. "The Emperor's Grave: A Sketch of St. Helena." *Oriental Herald*. 4, no. 23 (1839): 213–18.

Pringle, Thomas. "On the Demoralizing Influence of Slavery." *Anti-slavery Monthly Reporter* 2, no. 8 (January 1828): 161–74.

Prichard, John Cowles. *Researches into the Physical History of Man*. London: John and Arthur Arch, 1813.

———. *Researches into the Physical History of Mankind*. Vol. 2. London: Sherwood, Gilbert, and Piper, 1837.

Qureshi, Sadiah. *Peoples on Parade: Exhibitions, Empire, and Anthropology in Nineteenth-Century Britain*. Chicago: University of Chicago Press, 2011.

Rankin, F. Harrison. *The White Man's Grave: A Visit to Sierra Leone in 1834*. Vol. 1. London: Richard Bentley, 1836.

Reach, Angus B. "Chaka—King of the Zulus." *Bentley's Miscellany*, 1 January 1853, 292–99.

Reddy, William. "Against Constructionism: The Historical Ethnography of Emotions." *Current Anthropology* 38, no. 3 (June 1997): 327–51.

Rediker, Marcus. *The* Amistad *Rebellion: An Atlantic Odyssey of Slavery and Freedom*. New York: Penguin, 2013.

Reid, David Boswell. "Dr. Reid on the Ventilation of the Niger Steam Ships." *Friend of Africa*, 1 February 1841, 43–47; 24 March 1841, 65–73.

Reid, William Hamilton. [Sanhedrin Hadasha] *and Causes and Consequences of the French Emperor's conduct towards the Jews*. London: Day & Co. 1807.

Report and Proceedings: with Appendices of the Government Commission on Native Laws and Customs. Cape Town: W.A. Richards, 1883.

"Report of an Investigation into the Circumstances Attending the Loss of the Steam Ship *Forerunner*." In *Reports from Commissioners* (House of Lords). Vol. 35, 315–64. London: H.M. Stationery Office, 1855.

Report of the Parliamentary Select Committee on Aboriginal Tribes. London: William Ball, 1837.

"Report of Select Committee on West Coast of Africa." Appendix No. 19. *Reports from Committees: West Coast of Africa in Parliamentary Papers.* Vol. 12. London: H.M. Stationery Office, 1842.

The Report of the Wesleyan-Methodist Missionary Society for the Year Ending April, 1853. London: Wesleyan Missionary Society, 1853.

"Review [of Gardiner's *Narrative*]." *Athenaeum*, 7 May 1836, 326–28.

"Review [of Isaacs' *Travels and Adventures*]." *Athenaeum*, 18 June 1836, 425–26.

Robinson, Ronald, and John Gallagher. *Africa and the Victorians: The Official Mind of Imperialism.* London: Macmillan, 1989.

Rochlin, S.A. "Nathaniel Isaacs and Natal." *Transactions (Jewish Historical Society of England)* 13 (1932): 247–70.

Rose, Cowper. *Four Years in Southern Africa.* London: Henry Colburn & Richard Bentley, 1829.

Roth, Cecil. "The Membership of the Great Synagogue, London, to 1791." *Miscellanies* (Jewish Historical Society of England) 6 (1962): 175–85.

Routledge's Guide to the Great Exhibition; Containing a Description of Every Principal Object of Interest. London: George Routledge, 1851.

Rovner, Adam. *In the Shadow of Zion: Promised Lands before Israel.* New York: NYU Press, 2014.

Rowland Jr., Alexander. *An Essay on the Cultivation and Improvement of Human Hair; With Remarks on the Virtue and Cultivation of the Macassar Oil.* London: J. Stratford, 1809.

Royle, Stephen. *The Company's Island: St. Helena, Company Colonies and the Colonial Endeavour.* London: I.B. Tauris, 2007.

Samuelson, R.C.A. *Long Long Ago.* Durban: Knox, 1929.

Scanlan, Padraic X. *Freedom's Debtors: British Antislavery in Sierra Leone in the Age of Revolution.* New Haven, CT: Yale University Press 2017.

Schafer, Daniel L. "Family Ties That Bind: Anglo-African Slave Traders in Africa and Florida; John Fraser and His Descendants." *Slavery and Abolition* 20, no. 3 (1999): 1–21.

Schama, Simon. *Rough Crossings: Britain, the Slaves, and the American Revolution.* New York: Harper Collins, 2006.

Schwarz, Suzanne. "Commerce, Civilization and Christianity: The Development of the Sierra Leone Company." In *Liverpool and Transatlantic Slavery*, ed. D. Richardson, S. Schwarz, and A. Tibbles, 252–76. Liverpool: Liverpool University Press, 2007.

Schwarzfuchs, Simon. *Napoleon, the Jews and the Sanhedrin.* Boston: Routledge & Kegan Paul, 1979.

Semmel, Bernard. *The Rise of Free Trade Imperialism: Classical Political Economy and the Empire of Free Trade and Imperialism, 1750–1850.* New York: Cambridge University Press, 1970.

Shaw, William. "Extract of a Letter from Mr. W. Shaw, dated August 13, 1828." *Wesleyan Methodist Magazine* 8, no. 3 (1829): 130.

Shrewsbury, William. "Extract of a Letter from Mr. Shrewsbury, dated June 30, 1828." *Wesleyan Methodist Magazine.* 8, no. 3 (1829): 203–4.

———. "Extract of a Letter from Mr. Shrewsbury, dated Butterworth, Sept. 30, 1828." *Wesleyan Methodist Magazine* 8, no. 3 (1829): 268–70.

Simpson, William. *A Private Journal Kept during the Niger Expedition*. London: John F. Shaw, 1843.

Sipthorpe, A. B. C. *The History of Sierra Leone*. London: Frank Cass, [1868] 1970.

Smith, Adam. *An Inquiry into the Nature and Causes of the Wealth of Nations*. Edinburgh: Adam and Charles Black, [1776] 1863.

Solomon, Richard Allan. *The Solomons: The Genealogical Tree and a Short History of the Solomon Family in South Africa*. Johannesburg: R. A. Solomon, 1988.

Solomon, W. E. Gladstone. *Saul Solomon: The Member for Cape Town*. Cape Town: Oxford University Press, 1948.

Somerset, Charles Henry. "Proclamation by His Excellency General the Right Hon. Lord Charles Henry Somerset." 1 June 1824. In *Records of the Cape Colony*, vol. 25, ed. George McCall Theal, 203–4. Cape Town: Government of the Cape Colony, 1905.

"St. George's Gallery." *Standard*. 18 May 1853.

"St. George's Hall—the Zulu Kafirs." *Morning Chronicle* (London). 25 July 1853.

Stanton, Henry B. *Sketches of Reforms and Reformers, of Great Britain and Ireland*. New York: Baker & Scribner, 1850.

Stern, Philip J. *Empire, Incorporated: The Corporations That Built British Colonialism*. Cambridge, MA: Harvard University Press, 2023.

St. M. Watson, George Leo de, ed., *A Polish Exile with Napoleon: Embodying the Letters of Captain Piontkowski to General Sir Robert Wilson and Many Documents from the Lowe Papers*. London: Harper and Brothers, 1912.

Tales of the Wars: or, Naval and Military Chronicle. Supplementary volume. London: William Mark Clark, 1846.

Taylor, Thomas. *A Biographical Sketch of Thomas Clarkson*. London: Joseph Rickerby, 1839.

Temperley, Howard. *White Dreams, Black Africa: The Antislavery Expedition to the River Niger, 1841–1842*. New Haven, CT: Yale University Press, 1991.

Temple, Philip. *A Sort of Conscience: The Wakefields*. Auckland, NZ: Auckland University Press, 2002.

Theal, George M. *Records of the Cape Colony*. Vol. 19. London: William Clowes & Sons, 1904.

Thomas, Charles W. *Adventures and Observations on the West Coast of Africa, and Its Islands*. New York: Derby and Jackson, 1860.

Thomas, William. *The Philosophic Radicals: Nine Studies in Theory and Practice, 1817–1841*. New York: Oxford University Press, 1979.

Thompson, George. *The Palm Land, or West Africa, Illustrated: Being a History of Missionary Labors and Travels*. Cincinnati, OH: Moore, Wilstach, Keys, 1858.

———. *Travels and Adventures in Southern Africa*. London: Henry Colburn, 1827.

Tillotson, Kathleen. *Novels of the Eighteen-Forties*. New York: Clarendon Press, 1954.

Tombs, Robert. *The English and Their History*. New York: Knopf, 2014.

"Travels and Adventures in Eastern Africa" [Review of Isaacs' book]. *Court Magazine.* July 1836, 41.

"Travels in Eastern Africa" [Review of Isaacs' *Travels and Adventures*]." *Gentleman's Magazine.* December 1836. 630–31.

"Travels in Eastern Africa" [Review of Isaacs' *Travels and Adventures*]. *Metropolitan*, April 1837, 101–2.

"United States Exploring and Surveying Expedition." *Times.* 18 May 1853.

"University Intelligence." *Times.* 18 May 1853.

Unwin, Brian. *Terrible Exile: The Last Days of Napoleon on St Helena.* London: I. B. Tauris, 2010.

Valman, Nadia. "Muscular Jews: Young England, Gender and Jewishness in Disraeli's 'Political Trilogy.'" *Jewish History* 10, no. 2 (1996): 57–88.

Vick, Brian. *The Congress of Vienna: Power and Politics after Napoleon.* Cambridge, MA: Harvard University Press, 2014.

"Visit of the Zulu Kaffirs, by Command, at Buckingham Palace." *Standard.* 15 June 1853.

Volo, Dorothy D., and James Volo. *Daily Life in the Age of Sail.* Westport, CT: Greenwood Press, 2002.

Wade, John. *History of the Middle and Working Classes.* London: Effingham Wilson, 1835.

Wainaina, Binyavanga. "How to Write about Africa." *Granta* 92 (2005): 92–95.

Wahrman, Dror. *The Making of the Modern Self: Identity and Culture in Eighteenth-Century England.* New Haven, CT: Yale University Press, 2004.

Walker, James W. St. G. *The Black Loyalists: The Search for a Promised Land in Nova Scotia and Sierra Leone, 1783–1870.* Toronto: University of Toronto Press, 1999.

Wakefield, Edward Gibbon. "Notes on Chap. II Book 1." In *An Inquiry into the Causes and Nature of the Wealth of Nations*, 1:59–64, ed. E. G. Wakefield. London: Charles Knight, 1835.

———. *Plan of a Company to Be Established for the Purpose of Founding a Colony in Southern Australia.* London: Ridgway and Sons, 1832.

Walker, John Frederick. *Ivory's Ghosts: The White Gold of History and the Fate of Elephants.* New York: Atlantic Monthly Press, 2009.

Ware, E. Richmond. "Enoch Richmond Ware's Voyage to West Africa." In *New England Merchants in Africa: A History through Documents, 1802–1865*, ed. N. Bennett and G. Brooks, 298–313. Boston: Boston University Press, 1965.

———. "Health Hazards of the West African Trader, 1840–1870.". *American Neptune* 27 (1967): 81–97.

———. "West African Trading Voyage, Part 2." *Yachting.* 91, no. 4 (April 1952): 54–56, 102, 104.

Webb, L. A., Jr. "The Trade in Gum Arabic: Prelude to French Conquest in Senegal." *Journal of African History* 26. no. 2 (1985): 149–68.

"What They Are Doing in the Great Metropolis." *North Wales Chronicle and Advertiser for the Principality.* 27 May 1853.

Winterbottom, Thomas Masterman. *An Account of the Native Africans in the Neighbourhood of Sierra Leone.* London: C. Whittingham, 1803.

Wolf, Eric. "Aspects of Group Relations in a Complex Society: Mexico." *American Anthropologist* 58, no. 6 (December 1956): 1065–78.

"The Wreck of the African Mail Steamer Forerunner." Times. 21 November 1854.

Wright, John. "Henry Francis Fynn." *Natalia* 4 (1974): 14–17.

———. "Making Identities in the Thukela-Mzimvubu Region, c. 1770–c. 1940." In *Tribing and Untribing the Archiv,.* vol. 1, ed. C. Hamilton and J. Wright, 182–215. Pietermaritzburg, South Africa: University of Kwa-Zulu Natal Press, 2016.

———. "Reconstituting Shaka Zulu for the Twenty-First Century." *Southern African Humanities* 18, no. 2 (2006): 139–53.

Wylie, Dan. *Myth of Iron: Shaka in History.* Scottsville, South Africa: University of KwaZulu-Natal Press, 2006.

Zangwill, Israel. "A Land of Refuge." London: Jewish Territorial Organisation, 1908[?].

"The Zoolus of Eastern Africa [Review of *Travels and Adventures*]." *Mirror of Literature, Amusement, and Instruction,* 8 October 1836, 1–3.

"Zulu Kaffirs." *Illustrated London News.* 28 May 1853.

Index

Founded in 1893,
UNIVERSITY OF CALIFORNIA PRESS
publishes bold, progressive books and journals
on topics in the arts, humanities, social sciences,
and natural sciences—with a focus on social
justice issues—that inspire thought and action
among readers worldwide.

The UC PRESS FOUNDATION
raises funds to uphold the press's vital role
as an independent, nonprofit publisher, and
receives philanthropic support from a wide
range of individuals and institutions—and from
committed readers like you. To learn more, visit
ucpress.edu/supportus.

www.ingramcontent.com/pod-product-compliance
Lightning Source LLC
Chambersburg PA
CBHW030300160425
25215CB00018B/100/J